POLICY INTEGRATION FOR COMPLEX
ENVIRONMENTAL PROBLEMS

Policy Integration for Complex Environmental Problems

The Example of Mediterranean Desertification

Edited by
HELEN BRIASSOULIS
Department of Geography,
University of the Aegean, Greece

Routledge
Taylor & Francis Group

LONDON AND NEW YORK

First published 2005 by Ashgate Publishing

Published 2017 by Routledge
2 Park Square, Milton Park, Abingdon, Oxfordshire OX14 4RN
711 Third Avenue, New York, NY 10017, USA

First issued in paperback 2017

Routledge is an imprint of the Taylor & Francis Group, an informa business

British Library Cataloguing in Publication Data
Policy integration for complex environmental problems : the
    example of Mediterranean desertification. - (Ashgate
    studies in environmental policy and practice)
    1. Desertification - Control - Government policy -
    Mediterranean Region 2. Desertification - Control -
    Government policy - European Union countries
    3. Environmental policy - Mediterranean Region
    4. Environmental policy - European Union countries
    5. Desertification - Mediterranean Region 6. Mediterranean
    Region - Environmental conditions
    I. Briassoulis, Helen
    333.7'36'091822

Library of Congress Cataloging-in-Publication Data
Policy integration for complex environmental problems : the example of Mediterranean
    desertification / edited by Helen Briassoulis.
        p. cm. -- (Ashgate studies in environmental policy and practice)
    Includes index.
    ISBN 0-7546-4243-7
        1. Desertification--Government policy--Mediterranean Region. 2. Land use--
    Government policy--Mediterranean Region. 3. Desertification--Government policy--
    European Union countries. 4. Land use--Government policy-European Union countries.
    I. Briassoulis, Helen. II. Series.

    HD840.7.Z63P65 2005
    333.73'6'091822--dc22

                                                                            2004030114

    ISBN 13: 978-1-138-25903-4 (pbk)
    ISBN 13: 978-0-7546-4243-5 (hbk)

Contents

List of Figures and Tables

Figures

Tables

List of Abbreviations

AEMs	Agri-Environmental Measures
BAP	Biodiversity Action Plan
BEPGs	Broad Economic Policy Guidelines
BRAP	Better Regulation Action Plan
C&I	Criteria and Indicators
CAP	Common Agricultural Policy
CAS	Complex Adaptive Systems
CBD	Convention on Biological Diversity
CCD	Convention to Combat Desertification
CF	Cohesion Fund
CITES	Convention on International Trade in Endangered Species of Wild Fauna and Flora
CSF	Community Support Framework
CSG	Community Strategic Guidelines
CTP	Common Transport Policy
DG	Directorate General
EAFRD	European Agricultural Fund for Rural Development
EAGGF	European Agricultural Guarantee and Guidance Fund
EAP	Environment Action Program
EC	European Commission
ECB	European Central Bank
ECJ	European Court of Justice
ECOFIN	Economic and Financial Council (EU)
EEB	European Environmental Bureau
EEC	European Economic Community
EESC	European Economic and Social Committee
EFFIS	European Forest Fire Information System
EIA	Environmental Impact Assessment
EIB	European Investment Bank
EMAS	Environmental Management and Audit System
EMI	European Monetary Institute
EMU	Economic and Monetary Union
EPI	Environmental Policy Integration
ERDF	European Regional Development Funds
ESC	Economic and Social Council
ESDP	European Spatial Development Perspective
ESF	European Social Fund
ETUC	European Trade Union Confederation

EU	European Union
EUCFP	EU Common Forest Policy
EUFS	EU Forestry Strategy
EUROSTAT	Statistical Agency of the EU
EUSDS	European Union Sustainable Development Strategy
FAO	Food and Agriculture Organization
FCCC	Framework Convention on Climate Change
FFR	Forest Focus Regulation
FP	Framework Programme
FPI	Forest Policy Integration
GDP	Gross Domestic Product
HEP	Horizontal Environmental Policy
HICP	Harmonized Index of Consumer Prices
ICZM	Integrated Coastal Zone Management
IFC	Integration of Forest-Related Concerns
IFF	Intergovernmental Forum of Forests
IMF	International Monetary Fund
IMP	Integrated Mediterranean Programmes
IPCC	Intergovernmental Panel on Climate Change
IPF	Intergovernmental Panel of Forests
IPPC	Integrated Pollution Prevention Control
ISGF	Inter-Service Group on Forestry
ITTO	International Tropical Timber Organization
IUCN	International Union for the Conservation of Nature
LFAs	Less Favoured Areas
MCPFE	Ministerial Conference on the Protection of Forests in Europe
MS	Member State
NAP	National Action Programme
NEA	Network of Environmental Authorities
NFPs	National Forest Programmes
NGOs	Non Governmental Organizations
NSRF	National Strategic Reference Framework
NUTS	Nomenclature of Statistical Territorial Units
OECD	Organization for Economic Cooperation and Development
OMC	Open Method of Coordination
OP	Operational Programme
OSFA	One-Size-Fits-All
PDO	Protected Denomination of Origin
PI	Policy Integration
PIS	Policy Integration Schemes
RAP	Regional Action Programme
RBAs	River Basin Authorities
RDPs	Rural Development Programmes
RDR	Rural Development Regulation

RSPB	Royal Society for the Protection of Birds
SD	Sustainable Development
SDS	Sustainable Development Strategy
SEA	Strategic Environmental Assessment
SEAP	Sectoral Environmental Action Plan
SFM	Sustainable Forest Management
SFs	Structural Funds
SGP	Stability and Growth Pact
SIA	Sustainability Impact Assessment
SME	Small and Medium-Sized Enterprises
SPA	Social Policy Agenda
SPD	Single Programming Document
SPIS	Spatial Policy Integration Schemes
TEN	Trans-European Networks
TEN-T	Trans-European Transport Networks
UN	United Nations
UN/DESA	United Nations / Department of Economic and Social Affairs
UN/ECE	United Nations – Economic Commission for Europe
UN/ECOSOC	United Nations / Economic and Social Council
UNCCD	United Nations Convention to Combat Desertification
UNCSD	United Nations Commission on Sustainable Development
UNDP	United Nations Development Programme
UNEP	United Nations Environmental Programme
UNESC	United Nations Economic and Social Committee
UNFCCC	United Nations Framework Convention on Climate Change
UNFF	United Nations Forum on Forests
UNRISD	United Nations Research Institute for Social Development
WCED	World Commission on Environment and Development
WCS	World Conservation Strategy
WFD	Water Framework Directive
WTO	World Trade Organization
WWF	World Wildlife Fund
WWF EPO	World Wildlife Fund European Policy Office

List of Contributors

Helen Briassoulis, professor, Department of Geography, University of the Aegean, Mytilene, Lesvos, Greece.

Vassilis Detsis, lecturer, Department of Home Economics and Ecology, Charokopeio University, Athens, Greece.

Theodoros Iosifides, lecturer, Department of Geography, University of the Aegean, Mytilene, Lesvos, Greece.

Giorgos Kallis, research associate, Environment and Spatial Planning Laboratory, University of Thessaly, Volos, Greece.

Constantinos Liarikos, economist, project coordinator, WWF-Hellas, Athens, Greece.

Georgios Mantakas, research associate, Institute of Forest Research, Athens, Greece.

Apostolos G. Papadopoulos, assistant professor, Department of Geography, Charokopeio University, Athens, Greece.

Aristotelis C. Papageorgiou, assistant professor, Department of Forestry, Environmental Management and Natural Resources, Democritus University of Thrace, Orestiada, Greece.

Katerina Petkidi, geographer, research consultant, WWF-Hellas, Athens, Greece.

Foreword

Research projects start from a particular question but, in the process, it is not uncommon to discover that other, usually broader, questions have to be addressed first. This is more or less what happened when this editor was involved in a EU-funded research project[1] inquiring, among other topics, the policy aspects of combating desertification in Mediterranean Europe. The original purpose gradually but firmly broadened to cover the issue of policy integration, a necessary prerequisite for tackling complex socio-environmental problems, like desertification, and for facilitating the transition to sustainable development more generally. The search of the literature revealed that the systematic study of the subject has started only recently, mostly in the EU where there is related on-going policy activity. Moreover, the major thrust of the pertinent research is on Environmental Policy Integration (EPI) which is a narrower notion than that of policy integration. The idea of producing a book on policy integration based on the research experience gained from the afore-mentioned project was, thus, been born.

The current literature on EPI focuses on the procedural aspects of incorporating environmental concerns in sectoral policies mostly. This book interprets policy integration more broadly, as 'integration of policies', and aims to contribute to the discourse on the subject from the perspective of diverse EU policy areas by expanding on its meaning, diverse dimensions, importance in addressing complex policy problems, and its analysis. Desertification control is used as an illustrative example of a complex policy problem that is relevant for the Southern EU member states and for which few (if any) policy proposals at sub-global levels are available at present, not only in the EU but also internationally.

The present exploration of policy integration was far from easy; in fact it was very challenging. I would like to thank my colleagues who embarked with me on this journey and stood fast throughout the long, tiresome months of authoring their contributions. The preparation of the final camera-ready copy would not have been possible without the unflagging and tenacious efforts of Panagiotis Stratakis. Finally, the book may not have even seen the light of day without the unfaltering love, support and encouragement of my husband.

Helen Briassoulis
Mytilini, Lesvos, Greece

---

[1] MEDACTION (Policies for land use to combat desertification), Module 4. Contract No. ENVK2-CT-2000-00085.

Chapter 1

Complex Environmental Problems and the Quest for Policy Integration

Helen Briassoulis

Introduction

After a long gestation period, two ideas, sustainable development and complexity, matured in the 1990s and moved to the top of political and scientific agendas, stirring multifarious theoretical and methodological discourses on how to better and more effectively deal with pressing contemporary societal problems. The need for integrated, interdisciplinary approaches and policies to holistically address complex, crosscutting, 'wicked' socio-environmental problems and the derivative quest for policy integration to promote sustainable development feature prominently among them.

Contemporary socio-environmental problems are multifaceted, involving diverse and intricately related natural and human resources as well as individuals and organizations acting and interacting on multiple spatial and temporal scales. Numerous, frequently single-purpose and little coordinated, sectoral policies concern particular facets of these problems; their direct and spillover effects may either contribute minimally to problem resolution or produce overlaps, conflicts, new problems and waste of resources. After several decades of policy-making experience, it became evident that sectoralized, uni-dimensional, uni-disciplinary and uncoordinated policies do not serve well the cause of sustainable development.

Although interest in policy integration has a long history in several quarters, the recent renaissance of the subject is associated mainly with the environmental repercussions of economic activities that are not properly accounted for (if at all) by the policies impinging on these activities; hence, the proliferation of policy activity and research on environmental policy integration (EPI). The focus on EPI has overshadowed other concerns such as, for example, that, in addition to environmental, sectoral policies have spatial, social, cultural and other repercussions, and that neither environmental policies account adequately for their economic and social impacts nor social policies account for their environmental and economic impacts. The ensuing discussion will argue that EPI is a narrow and limited view compared to a broader conception of policy integration that ensure that policy making better fits the nature of contemporary problems and supports the transition to sustainable development.

The policy market faces the following situation. On the demand side, contemporary problems are complex and interrelated, defying treatment by means either of narrow, sectoral policies or of all-encompassing, super-policies. On the supply side, numerous policies, related to particular aspects of one or more of these problems, exist, making it unnecessary to devise new policies each time a problem arises. Therefore, it might be more prudent to properly combine extant policies to address these problems. Policy integration, in the broadest sense, 'adds value' to policies while economizing on resources (Sanderson, 2000).

The discourse on policy integration has met with the common problem of confusion over and differences in the meaning of overtly similar terms, namely 'policy', 'integration' and 'policy integration', in various policy contexts that carry over to differences in their operational expressions, proposed design, and so on. Often the terms are not defined or they are defined loosely, opening the way for multiple interpretations (and mis-interpretations). Proper and consistent analysis of policy integration and the design of policy integration schemes require clear definitions of 'integration of what, by whom, where, when, why and how'. Only then it can be judged how well policy integration facilitates the resolution of problems and contributes to sustainable development.

This book seeks to contribute to the discourse on policy integration from the perspective of diverse policy fields, focusing on policies of the European Union. The impetus came from EU-funded research,[1] concerned with the design of a policy framework for combating desertification, which is an exceptionally complex socio-environmental problem, relevant to the Southern EU member states, for which few (if any) policies are available at present, not only in the EU but also internationally. In the course of the research, policy integration emerged as the only viable approach to desertification-related policy making. Upon reflection, policy integration appeared to call for deeper analysis as it proved to touch on the whole genre of complex socio-environmental problems. The chapters of this book use Mediterranean desertification as an illustrative example of several issues related to the integration of EU policies.

This chapter aims to frame the discussion of the subject and introduce the book chapters. The next section discusses the complexity of socio-environmental problems and offers a concise account of desertification drawing attention to those features that make its combat a complex policy problem. The third section negotiates both Environmental Policy Integration (EPI) and Policy Integration (PI), reviews the evolution of the discourse, defines both concepts, elaborates on their object and dimensions, presents relevant operational expressions and measures proposed to promote EPI and PI, highlights factors affecting the success of EPI and PI and, lastly, argues for the necessity of PI, and not EPI, to promote the transition

---

[1] MEDACTION (Policies for land use to combat desertification), Module 4. Contract No. ENVK2-CT-2000-00085.

to sustainable development. The last section introduces the individual contributions.

Complex Socio-Environmental Problems – The Example of Mediterranean Desertification

The Complexity of Socio-Environmental Problems

The diffusion of Complexity and Chaos theory in the Natural, the Social and the Policy Sciences has enriched greatly the study of contemporary socio-environmental problems and has shed light on suitable policy making approaches to address them (Byrne, 1998; Berkes and Folke, 1998; Marion, 1999; True et al., 1999; Zahariadis, 1999). This section highlights the distinguishing characteristics of complex systems, and especially of human-environment systems, that help comprehend and negotiate the complexity of socio-environmental problems and its policy implications.

A voluminous literature documents the complexity of natural and social systems and, more importantly, of human-environment systems (Holling, 1986; Dryzek, 1987; Waldrop, 1992; Berkes and Folke, 1998; Byrne, 1998; Levin, 1999; Science, 1999; Gunderson and Holling, 2002). The present awareness is not new. For a long time now, researchers in diverse contexts, following different theoretical and methodological routes, have opined that the linear, equilibrium-centered view of nature and society does not fit the evidence.[2] What is new perhaps is that, after the 1980s, the growing popularity of Complexity and Chaos Theory coincided with and was reinforced by the coming to prominence of pressing socio-environmental problems and the understanding that their study as well as the design of effective policies could not be based on the dominant Newtonian-Cartesian, non-evolutionary scientific tradition that dissociated the environment from people, policies and politics (Berkes and Folke, 1998). Co-evolutionary, interdisciplinary, historical, and comparative systems approaches developed, adopting integrative modes of inquiry and using multiple sources of evidence, to study socio-environmental problems arising in the context of interlinked human-environment systems. Complexity-informed approaches, in particular, aim at characterizing the nature of a system "with reference to its constituent parts in a non-reductionist manner" (Manson, 2001, p.406). Table 1.1 presents selected, important differences between the Newtonian-Cartesian and the Complexity-Chaos approaches.

---

[2] Examples are the notion of creative destruction (Schumpeter, 1975), catastrophe theory applications in the social sciences and in planning (Poston and Stewart, 1981; Dendrinos and Mullaly, 1985), path dependence and increasing returns in economics (Arthur, 1989; Liebowitz and Margolis, 1995).

Table 1.1   Selected differences between the Newtonian-Cartesian and the Complexity-Chaos approaches (*)

Newtonian-Cartesian Approach	*Complexity- Chaos approach*
Static, reductionist, based on 19th century Physics; deterministic causality	Dynamic, (co)-evolutionary, holist, living (self-organizing) systems model-based; indeterminate causality
Complete rationality; perfect knowledge and information	Bounded rationality; limits to knowledge; incomplete information
Structurally simple systems; linear or quasi-linear relationships among variables	Inherently complex systems; non-linear relationships among variables prevail
Certainty and predictability	Limited certainty; unpredictability
Systems tend towards equilibrium through negative feedbacks	Inherently unstable systems; positive feedbacks are more common; multi-equilibrium; surprises integral part of anticipated adaptive responses
Individual differences, externalities and exogenous influences deviating from the norm are considered exceptional and treated as noise	Individual differences and random externalities are considered normal events and the driving forces of variety, adaptation and complexity

(*) Based on Berkes and Folke (1998); Russell and Faulkner (1999); Geyer (2001); Holling (2001)

Complex systems are difficult to describe succinctly because they are dynamic, indeterminate, characterized by novelty and surprise, and generating new behaviors while maintaining their structure and coherence through adaptation (Batty and Torrens, 2001; Holland, 1995). 'Complex adaptive systems' (CAS) is the term commonly used to refer comprehensively to complex systems that are constantly adapting to their environment (Janssen et al., 2000; Finnigan, 2003; Gunderson and Holling, 2002; Holling, 2001). These include weather systems, immune systems, ecosystems, economic systems, social systems, cultures, traffic, and many more.

Like all systems, CAS comprise components, individual, self-interested agents (Levin, 1999)[3] with different characteristics; for example, flora and fauna in ecosystems, individuals and organizations in social systems. Human agents, in particular, differ in their socio-economic features, viewpoints, preferences, goals, aspirations, emotions, future outlooks, and amounts of resources they possess[4] (Detombe, 2001). Agents interact in parallel, on the basis of simple rules, among them and with their environment through flows of various types of resources in and

---

[3] Semi-autonomous units seeking to maximize some measure of goodness, or fitness, by evolving over time (Dooley, 1997).

[4] Money, know-how, power, etc.

out of the system boundaries (Levin, 1999; Limburg et al., 2002). In socio-ecological systems, formal and informal institutions govern the multiple, context-specific, frequently unknown interactions among actors and between them and resource systems (Ostrom, 1990; Gunderson and Holling, 2002). The connectivity of a CAS, the particular ways in which its agents connect and relate to one another, is critical to its evolution and survival.

In fact, one of the most prominent features of CAS is that their properties are explained by an understanding of the *relationships* among their parts than by an understanding of these parts separately (Gallagher and Appenzeller, 1999; Manson, 2001; Limburg et al., 2002). The *non-linear* nature of the relationships among their components distinguishes complex from simple, linear systems and defines their internal structure, behaviour and mode of change (Manson, 2001). Linear systems have a single equilibrium state and their evolution is smooth and continuous. Small disturbances produce equally small changes in system state attributes because *negative feedback* mechanisms bring the system back to the initial equilibrium state. On the contrary, non-linear systems possess multiple equilibria states. The transition between them may be abrupt and discontinuous because mutually reinforcing *positive feedback* mechanisms are at work (White, 2001). These amplify microscopic heterogeneity hidden within complex systems that, consequently, exhibit a characteristic Sensitive Dependence on Initial Conditions (SDIC) (Glasner and Weiss, 1993).[5] Minor, random, and sometimes overtly insignificant, changes may precipitate an avalanche of changes that bring the system to a new equilibrium state, a fundamental shift in the structure of the system or a large-scale event; this is popularly known as the 'butterfly effect' (Gleick, 1994; Allen, 2001).

The nonlinear nature of natural, social and economic systems[6] and the central importance of random, significant or not, historical events in influencing *selection* among the multiple equilibria of a system and, thus, determining its state and evolution have been frequently pointed out. Holling (1978, 1995) has illustrated how small disturbances, such as pest infestations or fire, can trigger large-scale redistribution of resources or connectivity within the internal structure of an ecosystem (Manson, 2001, p.410). Increasing returns in economics (Arthur, 1989) is a widely publicized kind of positive feedback mechanisms, magnifying chance events, which explain several economic and spatial phenomena[7] and the path dependence of socio-economic and spatial development (Anderson et al., 1988; Arthur et al., 1997). Path dependence, the dependence of current outcomes on the path of previous outcomes and, thus, the persistence of change along well-defined, historically determined pathways (Berkhout, 2002; Puffert, 2003), is a

---

[5] In other words, small disturbances tend to be compounded, growing exponentially and producing discontinuous, abrupt and, thus, unpredictable changes (surprises) in state variables.

[6] And of the bio-sphere as a whole (Abel and Stepp, 2003).

[7] Such as technical standards (the 'QWERTY' standard typewriter and computer keyboard), urban sprawl and urban concentrations, clustering of economic activities, and so on.

characteristic of biological, economic and social systems, underscoring the contingent nature of their evolution and the influence of the institutional settings within which they are embedded (Liebowitz and Margolis, 1995; Henderson, 2001; Berkhout, 2002).

Some systems may exhibit inertia and insensitivity to external changes if, because of a particular event, they are *'locked in'* a particular state where change is irreversible (Arthur, 1989; Liebowitz and Margolis, 1995). For example, overgrazing or climate change can push vegetated systems into a new stability domain (desertification) that is reinforced by feedback loops that maintain high temperatures and low water and nutrients (Limburg et al., 2002), thus, 'locking' the system in an undesirable state.

CAS change[8] and *adapt* to changing external conditions and 'shocks' through *self-organization*; i.e. spontaneously re-arranging their elements[9] in different patterns, behavior and structure to better interact with their environment (Manson, 2001). However, their changes are not permanent because CAS are in a state of tenuous equilibrium, at the 'edge of chaos' as it is commonly known,[10] tending to collapse into a rapidly changing state of dynamic evolution. Their property of *self-organized criticality*[11] means that they reach an equilibrium state that is not stable but which is the most productive and creative, leading to new possibilities.

The evolution of CAS is characterized by the co-existence and complementarity of long periods of relative stasis (phases), where change is gradual (slow processes) and the system functions smoothly, punctuated by bursts of evolutionary change (phase shifts), discontinuities and rapid transformations (fast processes) (Berkes and Folke, 1998; Wollin, 1999). Turbulence is triggered by either minor disturbances or external events (e.g. a small price change, a new subsidy scheme, an earthquake) (Wollin, 1999; Holling, 2001; Gunderson and Holling, 2002).

Natural and human CAS exhibit *dissipative behavior*,[12] i.e. when external forces or internal perturbations drive a system far from equilibrium, to a highly unorganized state where its structure is characterized by irregular patterns, it self-organizes by the dissipation of energy, according to the second Law of Thermodynamics, moving to a state with more organization (Schieve and Allen,

---

[8] I.e. move between equilibrium points.

[9] Through positive and negative feedback mechanisms.

[10] A notion introduced by Langton (1990) and denoting the balance point where the components of the system never lock in place and yet never quite dissolve into turbulence either (Waldrop, 1992). In a state of equilibrium, a system does not have the internal dynamics enabling it to respond to its environment; eventually it will die. In a state of chaos, on the other hand, a system ceases to function as a system; hence, the importance of the system being at the 'edge of chaos'.

[11] Their ability to balance between randomness and stasis.

[12] Originating in the work of Prigogine (see, e.g. Prigogine, 1980) on non-equilibrium Thermodynamics, who coined the terms, dissipation and dissipative structures are still contested notions among scientists (see, for example, ISCID, 2004).

1982 cited in Manson, 2001, p.410). The 'dissipative structures' formed may appear spontaneously on various levels of the spatio-temporal hierarchy (Gunderson and Holling, 2002), feeding back to capture and dissipate more energy; thus, they can be characterized as evolutionary because they generate variation (Abel and Stepp, 2003).[13]

CAS co-evolve with their environment as they exist within it, being are also part of it. Therefore, following a disturbance, i.e. a change in internal or external conditions, *adaptation* occurs through reorganization, redistribution and restructuring of the CAS components (Holling 1986, 2001), a process based on *learning*[14] (Lee, 1993; Berkes and Folke, 1998; Manson, 2001; Wilson, 2002). Through continuous exchanges of information with their environment (feedbacks), the system's agents are able to anticipate the results of their actions as well as to adapt to changing conditions. But because they are part of the environment, their changes modify the environment too, triggering new rounds in a constant process of change.

CAS are open systems. Their boundaries are difficult to delineate precisely because they are a function of the problem studied and, also, system components and their relationships are continuously changing. At a given spatio-temporal level, a CAS consists of smaller, semi-autonomous systems, nested (embedded) within larger aggregate systems (Berkes and Folke, 1998; Gibson et al., 1998; Gunderson and Holling, 2002). The hierarchical organization characteristic of CAS differs from top-down, serial, command-and-control authoritative structures.[15] Because functions and control are decentralized, in a parallel mode of operation with numerous, non-linear feedbacks linking system components, CAS possess an inherent variety[16] and flexibility that allow them to respond fast to unforeseeable events and to try multiple options simultaneously.

System agents may belong to various hierarchies (Manson, 2001). In social systems, human agents[17] are members of diverse socio-cultural groups and organizations, occupying different positions in various relational webs spanning the whole spatial/organizational hierarchy (Healey, 1997). The same is true for resource systems which, depending on the use and ecological status of resources, may participate in different production, consumption and ecological systems at multiple hierarchical levels. Causation is, thus, difficult to establish in human-environment systems where the number and complexity of multifarious, non-linearly interacting hierarchies are rather impossible to observe and record satisfactorily. Moreover, the multiple memberships of their agents may generate

---

[13] For the case of economic systems, see Harvey and Reed (1994).

[14] "A system 'remembers' through the persistence of internal structure (Holland, 1992)" (Manson, 2001, p.410).

[15] Simon (1969) used the Russian dolls metaphor to describe these hierarchies, which some authors do not find very appropriate.

[16] That owes, among others, to the combinatorial explosion of interactions among systems components.

[17] Individuals, households, public and private formal and informal organizations.

redundancy as well as contradictions[18] but they increase the strength and resilience of CAS to external and internal shocks, while being a source of creativity and creation of new possibilities for system survival. Polycentric governance systems, that possess these characteristics, have proven less vulnerable to unexpected contingencies than rigidly organized, centralized systems (Ostrom, 1990, 1998).

A much-celebrated feature of CAS is *emergence*. Instead of being planned and pre-determined, order and control emerge from the bottom up as local interactions, based on simple rules, among individual agents produce over time regular patterns that feedback on the system, informing the behavior of agents (Dooley, 1997; Levin, 1999). *Emergent* qualities are system-wide characteristics, not features of individual components.[19] They are a function of synergism, not superposition, among system's parts at a particular level, rendering CAS unpredictable and difficult to control (Baas and Emmeche, 1997; Manson 2001).

Contemporary socio-environmental problems, originating in complex human-environment systems, are similarly complex involving a multitude of diverse actors and resources interacting over and across different spatial and temporal levels. Greater numbers of actors and resources imply more interactions and greater differentiation among them that reduce the chances of understanding one another's functions and increase the incidence of problems (Zahariadis, 2003). Problem definitions and the associated goal-setting are contingent and contextual, a function of the initial system conditions, i.e. of the actors that defined the problem and the state of the human-environment system. These evolve in the course of problem solving as actors, resources and their relationships change. This is why these problems are often ill- or multi-defined,[20] definitions and goals may be conflicting, and causation difficult to establish as it depends on the scale of analysis and the amount of available information, which is usually incomplete. It is uncertain when, why, and where the problems started, as well as where they lead and when they will end. Consequently, these problems are unpredictable and hard to analyze and handle with ready-made solutions (DeTombe, 2001). Problem solving is a continuous and fluid process. Seldom are these problems 'solved'; at best, they are 'resolved' (Patton and Sawicki, 1986) for their 'owners' and over a finite time period. Solutions to problems cannot be imposed; rather, they *emerge* from the interactions among system components following the rules set.

The implications of socio-environmental problem complexity for policy making and management are crucial (Dryzek, 1987; Wiman, 1991; Wollin, 1999; Zahariadis, 2003). Resource-related public policies are collective choice institutions that mediate "the relationship between a social group and the life-support ecosystem on which it depends" (Berkes and Folke 1998, p.9). Policy

---

[18] As in, e.g. a democracy that builds on and thrives amidst opposing points of view.
[19] For example, land degradation at the regional level cannot be explained as a function of each individual engaging in land and water degrading activities and of the physico-chemical properties of individual land parcels.
[20] I.e. definitions are observer-dependent.

interventions have generally unpredictable and uncertain impacts at various spatial/organizational levels, which depend on the scale at which the policy problem is defined and treated. "Policies may result in problem displacement, across time, space and medium, rather than amelioration" (Brown, 2000, p.577). Policies that ignore system complexity and variously reduce it frequently give rise to undesirable 'surprises' in both the short and the long run[21] or they prove to be perverse; i.e. they do not produce economic benefits while they generate environmental costs ((Berkes and Folke, 1998; Ascher, 2001).

Formulation of responsive and effective policies necessitates an understanding of the complexity of linked human-environment systems and the inherent uncertainty of policy problems as well as recognition of the fact that "there is no single, universally accepted way of formulating the linkage between social systems and natural systems" (Berkes and Folke, 1998, p.9). Contemporary policy approaches gradually abandon the Newtonian-Cartesian, linear worldview and espouse the complex, non-linear world model. The adoption of the precautionary principle, of learning-based and strategic approaches and of the adaptive management paradigm, all signify the influence of complexity-thinking in policy making (Wiman, 1991; Lee, 1993; Healey, 1997; Brown, 2000; Ascher, 2001; White, 2001).

However, because the policy market already supplies a multifarious, disjointed and largely uncoordinated policy basket, unable to come to terms with the complexity of contemporary, cross-cutting problems, policy integration appears to be another avenue worth-exploring to manage this complexity, capitalizing on the flexibility of complex human-environment systems[22] (Zahariadis, 2003). At present, most policy systems, at all levels, are complicated rather than complex, comprising policies built on the linear world assumption while reality is nonlinear (Figure 1.1). Because of vertical administrative organization and compartmentalization and institutional fragmentation, such systems cannot cope effectively with policy externalities[23] and unexpected changes.[24] Hence, they are

---

[21] "Policies that assume smoothly changing and reversible conditions, and limitless ability of the economy to adapt and substitute, lead to reduced options, limited potential and perpetual surprise. The political window that drives 'quick fixes' for quick solutions simply leads to more unforgiving conditions for decisions, more fragile natural systems and more dependent and distrustful citizens."(Holling et al., 1998, p.354).

[22] "A complex system can deal with truly novel situations because it has a wide array of internal components and subsystems linked by complex relationships. Some subset of these components may have some ability to accommodate a novel relationship. In the rare cases when no suitable components or sub-systems exist, the system cannot respond to new relationships with the environment, with potentially catastrophic results.... The destruction of complex, diverse internal relationships may lead to a lack of resilience and adaptability in ecosystems" (Manson, 2001, p.410).

[23] I.e. changes in one policy area produced by changes in another area.

[24] "Changes in one variable are more likely to trigger significant changes in other than the local subsystem, changes less readily apparent to the responsible organizational unit. This is because the response of its own task environment would reflect the impact of implemented

not *fit* to the complexity of socio-environmental systems. Policy integration may help, under certain conditions, to fix the policy system, by better coupling policy supply to the features of the complex world that generates policy demand. It can reduce redundant and conflicting, while preserving useful, overlaps and linkages, and turn the policy system from complicated to complex, thus, providing for better *institutional fit*. Before turning to the subject of policy integration, desertification, a complex socio-environmental problem that poses a real challenge to policy integration, is presented in the next section.

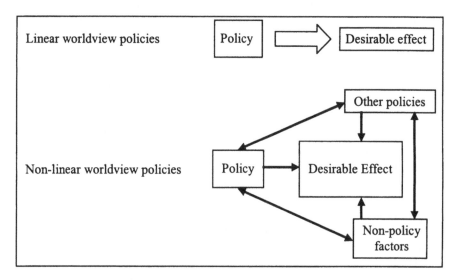

Figure 1.1   Causality structures of linear vs. non-linear worldview policies

Desertification – A Complex Socio-Environmental Problem

Nearly half a century after the French researcher Aubreville first coined the term 'desertification' (Aubreville, 1949, cited in Reynolds and Stafford-Smith, 2002b) and 17 years after the 1977 United Nations Conference on Desertification took place in Nairobi, the United Nations Convention to Combat Desertification (UNCCD) was signed in 1994. Desertification was no longer considered a problem of the drylands of Africa but a problem relevant to several bioclimatic world regions (except Antarctica!).

Desertification is "land degradation in arid, semiarid and subhumid tropics caused by a combination of climatic factors and human activities" (UNCCD, 1994). Land degradation means reduction or loss of the biological or economic productivity and complexity of rainfed cropland, irrigated cropland, or range,

---

politics less accurately than if the variables it manipulated operated more independently from 'distant' subsystems" (Lustick, 1980, pp.346-7 cited in Zahariadis 2003, p.292).

pasture, forest and woodlands resulting from land uses or from a process or combination of processes, including those arising from human activities and habitation patterns, such as: (a) soil erosion caused by wind and/or water, (b) deterioration of the physical, chemical, biological and economic properties of soils and (c) long-term loss of natural vegetation (UNCCD, 1994).

This definition makes clear that the term 'desertification' refers to the *formation and expansion of degraded soil and land* and not to the expansion or advance of the current deserts. It results from complex interactions between biophysical and societal factors and processes at various spatial and temporal scales (Williams, 2000, Reynolds and Stafford-Smith 2002b, Prince, 2003) that are briefly discussed below focusing on Mediterranean desertification.

Biophysical determinants of desertification Climate, soils and geology, surface and ground water, topography and vegetation are influential factors in the process of land degradation. In the arid, semi-arid and dry sub-humid zones[25] of the Mediterranean, adverse climatic conditions such as low as well as uneven annual and interannual distribution of rainfall, extreme weather events and the out-of-phase nature of the rainy and vegetative seasons favour overland flow and erosion of bare soils that greatly reduce the potential for biomass production, ultimately leading to desertification (Yassoglou, 2000). Global climate change is expected to widen the present geography of the vulnerable zones in the Mediterranean.

Soil depth, structure and stability, organic content and soil-water balance are critical determinants of a satisfactory land cover and the reversibility of degradation processes.[26] Human actions that modify negatively these soil parameters, such as forest fires, land use changes, intensive cultivation and overgrazing, contribute to the acceleration of the erosional processes in already sensitive bioclimatic regions. Soils derived from limestone and acid igneous *parent materials* are shallow with a relatively dry moisture regime, and high erodibility and desertification risk.[27] Under Mediterranean climatic conditions, regeneration of these soils and vegetation may be slow or even impossible, and desertification irreversible. The stoniness of land has a great although variable effect on run-off, soil erosion, soil moisture conservation and biomass production influencing, thus, land protection in the Mediterranean (Poesen and Lavee, 1994).

Important topographic determinants of soil erosion are the *slope gradient* and *slope aspect*. Severely eroded soils are commonly found on moderately to very steep slopes (greater than 12 per cent) in the semi-arid zone of the Mediterranean.

---

[25] These are areas, excluding polar and sub-polar regions, where the ratio of annual precipitation to dynamic evapotranspiration ranges between 0.5 and 0.65 (UNCCD, 1994).

[26] Desertification proceeds when soil cannot provide rooting space, water and nutrients to plants.

[27] Also, soils formed on marl are very susceptible to desertification because they cannot support any annual vegetation in particularly dry years, despite their considerable depth and high productivity in normal and wet years (Kosmas et al., 1993).

Recovery of vegetation is slower and erosion rates higher in warmer southern and western rather than in northern and eastern aspects.

The availability of *surface and ground water*, a function of the previous factors, is a critical determinant of desertification. It influences mainly the course of soil-water balance throughout the year and affects the type and conditions of land cover. The arid and semi-arid zones of Mediterranean Europe have large moisture deficits due to scarcity of surface and ground water.

Lastly, the *vegetative cover*, a function of land use and its change, among others, is the dominant biotic determinant of desertification affecting both run-off and sediment loss under unfavourable soil and climatic conditions. Land is considered desertified when biomass productivity drops below a certain threshold value. A value of 40 per cent vegetative cover is considered critical below which accelerated erosion dominates in sloping landscapes (Thornes, 1988). Mediterranean ecosystems have a great adaptation potential and resistance to aridity. Among Mediterranean perennial crops, olive trees are particularly adaptable and resistant to long term droughts, supporting a remarkable diversity of flora and fauna in the understorey, even higher than in some natural ecosystems (Margaris et al., 1995).

Both slow and fast physical and/or chemical processes contribute to degradation and desertification. Slow processes include *soil erosion*, which is activated by the destruction of the vegetative cover and affects marginal sloping lands, and *soil salinization* and *nitrification*, which are localized, resulting from irrational irrigation practices, but affect valuable low lands (Yassoglou and Kosmas, 2000). Fast processes include droughts and extreme weather events.

Societal determinants of desertification   The human determinants of land degradation and desertification encompass diverse and interdependent socio-economic, cultural, political and institutional factors and processes that can be broadly categorized into driving and mitigating forces[28] and proximate causes of the phenomenon.[29] Human driving and mitigating forces, operating on various

---

[28] The distinction between driving and mitigating forces is not always clear as they exchange roles depending on the historical and geographic context.

[29] *Human driving forces*, or macroforces, are fundamental societal forces that, in a causal sense, link humans to nature and bring about global environmental changes. They include: population change, technological change, sociocultural/socioeconomic organization and change (economic institutions and the market, political economy, political institutions).

Human mitigating forces are those forces that impede, alter or counteract human driving forces. They include: local to international regulation (policies), market adjustments, technological innovations, and informal social regulation through norms and values.

Proximate driving sources or *proximate causes* are the aggregate, human activities that, under the influence of driving and mitigating forces, directly cause environmental transformations, either through the use of natural resources, the use of space, the output of waste or through the output of products that in themselves affect the environment. Other examples are: biomass burning, fertilizer application, species transfer, plowing, irrigation, drainage, livestock pasturing, pasture improvement, deforestation and site abandonment,

spatio-temporal scales, affect the decisions of individuals to change the use of their land. Human activities and practices used in the process, the proximate causes of change, produce *land cover change*, which, under adverse biophysical conditions, may lead to land degradation and, eventually to desertification. The most influential societal factors are briefly presented below.

Population structure and dynamics (mobility and migration) relate to land degradation and desertification in complex, non-linear and context-dependent ways (Blaikie and Brookfield, 1987; Perez-Trejo, 1994; UNSO, 1994; Reynolds and Stafford-Smith, 2002a). No simple and clear causal connections seem to exist, or a value of land sensitivity, after which problems worsen. Intense *outmigration* causes population decline and ageing, land abandonment and low land maintenance. Unfavourable biophysical conditions increase the chances for land degradation while favourable conditions facilitate their gradual restoration. In-migration may exacerbate existing land degradation but may also provide the necessary labour force for the maintenance of local land resources.

Non-linear, context-dependent relationships develop also between *poverty* and *social inequality* with land degradation and desertification. One point of view sees a vicious spiral of land degradation causing poverty and vice versa, ignoring several intervening factors. Another point of view sees poverty as accelerating land degradation and desertification, and vice versa, in the presence of unfavourable factors such as weak public policies, inefficient markets, and problematic institutions (UNSO, 1994).

Degradation of local resources may result also from non-local forces. Global socio-economic restructuring and technological change produce numerous, intricately related changes in *modes of production, social values, consumption patterns, life styles, family structure, employment composition*, which bring about changes in the valuation and modes of utilization and management of land resources following direct and indirect pathways (Reynolds and Stafford-Smith, 2002a).

Agricultural product price changes, market and/or public policy-induced, *capital availability* and *competition among economic activities* are influential economic determinants of land use change[30] and of changes in the mode of land resources utilization. Land degradation and desertification may be the long-term consequence of these changes under adverse bio-climatic conditions and unfavourable sociopolitical circumstances. However, the links between economic forces and the phenomenon are non-linear and context dependent making assessments uncertain and risky (Reynolds and Stafford-Smith, 2002a).

Legal, institutional and *administrative factors*, including inappropriate or inexistent environmental and planning legislation, problematic policy

---

breaking up of large tracts of grassland, expansion of cultures which promote erosion, farming of fields in the fall line, urbanization, suburbanization, urban fringe development, fire (Briassoulis, 2000; Turner et al., 1990).

[30] Modification of existing uses or conversion to other uses.

implementation, unclear or inexistent systems *resource rights* for critical resources, such as water and soils, administrative compartmentalization and lack of coordination, are often important influences on land degradation and desertification.[31] More generally, lack of planning coupled with resource-depleting practices impedes the wise management of land and environmental resources (Reynolds and Stafford-Smith, 2002a).

Critical local level institutional influences on land and resource use decisions is *land tenure and ownership*. Rural land rental, combined with absentee ownership, land fragmentation, and vague and incompatible resource regimes often lead to inappropriate land management and degradation and render the implementation of formal policies problematic (Reynolds and Stafford-Smith, 2002a).

National or supranational *policies* act either as driving or as mitigating forces of the phenomenon. In Mediterranean Europe, the negative impacts of the Common Agricultural Policy (CAP), transport policy and the Structural Funds (SFs) on water and soil resources have been extensively analyzed (Buller, 2002). Revisions of these and other policies aimed, among others, at avoiding their negative environmental impacts (such as the Agri-Environmental Regulation and the requirement for EIA in the SFs). The same is true for several national policies that disturb bioclimatically sensitive regions, setting the stage for future degradation (Reynolds and Stafford-Smith, 2002a).

Political regimes and their change are deeper underlying forces of land degradation through their influence on the 'philosophy' and approach to land management and on socio-economic organization. *Natural and social historical events*, such as wars, famines, natural disasters, new technologies, price shocks, resource crises, and the like, may trigger processes leading to land degradation in sensitive regions, at least in the long-term (Reynolds and Stafford-Smith, 2002a).

Technology is not a driving force *per se* but its irrational use may reinforce the negative impacts of other socio-economic drivers of land degradation. The intensive use of agricultural machinery and road vehicles, especially on steep slopes or erosion-prone land, the application of agrochemicals, and other practices change the soil structure and composition, accelerating degradation especially under favourable climatic conditions (Yassoglou and Kosmas, 2000).

The previous discussion reveals the complex and context-dependent relationships between land degradation and its determinants. In fact, *geographic location, accessibility* and the *spatial distribution* of economic activities, uses of land, population and infrastructure determine critically the level of socio-economic development and, consequently, the degradation and desertification prospects. The urban-rural dynamics is the broader explanatory schema that should be used to

---

[31] The legal framework governing landed property, in particular, has proven incapable of controlling the abuse of public property and the fragmentation of private property that creates numerous landowners who act in uncoordinated fashion contributing to, and at the same time being affected by, land degradation.

frame the meaningful analysis and policy making on desertification (van der Leeuw, 1999).

The aforementioned societal factors act on multiple spatio-temporal scales, some producing short-term, localized impacts (fast processes) while others producing longer-term, large-scale effects (slow processes). This is the most serious source of uncertainty as it concerns both the definition and identification of desertification in an area and the proper measures to combat it (Reynolds and Stafford-Smith, 2002a).

More certainty and scientific agreement exists as regards the proximate causes of the phenomenon that are common to most world regions including the Mediterranean Europe. Human activities that degrade land and water resources and increase the desertification risk include deforestation, forest fires, overgrazing, as well as inappropriate land management practices such as intensive cultivation,[32] monocultures, abandonment of traditional practices (such as terracing), unsatisfactory (or absent) maintenance of rural holdings, surface and groundwater overdrafts,[33] drainage of wetlands, and large construction works (MEDACTION, 2004b).

Desertification-related policy activity   The UNCCD is the international regime providing the broad frame of actions to combat desertification. Signatory countries to the Convention are obliged to prepare National Action Programmes (NAPs) as well as Regional Action Programmes (RAPs) to protect their affected regions. In Mediterranean Europe, Italy, Greece, Portugal and Spain have drafted NAPs while the Mediterranean RAP is under preparation in cooperation with other countries bordering the Mediterranean.

Because of the low level of public awareness and priority among other pressing issues and the complexity of the problem, actions to combat desertification are still fragmented, sectorally and spatially uncoordinated, concerning mostly the proximate causes rather than the driving forces of the phenomenon. The European Union has supported the fight against desertification through research funding, beginning with the First Framework Programme (FP1) in 1989 up to FP6 at present (2004), specific projects (e.g. INTERREG, LIFE), research at the Joint Research Centre at Ispra, Italy, technical and information support provided by the European Environment Agency, and specific measures included in the CAP and the Structural Funds. Recently, the European Commission published Communication COM (2002) 179, "Towards a Thematic Strategy for Soil Protection", to serve one of the objectives of the 6th Environmental Action Programme, namely soil protection against erosion and pollution (MEDACTION, 2004b). At the national level, policy measures vary by country and do not always target explicitly desertification. Soil protection, afforestation, fire protection, water resources

---

[32] With the application of heavy machinery, agro-chemicals, etc. (especially on marginal and sensitive lands).
[33] To support agricultural, tourism and industrial uses.

conservation and other measures indirectly contribute to mitigating the longer term occurrence of the phenomenon.

Epilogue: the complexity of desertification   Desertification is a contentious issue (Thomas, 1997) and a 'wicked' policy problem. It results from complex, non-linear, context- and scale-dependent interactions among numerous natural resources and human activities that are mediated by a considerable number of nature-society institutions. Its occurrence is path-dependent, sensitive to the initial conditions prevailing in particular spatio-temporal contexts. It is, thus, difficult to disentangle its multi-scale biophysical and societal causes and predict its consequences and its reversibility.[34] It is an emergent phenomenon as it is a higher-level land feature, the cumulative outcome of numerous individual, local level inappropriate land management practices induced by the intricate interplay of multi-scaled biophysical and societal forces. Once unsustainable conditions set in an area, positive feedback mechanisms usually intensify degradation leading the land to several states of equilibrium. It cannot be known with certainty, however, whether and when an area will be 'locked' in an irreversibly desertified state.

As a result of this complexity, considerable uncertainty surrounds the phenomenon, leading to controversies over its definition, the relative importance of its anthropogenic causes, the assessment of land affected or at risk, its reversibility and the importance of its impacts that vary with the spatial and temporal scale of analysis (Reynolds and Stafford-Smith, 2002b). Moreover, desertification is a socio-culturally defined and determined construct. The *meaning and interpretation* of its determinants and impacts have influenced the collective sense of urgency to combat it. This explains the preference for particular abatement approaches,[35] the slow progress towards developing integrated approaches, the relative inaction and the partial, fragmented, and uncoordinated efforts to address it.

Policy making to combat desertification is a complex enterprise owing to the multiplicity and diversity of individual and collective actors involved and the numerous formal and informal institutions implicated[36] at multiple spatial and temporal scales (Briassoulis, 2004).   Managing this complexity requires the creation of the necessary *institutional capacity* (or, *institutional capital*), the adoption of integrated spatial planning approaches and appropriate local resource management schemes, such as co-management (Berkes and Folke, 1998) and the

---

[34] Controversy centers on its causes and consequences. The problem is two-fold: "(a) whereas desertification is most often attributed to a myriad of human activities, ... it may be triggered or exacerbated by climate variability, ... so that the causes are not necessarily solely anthropogenic (at least at the local land use level), (b) not all such ecological, biogeochemical and hydrological changes have an immediate or direct economic impact on human activities" (Reynolds and Stafford-Smith, 2002b, p.3)

[35] As regards the choice between social and institutional vs. scientific and technological solutions.

[36] I.e. the rules governing the interactions among actors and their relationships with the natural resources involved.

continuous top down-bottom up communication and coordination of local level interventions to alleviate the larger problem.

Environmental Policy Integration or Policy Integration?

Introduction

Policy integration became the subject of academic research and of policy making after the early 1970s when policy activity in several areas was growing rapidly, environmental concerns were gaining prominence and sustainable development was politically accepted as the ultimate societal goal, demanding holistic and integrated approaches to address pressing development problems. Beginning in the 1990s, concerns were voiced also that the departmentalization of policymaking and the lack of coordination among policies, developing along narrow sectoral and interest-centered lines, was frequently generating negative policy externalities. Implementation of one policy caused unwanted impacts on the object of another, the most prominent case being that of the environmental impacts of sectoral policies. This is why the bulk of the post-1990 literature on policy integration (PI) is devoted mostly to Environmental Policy Integration (EPI) although PI more generally received attention on several occasions as it concerns not only the environmental but also the agricultural, social, welfare, economic and other policy spheres (O' Riordan and Voisey, 1998; Ardy and Begg, 2001; Avery, 2001; Eggenberger and Partidario, 2000; ILO, 2001; Persson, 2002; Shannon, 2002).

This section probes into the question of whether EPI is sufficient or whether deeper PI should be sought to support the transition to sustainable development. The discussion is inevitably dominated by the EPI literature that is exploited, however, to address the more general PI notion. A historic overview is offered first of landmark political, and parallel scientific, developments on EPI and PI since the 1970s. The current thinking on the object, dimensions and operational measures of EPI and PI, of policy measures to promote them and of factors affecting their success is presented next. Lastly, arguments in support of PI are offered as the basis of elaborating an analytical methodology in Chapter 2 of this volume.

Landmark Developments on EPI and PI since the 1970s

In the European Union (EU), the evolution of the EPI notion has gone roughly through three periods – the pre-1987 (or, pre-Brundtland), the 1987-1998 and the post-1998 (or, post-Cardiff) period – being influenced by international and European developments. The official starting date of the first period could be 1972, when the United Nations Conference on the Human Environment in Stockholm introduced the notion of 'eco-development' to jointly promote the interdependent goals of environmental protection and economic development (Nelissen et al., 1997), and a number of landmark studies were published in that and subsequent years. Prominently among them figure *The Limits to Growth* (Meadows, 1972), *A Blueprint for Survival* (Goldsmith, 1972), *Towards a Steady-state Economy* (Daly, 1973) and many more constituting classic environmental texts by now.

The First Environmental Action Programme (EAP) of the EU (1973-76) established that effective environmental protection required the consideration of environmental consequences in decision-making and that comprehensive assessment of the impacts of policies to avoid any damaging activities, with special reference to agriculture and spatial planning, was necessary. The Third EAP (1982-1986) went further asking that concern for the environment be integrated in certain policy areas, namely, agriculture, energy, industry, transport and tourism (Lenschow, 2002). At the international level, the 1980 World Conservation Strategy (WCS) of the International Union for the Conservation of Nature (IUCN) emphasized also the integration of the ecological dimension in development.[37]

This was an intensive period of conceptual developments worldwide on the relationships among the environmental, social and economic dimensions of societal problems at various spatial scales that laid the foundations and produced pioneering theoretical and methodological works on integrated approaches to development issues. An emphasis on interdisciplinarity replaced the original focus on multi-disciplinarity. 'Integration' and 'integrated approach', as well as various synonyms such as 'harmonization', 'synthesis', 'coordination', had become established terms in the academic literature and in policy quarters by the mid-1980s. Important areas of applications included the economic-environmental analysis of urban and regional systems and related decision-making, integrated land use-transportation models and integrated resources management (Isard 1972, Putman, 1983; Batey and Madden, 1986; Briassoulis, 1986; Nijkamp, 1986; Braat and van Lierop, 1987).

The next period began with the publication of *Our Common Future* in 1987 (WCED, 1987) that established the pursuit of sustainable development as the superior political goal worldwide. 'Integration', in general, and EPI, in particular, automatically came to centre stage in policy and academic circles as the backbone of sustainable development. The Single European Act of 1987 formally incorporated the integration principle into the EEC Treaty. Article 130r demanded that 'environmental protection requirements shall be a component of the Community's other policies' (Hertin and Berkhout, 2003). In the same year, the Fourth EAP (1987-1992) devoted a subsection to integration of the environmental with other Community policies (Lenschow, 2002).

Five years later, three developments coincided. In March 1992, the Fifth EAP "Towards Sustainability" (1993-2000) promoted EPI in five sectors: agriculture, energy, industry, transport and tourism. In June 1992 the Rio Summit produced Agenda 21 that devoted Chapter 8 to integrating environment and development in decision-making. Lastly, the 1992 Treaty of the European Union (Articles 2 and

---

[37] The term 'cross-sectoral' conservation policy appeared for the first time in the WCS (Lafferty and Hovden, 2002).

130r(2)) required the integration of environmental protection requirements in the definition and implementation of Community policies (Lenschow, 2002).

During this second period, until 1997, research in the natural, social and policy sciences promoted the notion of integration on several fronts, from theories and philosophical treatments of the subject to methodological and modeling tools. Integrated models, multi-criteria decision making methods, sustainable development indicators, policy integration indicators, sustainability planning approaches, integrated pollution control and many more, all explicitly or implicitly acknowledged the necessity of integrated approaches to development (Jansen, 1991; Meyer and Turner, 1994; OECD 1994). In that period, Complexity theoretical ideas and tools started to find applications in the development of holistic, conceptual and applied frameworks for sustainable development planning and management (Berkes and Folke, 1998).

The signing of the Amsterdam Treaty in 1997 marked the beginning of the third period. Article 6 of the Treaty gave EPI its current political significance, constituting the official statement for the integration of environmental concerns in sectoral policies: "Environmental protection requirements must be integrated into the definition and implementation of the Community policies and activities referred to in Article 3, in particular with a view to promoting sustainable development" (Lenschow, 2002, p.14). The developments that followed aimed at providing the necessary procedures and rules to materialize the injunction of Article 6.

The decision of the Heads of Government and State, at the 1997 Luxembourg Summit, that sustainable development can be implemented only by means of policy integration, was the basis for launching the *Cardiff Integration process* at the 1998 Cardiff Summit. Following the Commission's Communication "Partnership for Integration" (CEC, 1998), the European Council invited the Councils of Transport, Energy and Agriculture to start preparing their own environmental strategies. This first 'wave' of Council formations was followed by the second 'wave' involving the Councils of Development, Internal Market and Industry that were called to prepare environmental strategies too at the 1998 Vienna Summit. The European Council invited the Commission to submit a progress report on the mainstreaming of environmental policy.[38] Finally, at the 1999 Cologne Summit, the third 'wave' of Council formations, Fisheries, ECOFIN and General Affairs, joined the Cardiff Integration process (Fergusson et al., 2001). Following the adoption of the EU Sustainable Development Strategy by the Gothenburg European Council in 2001,[39] the Council was invited "to finalize and further develop sector strategies for integrating environment into all relevant

---

[38] The Parliament's opinion of May 28th, 1998 is worth mentioning here: "...the Commission [...] has immediate responsibility for and the opportunity of improving the complementarity and consistency of Community policies, in particular by establishing the internal mechanisms for co-ordination between its various departments...".

[39] When the environmental pillar was added to the Lisbon Strategy also.

Community policy areas with a view to implementing them as soon as possible [...].[40] In October 2002, the Environment Council asked the European Council to invite the Council formations responsible for education, health, consumer affairs, tourism, research, employment and social policies to develop environmental integration strategies (CEC, 2004).

Since 1999, at the EC Summits, the Councils have reported on progress made with respect to their environmental strategies and the Commission periodically has issued Communications and reports on related matters.[41] By the end of 2003 all nine sectors had adopted integration strategies that are now at various phases of a review process. Some sectors have made commitments to implement their strategies within given time horizons. At the 2003 Spring European Council, it was decided that the European Commission will carry out an annual stocktaking of environmental integration as a complement to the Environment Policy Review, which will feed into the Commission's Spring Report and the Spring European Council debate (CEC, 2004). The first stocktaking was issued in June 2004 which opined that the Cardiff process has produced positive results, such as raising the profile of environmental integration and concrete improvements in some sectors, but is suffers also from several drawbacks. These include a general lack of consistency, insufficient political commitment, need for improved delivery, implementation and review mechanisms as well as clearer priorities and focus. Its suggestions emphasize the need to improve the consistency of strategies across Council formations and to place greater emphasis on good practice in terms of content and implementation. Community and national level measures are proposed to support sectoral Councils in their environmental integration efforts and help maximize the benefits of these efforts in terms of concrete environmental improvements (CEC, 2004).

The official developments on EPI ran in parallel, or, in certain cases, in conjunction with intense and maturing research activity, continuing from the past, related directly or indirectly, both to EPI and to policy integration more broadly from the level of the individual firm to that of the globe. Besides the proliferation of 'integration indicators' (CEC, 1999a; OECD, 1999; EC, 1999; 2000; EEA, 1999b; 2000), theoretical and methodological developments on integrated approaches to (sustainable) development gained ground informed by Complexity theory and informing at times the policy process (Detombe, 2001; Gunderson and Holling, 2002). Some of them take a broader look at PI as it is discussed below.

---

[40] Paragraph 32, Presidency Conclusions to the Gothenburg European Council (15-16 June 2001).
[41] Such as the *Report on Environment and Integration Indicators to Helsinki Summit* (CEC, 1999a), *From Cardiff to Helsinki and Beyond* (CEC, 1999b) and *Bringing our needs and responsibilities together - Integrating environmental issues with economic policy* (CEC, 2000).

Conceptualizing and Defining Environmental Policy Integration and Policy Integration

Different conceptualizations and meanings of EPI and PI, varying in clarity, certitude and rigour, have led to different analyses over time and in various contexts. The meaning of policy integration depends on how 'policy' and 'integration' are conceptualized. According to Merriam-Webster's Collegiate Dictionary, 'integrate' can mean either "to form, coordinate, or blend into a functioning or unified whole" or "to unite with something else" or "to incorporate into a larger unit" (cited in Persson, 2002, p.9).[42] The fine difference between 'blending into a unified whole' or 'uniting with something else' and 'incorporating into a larger unit' is schematically depicted in Figure 1.2 below (inspired from Persson, 2002).

In discussing PI for sustainable development, Thomas (2003) suggests four conceptions of the term 'integration' – integration as efficiency, as mindfulness, as institutional coordination and as 'compatibility-within-a-framework'. The first conception refers to coordinating sectoral action to maximize the use of resources; i.e. achieving more than one policy objective. The second conception, mindfulness, is about incorporating environmental (or economic or social) considerations in policies that are not environmental (or, economic or social, respectively). The third conception refers to administrative coordination being the prerequisite for maximizing environmental (or social or economic) achievement. The fourth conception, 'compatibility-within-a-framework', aims at striking a compatible relationship among various sectoral goals within an overarching framework, allowing for compromises among them. One step further, 'goal integration' involves the integration of environmental, economic and social dimensions in ways that secure a simultaneous realization of the goals of each in a single policy, programme or project intervention (Thomas, 2003). This ideal integrationist conception assumes that "all apparent irreconcilability can be overcome through attaining an underlying unity of purpose" (Thomas 2003, p.203).

Turning to the meaning of 'policy', several authoritative sources concur that public policies are purposeful courses of action, comprising a long series of more-or-less related activities, which governments pursue to reach goals and objectives related to a problem or matter of concern and to produce certain results (Friedrich, 1963; Lowi, 1964; Anderson, 1984; Pressman and Wildavski, 1992). Hence, a policy is not a single, discrete, unitary, disembodied phenomenon, but a series of decisions. It concerns what is actually done (or not done) as opposed to what is proposed or intended, which is the case of decisions; policy implementation and enforcement complete the actual policy process. Essential constituent elements of a policy are its object (the characteristics of the problem considered and the theory

---

[42] The *Oxford Combined Dictionary of Current English & Modern English Usage* (1982, p.129) leans towards the latter interpretation ("To *integrate* is to combine components into a single congruous whole").

about it), interested and/or involved actors, their goals (reflecting their value systems), the resources and means available, the instruments used to achieve the goals set and the implementation mechanisms.

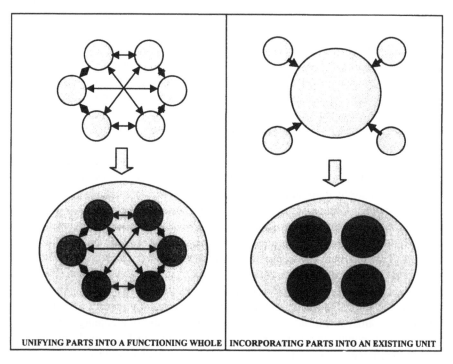

| UNIFYING PARTS INTO A FUNCTIONING WHOLE | INCORPORATING PARTS INTO AN EXISTING UNIT |

Figure 1.2   The difference between unifying and incorporating parts

Policy integration, then, can be conceptualized as a process either of coordinating and blending policies into a unified whole, or of incorporating concerns of one policy into another. Leaving aside for the moment the specifics of how integration can be achieved, obviously, the key, fundamental requirement of this process that move it forward and the key features of its product, an integrated policy, are *communication* and *collaboration* among the actors involved[43] who work *jointly*, on the basis of *joint* procedures, to design solutions to *shared* (or common) problems, *sharing* resources,[44] and building new relationships as needs and problems arise (Shannon, 2002).

In the discourse on PI and EPI, the intertwined 'normative vs. rational concept' and 'output (substance) vs. process (procedure)' questions arise (Persson, 2002). Considered as a normative concept, the concern is with the relative weights given to the policies, or the aspects of reality, that are being integrated (e.g.

---

[43] Some of which may be common to several policies.
[44] Staff, budgets, know-how, data and information, etc.

environmental, social, or economic). Thus, the emphasis is on the output of the integration process; i.e. what an integrated policy should look like. If absolute criteria (weights) are adopted, then a policy is or is not integrated. Looser criteria allow for alternative forms of an integrated policy. As a rational concept, PI is seen as a process of integrating various aspects of reality into (sectoral) policy making, that takes place at various levels and in diverse contexts.[45] Upon reflection, however, it turns out that the two questions are interrelated because an integrated policy results only through a process of integration (Lenschow, 2002).

The literature contains very few definitions of PI. Underdal (1980), adopting a normative stance, focused on the output side of the process defining an integrated policy as one in which "all significant consequences of policy decisions are recognized as decision premises, where policy options are evaluated on the basis of their effects on some aggregate measure of utility, and where the different policy elements are in accord with each other" (Underdal, 1980, p.162).

Peters (1998), from a public policy and administration perspective, considers PI as an organizational issue that is necessary so that "various organizations... charged with delivering public policy work together and do not produce either redundancy or gaps in services" (p.5). He distinguished between 'policy coordination' that is more pertinent to policy formulation and 'administrative coordination' that concerns policy implementation mostly.

Eggenberger and Partidario (2000), considering integration more broadly, both as an output and as a process, argued that:

> ...whenever there are two professionals with different backgrounds looking at the same problem with similar objectives, they are integrating. Whenever there are two different topics that need to be tackled together, there is integration.... Integrating, in fact, means a new entity that is created where new relationships are established, bearing on individual entities that have specific characteristics and specific dynamics but in combination they act in a different way (p.204).

Without an adjective, no priority among the objectives of PI is assumed; the task is delegated to the political (democratic) process. The adjective 'environmental', or in the same spirit, 'social' or 'economic', denotes a particular point of view and priority in integrating policies.[46] Since the 1980s, most definitions found in the literature refer to EPI (even when they omit the adjective 'environmental') as the impetus for intensively researching PI came with the heightened interest in the environmental repercussions of economic activity and the pursuit of sustainable development.

---

[45] From the governmental to the street level.

[46] The economic policy domain exemplifies the integration of one policy's goals into other policies. All policies take into account economic factors from the start of the policy making process – from agenda setting and policy formulation (budgeting), to policy implementation (following budgets) to policy evaluation (accounts and auditing) (Lafferty and Hovden, 2002).

Collier (1994) defines policy integration as aiming at (a) achieving sustainable development and preventing environmental damage; (b) removing contradictions between as well as within policies; and (c) realizing mutual benefits and the goal of making policies mutually supportive. She highlights that EPI involves different requirements at different stages of the policy process and considers that a set of criteria for determining goal trade-offs is necessary to guide the EPI process.

The OECD (1996a), focusing on the process side of EPI, defined it as: "Early co-ordination between sector and environmental objectives, in order to find synergy between the two or to set priorities for the environment, where necessary." The European Environment Agency sees EPI as a process of shifting the focus of environmental policy away "from the environmental problems themselves to their causes ... [and] ... from 'end-of-pipe' ministries to 'driving force' sector ministries" (EEA, 1998, p.283).

For Lafferty and Hovden (2002, p.15), EPI implies:

(a) the incorporation of environmental objectives into all stages of policy making in non-environmental policy sectors, with a specific recognition of this goal as a guiding principle for the planning and execution of a policy, (b) accompanied by an attempt to aggregate presumed environmental consequences into an overall evaluation of policy, and a commitment to minimize contradictions between environmental and sectoral policies by giving principles priority to the former over the latter.

Shannon (2002) emphasizes the communicative dimension of EPI; intersectoral policy integration concerns certain kinds of cooperative behavior, forms of institutions, and kinds of communicative action. Other authors concur on this point too (O' Riordan and Voisey, 1998; Hertin and Berkhout, 2003).

Finally, according to EEB (2003, p.10, 14):

Environmental Policy Integration is a long-term process that requires changes in administrative practice and government culture, institutional adaptation and also specific tools... The integration of environmental aspects into other policy areas must contribute to policies that effectively lead us to higher environment protection and greater sustainability.

This is a holistic definition of EPI as a process leading to policies sufficiently and effectively embodying environmental and sustainability concerns.

The Object and Dimensions of PI and EPI

To develop operational expressions for EPI and PI as well as measures to achieve them, it is necessary to clarify (a) what should be integrated and (b) in what sense, along which dimensions. The answers are partly conditioned by the focus on PI as process, output or both, and by the stage of the policy making process where PI takes place.[47] Two generic approaches to these intertwined questions are found in the literature, the vertical and intrasectoral and the horizontal and intersectoral (Lafferty and Hovden, 2002; Persson, 2002; Hertin and Berkhout, 2003).

The *vertical and intrasectoral approach* has dominated thinking on PI given the emphasis on EPI. In a rather narrow sense, the object of PI is to incorporate environmental concerns into a sectoral policy through proper procedures (Lafferty and Hovden, 2002).[48] This can happen in all stages of the sector policy process. In practice, it is encountered mostly during agenda (and goal) setting where environmental goals are added to the sector policy's set of goals (e.g. reduce air pollution from transport). Lafferty and Hovden (2002) state: "Vertical Environmental Policy Integration (VEPI) involves the degree to which sectoral governance has been 'greened'" (p.19). Naturally, the critical question is how one moves from proper goals to procedures to realize these goals. The literature leans heavily towards procedural measures and instruments for PI and EPI (Peters, 1998; Persson, 2002; EEB, 2003; Hertin and Berkhout, 2003). Underdal (1980) suggested the 'vertical consistency' criterion for PI; a policy should be consistent throughout all its levels, from policy goals to more detailed guidelines. Although often it is not explicitly stated, evidently, the result of a vertical, intrasectoral integration process is an integrated policy (in the sense reflected in its goals).

Vertical PI has a spatial dimension that is mostly implicit in the literature with some exceptions (to this author's knowledge so far). Because the policy process takes place at several organizational levels, the spatial dimension is implied when considering PI at various stages of the process. The importance of the spatial dimension, both for the analysis of PI and the design of appropriate measures, owes to the fact that different actors are involved at each level and stage, influencing the ease and success of PI. At higher levels, PI in terms of goals is easily achieved (Lenschow, 2002; Persson, 2002) while getting formal and informal policy actors comply with 'integrating' procedures and rules[49] at lower levels proves difficult if not infeasible (O' Riordan and Voisey, 1998). If linkages among the relevant spatial/organizational levels do not exist or are not fully functional, the initial PI intentions never materialize on the ground.

Buller (2002) makes direct reference to the spatial dimension of EPI, in the context of the CAP:

EPI needs to take place at a series of different levels in order to be effective – from that of policy formulation at the EU level and policy implementation at the national and sub-national level, down to the actions and attitudes of farmers and farming communities (p.122).

---

[47] Mainly, agenda setting, policy formulation, policy implementation.
[48] Obviously, this may apply to any other kind of concerns such as social (equity, gender equality, etc.), and so on.
[49] E.g. implement the EIA requirement.

Similarly, EEB (2003, p.15) considers that "vertical integration relates to an integration strategy coordinated at every level, from international to national, to regional and local level" suggesting to "establish and improve vertical integration: EPI at one level only is ineffective and it needs to be implemented in coordination at every level possible (European, national, regional, local and even international)". It is not clear, however, if by 'vertical' the EEB refers to one sectoral policy or to a broader and more abstract notion of vertical policy integration across spatial scales.

The *horizontal and intersectoral approach* to PI is increasingly recognized as the most appropriate for effective PI, requiring substantive and procedural cross-sectoral communication, cooperation and coordination. A vertically integrated sectoral policy has limited possibilities to bring about even narrow positive effects lest to contribute to sustainable development. Most studies again concern the environmental version of horizontal policy integration (HEPI); i.e. ensuring that the environmental dimension is integrated in all other policies as a cross-cutting concern. Lafferty and Hovden (2002) consider that HEPI concerns the extent to which a central authority[50] has developed a cross-sectoral strategy for EPI. Until now they could not find evidence of HEPI as it requires negotiation of trade-offs among environmental and other objectives.

Essential requirements for effective EPI include horizontal communication and networking between environmental and non-environmental sectors, joint responsibilities, lack of administrative fragmentation (including authorization procedures), and constructive inter-departmental cooperation (Lenschow, 2002b; Shannon, 2002; EEB, 2003; Hertin and Berkhout, 2003). Shannon's (2002) discussion of intersectoral policy integration[51] from an environmental perspective, as the main requirement to address the demands of contemporary ecologically, socially, politically, administratively, and legally crosscutting policy problems can be extended to the general case of PI. Finally, Peters' (1998) more general discussion of policy coordination[52] has also a horizontal rather than a vertical orientation.

Horizontal policy integration has a spatial dimension that is implicit in most of the literature. Liberatore's (1997) discussion of the six dimensions of EPI is relevant here. The 'space and time' dimension requires paying attention to:

> ...the interactions between different spaces: the *geographical space* of the affected environment, the *economic space of the activities* that have an impact on the environment, the *institutional space of the relevant authorities and policy instruments*, and the *cultural space of values* (p.117).

---

[50] The state itself or any body with horizontal competences and mandate.

[51] I.e. developing intersectoral policies that link policy networks, policy purposes, and affect desired changes in policy outcomes (Lee, 1993, cited in Shannon, 2002).

[52] That concerns the implementation aspects of PI.

Also, the 'organization' dimension concerns the mismatch between the territorial competences of environmental authorities with the affected environment.

Although analytically useful, the distinction between a vertical and a horizontal approach to EPI and PI is not unambiguous or straightforward to make in practice (Persson, 2002). The definition and usage of the terms 'vertical' and 'horizontal' vary with the analyst and determine the content of 'policy integration'. The vertical approach concerns a single policy that incorporates several concerns that were not included in its original design and VPI means the production of an integrated policy. On the other hand, the horizontal approach concerns *relationships* among policies with respect to a given issue (e.g. environmental) or to several interlinked issues and HPI may not necessarily lead to a unitary integrated policy. In the perspective of promoting sustainable development, both approaches should be merged through appropriate procedures for complete and effective PI extending beyond the environmental field.

Operational Measures of EPI and PI

Operational expressions for EPI and PI have been developed, variously called variables, criteria, or indicators, reflecting their conceptualization (process vs. output) and direction (vertical, horizontal or both) from which they are approached (Liberatore, 1997; OECD, 1997; OECD, 1999; CEC, 1999a; EEA, 1999b, 2000; CEC, 2000; OECD, 2002). With a few exceptions,[53] most of them relate to EPI although several can be generalized to the case of PI also. They can be used to identify the presence or absence of some form of PI or of sufficient conditions enabling or favouring some form of PI, as well as to develop measures to promote PI. The question of measures to gauge the *degree* of EPI or PI has not been addressed yet with the exception of the Metcalfe scale discussed at the end of this section. Naturally, the more criteria are satisfied, the higher the degree of PI achieved.

Although integration among policies can occur spontaneously and informally[54] (Peters, 1998; Persson, 2002), the proposed measures reflect PI either through imposition (direct government intervention) or through bargaining (market-based, indirect government intervention) or through a combination of the two approaches (Peters, 1998; Lenschow, 2002). Their deeper aim is to induce behaviour change. Adapting the measures/criteria proposed in the literature[55] to the general case of PI, three broad groups, generic, substantive and procedural measures/criteria, are presented below. The *generic* measures/criteria concern either enabling conditions

---

[53] For example, Peters (1998)

[54] This does not mean arbitrarily but through non-institutionalized actions of particular actors.

[55] Among the main sources used are: OECD (1997, 1999, 2002), Potier (1997), Peters (1998), UNECE (1999), CEC (2000), Lafferty and Hovden (2002), Shannon (2002), Lenschow (2002a), Persson (2002), EEB (2003), Hertin and Berkhout (2003), Jacob and Volkery (2003).

for the realization of of EPI or PI or general properties of either the output or the process of integration, or both. These include:

- Political commitment and leadership for EPI/PI in general;
- Need for compliance with international and EU commitments;[56]
- Existence of long term SD strategy (or a relevant Report or Forum);
- The environmental, social, economic agendas of different sectors form a consistent overall strategy (perhaps guided by a SD strategy);
- Favourable policy tradition and administrative culture (open, participatory, flexible);
- Core belief systems shared across policy sectors and communication processes fostering them;
- Intra-governmental power relations hindering EPI/PI; presence of vertical alliances hindering horizontal networking.

The *substantive measures/criteria* include Underdal's (1980) generic criteria characterizing an integrated policy: (a) *comprehensiveness* (inclusiveness of space, time, actors and issues), (b) *consistency* (all the components of the policy are in agreement) and (c) *aggregation* (an overarching criterion is used to evaluate different policy elements). These criteria are not easily operationalized, for theoretical and practical reasons, but their value as guiding principles for PI should not be underestimated.

The substantive measures/criteria refer to the object of integration, require knowledge about economy-environment-society interactions and are subject- and context-specific (Persson, 2002). Although they are mostly suited to an output view of PI, the properties of an integrated policy, they may be adapted to identify 'ideal processes of integration'. These measures/criteria are difficult to specify; hence, they are more challenging and controversial. They are unevenly distributed across sectors with criteria being available for agriculture, transport and energy[57] where groundwork already existed (CEC 1999a; EC, 1999, 2000; EEA, 1999a, 1999b). The development of related measures/criteria for other sectors has begun rather recently (Hertin et al., 2001). General environmental headline indicators and sustainable development indicators have been proposed also (EEA, 1999; OECD, 2001). However, all available substantive measures/criteria still have to be evaluated as to whether they are real *policy integration* criteria.[58]

The *procedural measures/criteria* represent the bulk of available EPI and PI measures/criteria. They concern mostly the conduct of the integration process (principles, prerequisites, mechanisms, procedures, instruments), deriving from theories of rational decision-making, administrative/organizational theory, and of the policy process. Their diversity owes to the fact that they have been developed

---

[56] Agenda 21, EU SDS, Article 6 of the 1999 Amsterdam Treaty, etc.

[57] Developed in the context of the Cardiff Integration Process.

[58] Persson (2002) found no substantive criteria for specific policies in his literature review.

in various contexts, from different perspectives, and in different time periods. The following is a list of relevant measures/criteria grouped into institutional, legislative, administrative, fiscal/financial, economic, technical, practical/ communication and hybrid.

Institutional measures/criteria

    Institutionalizing PI;[59] constitutional provisions for PI

    Existence of a formal overall *policy framework* for EPI/PI ('positive framing' of PI)

    Cardiff Integration Process and similar broader processes

    Establishment of required systems of rights for PI (agenda setting, participation, etc.)

    Sectoral strategies for PI (internalizing PI)

    Formulation of sectoral environmental action plan (SEAP)[60]

    Common or coordinated/compatible Action Plans[61]

    National environmental planning

Legislative measures/criteria

    Legal/institutional reforms in favour of EPI/PI

    Existence and adequacy of a legal framework and instruments for EPI/PI

    Streamlining/coordinating sectoral legislation

    Consistent, regular and monitored use of Environmental Impact Assessment (EIA), Strategic Environmental Assessment (SEA), or Sustainability Assessment in all sectoral and government policy decision

Administrative measures/criteria

    Favourable government architecture[62]

    Administrative capacity for EPI/PI; this includes:

- Organization in charge of EPI/PI; existence of central unit specifically entrusted with supervision, coordination and implementation of the integration process;
- *Horizontal administrative structures* (e.g. inter-ministerial committees and task forces, interdepartmental working groups and committees,

---

[59] The specific form depends on the political philosophy of the state – ranging from regulating PI to using persuasion and voluntary measures.

[60] Or, in general comprehensive sectoral plan accounting for relationships with other policy sectors.

[61] E.g. forest, biodiversity, desertification, transport, spatial development.

[62] Fostering *horizontal cooperation* and integration through various mechanisms and instruments; a mix of centralized and decentralized decision making; relevant questions include:

- Who has the right to elaborate a PI proposal and to set it on an interministerial agenda for negotiation and coordination?
- Who is entitled to have control regarding the procedures for coordination, the scope of the negotiation and the timing of the negotiating process?
- At what level is a decision taken to solve controversial issues?

coordination boards, networking schemes, joint responsibilities of different policy sectors, issue-specific *joint* working groups, etc.);
- Environment (or social, or economic) unit in a sectoral agency;
- Coordination units within agencies;
- Assigning existing institutions a new mandate, responsibility and accountability for EPI/PI (for within and between policy integration);
- Strengthening the Ministry of Environment and its administration with regard to procedural rights and rules relevant for the interministerial coordination and problem-solving processes;
- Officials (or, even a high-level official) charged with environment (or social or economic) tasks; 'integration correspondents';
- *Horizontal administrative procedures* (horizontal and vertical networking; early participation by environmental and other departments or agencies in decision-making, regular circulation of staff between sectoral departments, consultation processes, frequent *ad hoc* meetings and informal discussions;
- Coordination/integration of sector approval/licensing/authorization procedures, spatial planning, EIA, etc.;
- Consistent and compatible rules of decision-making in competent organizations (right to set formal agendas and develop EPI/PI proposals, relatively clear designations as to sectoral responsibility for EPI/PI goals);
- Balance of power; *joint decision making* and shared resources (budgets, staff, etc.) between environmental and sector stakeholders and authorities in sector decision-making;

 Administrative reform (restructuring) in favour of EPI/PI (e.g. merging ministries, integrating departments and functions, creating new organizations)

EPI/PI regulatory review procedures

Compliance, enforcement and accountability mechanisms for EPI/PI

Fiscal/Financial measures

Favourable (EU and national) budgetary process and reform (e.g. greening EU and national budgets; balanced budgeting)

 The Broad Economic Policy Guidelines (should) ensure that consistent policies for integration are proposed across different sectors of the economy

Flexible general taxation and special taxation

Environmental Fiscal Reform; use of fiscal incentives[63] and mechanisms

Subsidy reform[64]

---

[63] Fiscal incentives include income tax deductions for environmental investments, investment tax credit, accelerated depreciation, green loans, green funds, debt-for-nature swaps, etc.

[64] Eliminating perverse (environmentally and economically or socially harmful) subsidies.

Economic (market-based) measures

Market-based integration; use of economic instruments to internalize environmental costs of economic activity (e.g. resource and product pricing, user charges, fees, etc.)

Technical measures

Environmental management measures within sectors[65]

Agri-environmental programmes

Good Practice Codes

Integrated pollution prevention control measures

Practical and communication measures/criteria

Common (or compatible and consistent) assessment and evaluation methodologies, techniques and tools; development and use of policy integration indicators

Common data and information bases and services

Common (or compatible and consistent) monitoring programmes and infrastructure

Target-setting for PI; concrete measures, practical advice, implementation timetables and commitments to substantive results for EPI/PI

Periodic reporting of progress with respect to targets at both the general and the sectoral levels (within and between policies)

Regular, indicator-based reporting of the state of environmental, social and economic policies within the sector

Policy review procedures

Evaluation procedures

Sectoral conferences

Education and training services for civil servants, bureaucrats, etc. on EPI/PI issues

Hybrid measures

Demand management (in one sector to control its impacts on others)

Public-Private Partnerships (PPPs)

Tradable permits (for development purposes, pollution control, etc.)

Negotiated agreements

Lastly, measuring the strength of EPI/PI is not an easy task. Peters (1998) suggests two extremes of the PI continuum: (a) minimalist PI, where there is just recognition of cross-policy relationships and efforts to avoid conflicts may be made and (b) maximalist PI, where there exist regulations for coordination, cooperation, and other PI provisions, mechanisms and instruments. Metcalfe's

---

[65] E.g. use of eco-labels, Environmental Management Schemes (EMAS), etc.

(1994) policy co-ordination scale[66] is a general guide for gauging the degree of PI (OECD, 1996b; Jordan, 2002). A modified version is shown in Table 1.2 below.

Table 1.2   Policy coordination scale

Step 1: Independent decision making by state agencies (absence of EPI/PI);
Step 2: Information exchange among agencies; existence of reliable regular; communication channels;
Step 3: Two-way consultation among agencies, especially during agenda setting and policy formulation;
Step 4: Avoiding divergent positions among agencies; government 'speaks with one voice';
Step 5: Consensus seeking among agencies; joint problem solving;
Step 6: Central state mechanisms for arbitration of differences among agencies;
Step 7: Setting parameters on the discretionary power of organizations (what they must not do);
Step 8: Establishing government priorities;
Step 9: Overall government strategy (state of perfect EPI/PI).

Source: Adapted from Metcalfe (1994)

Factors Affecting the Success of EPI and PI

Drawing on theoretical and empirical evidence, the literature suggests several interdependent factors that contribute to or detract from the success of PI efforts. Although these refer mostly to EPI, several of them are relevant for PI in general. Some of them can be influenced (at least partially) through policy measures while others lie beyond the reach and ability of policy makers to influence. Several policy-related factors are reflected in the PI criteria presented previously. Selected important explanatory factors are discussed in the following, grouped into certain categories for ease of presentation, although in reality they interact dynamically in complex context, time-, sector- and policy-specific combinations. Some of them are of a local and short-term importance while others are macro-factors acting over the longer term.

Normative, ideological, and cultural factors are deep structure factors, conditioning EPI and PI success (see, Wollin, 1999). Societal values, belief systems, norms, and traditions influence first of all the perception of the problems created from lack of PI and, consequently, social acceptance and level of target group support. A tradition of cooperation and trust in government facilitates the adoption of related measures while shared values, such as environmental protection

---

[66] The scale was originally developed as a comparative tool to assess the extent to which countries are co-ordinated at a national level for effective participation at the international level. The scale was defined to compare member states of the EU.

and social equity, act as integrative mechanisms across policy sectors (Lenschow, 2002a, 2002c; Persson, 2002; Shannon, 2002).

Government and administrative culture, policy traditions[67] and preferences affect the priorities among policy goals, the adoption of the sustainability vision and of PI, and determine how they are addressed and acted upon. The degree of clientelism in government (Muller, 2002) and in state-society relationships[68] conditions the ease of pursuing PI efforts.[69] All analysts concur that political commitment and leadership remain the ultimate, critical factor for the success of EPI/PI (Peters, 1998; Hey, 2002; Lenschow, 2002a, 2002c; Muller, 2002; Persson, 2002).

These factors are, however, elusive, deeply embedded in particular locales, organizations and societies, changing through slow, social learning processes (Sabatier and Jenkins-Smith, 1999; Pressman and Wildavsky, 1992). Hence, it is difficult to trace their influence,[70] predict their impacts and manipulate them accordingly.

Legal, institutional, and organizational factors influence the structural and procedural prerequisites for the success of EPI and PI. Government architecture, the degree of centralization, the responsibilities and legal competencies of state agencies, and the breadth and distribution of formal mandates, accountabilities and resources among them determine the degree of administrative and regulatory capacity available for EPI and PI. Sectoral specialization, departmental pluralism, institutional fragmentation and insulation, narrow agency mandates and the functional differentiation of environmental and sectoral policy making, that characterize EU and national structures, present strong barriers to effective PI (Avery, 2001; Persson, 2002; Muller, 2002; Hey, 2002; Lewanski, 2002; Lenschow, 2002a, 2002c; Hertin and Berkhout, 2003). These conditions foster turf mentality, the development of vertical alliances of formal and informal actors, and keen competition among sectoral departments that hinder horizontal networking and administrative reforms necessary for policy integration.

The legal and institutional framework, defining rights, rules, mechanisms, procedures and tools related to political and administrative decisions,[71] inter-departmental coordination and collaboration[72] and resource allocation, is decisive

---

[67] Including the tradition of using knowledge and science in policy-making.

[68] Such as the 'Mediterranean syndrome' characterizing Southern European societies (La Spina and Sciortino, 1997).

[69] Shannon (2002) argues that if policies are tightly held in place by beneficiaries (interests, political alignments, and agencies), then working across policy sectors can be very difficult than if policies are loosely related to the interests of beneficiaries, the structural and ideological preferences of organizations and agencies, and the shifting alignment of political interests.

[70] As, for example, on the lack of clear objectives, quantifiable targets and timetables for individual policies and for policy integration.

[71] Such as sustainable development strategies.

[72] Such as, for example, the rights and standing to intervene in other agencies.

in the development of the required horizontal and vertical cooperative relationships and joint problem-solving mentality and capacity to address shared problems (Lenschow, 2002a; Lewanski, 2002; Hey, 2002; Shannon, 2002; EEB, 2003). Rules favouring narrow, sectoral policies compared to cross-cutting intersectoral issues, and fragmented authorization procedures have frequently impeded successful EPI and PI initiatives. Given the complexity of the undertaking and the numerous interdependent actors involved, assigning accountability for PI is another serious consideration (Peters, 1998). Similarly, the limited reach of several policy instruments, the inadequate provision and unsatisfactory implementation of integrative instruments,[73] the incomplete provision of knowledge, information, strategic assessment tools, indicators and monitoring systems all detract from the success of integration efforts.

Societal macro-factors and historical events are broader socio-political and economic influences on the acceptance, adoption and success of EPI and PI efforts. Globalization of economic and financial markets, international and EU commitments, socio-economic structure and organization, distribution of political power, and technological change may facilitate or detract from PI, acting in inconspicuous or difficult to uncover ways (Peters, 1998; Buller, 2002; Lewanski, 2002; Muller, 2002; Persson, 2002). Finally, case studies reveal that the success of PI depends decisively on sector-, policy-, context- and actor-specific characteristics, which usually act on lower levels, as well as on the gradual accumulation of knowledge on cross-policy undesirable impacts (Buller, 2002; Hey, 2002; Lenschow, 2002a, 2002c). The broad conclusion is that the success of PI provisions depends on the degree of institutional fit with the context and contingencies of their application (Lenschow, 2002b).

EPI or PI for Sustainable Development?

Integration, the centerpiece of sustainable development and a goal enshrined in several international conventions and in the Treaty of the European Union, has proven a controversial and contested concept that is interpreted variously in different contexts. Given the currently available supply of multi-level, disjointed and little coordinated institutional and administrative arrangements, policy integration appears as an absolute necessity to meet satisfactorily the demand for the management of contemporary, complex socio-environmental problems and the achievement of sustainable development. EPI, its prevailing variant, is a narrow, partial and vertical view of PI with a predominantly procedural orientation that downplays the role of several deeper and more essential preconditions in the process, assuming that, once proper structures and procedures are in place, integration will materialize. The big question then is whether a vertical orientation to PI suffices to promote sustainable development and to realize even the EPI goal, or whether a more complete interpretation of PI in both a vertical and a horizontal

---

[73] Such as EIA, SEA, budgeting, 'green' taxes and subsidies, etc.

sense, should be promoted. The ensuing discussion synthesizes and expands on the available literature to present the demand and supply factors that corroborate to make the latter interpretation the preferable option for the analysis and practice of PI.

The nature of contemporary socio-environmental problems, globalization, the Europeanization of policies, the emergence of multi-level governance, and the requirements of planning for the needs of client groups and affected regions generate demand for more encompassing, multi-level, spatially and temporally integrated policy approaches. Contemporary policy problems result from the intricate interplay of biophysical and human driving forces, cut across ecological, social, economic, administrative and political boundaries and spatio-temporal levels and transgress the functional specialization of most current political and administrative systems (Peters, 1998; Robert et al., 2001; Shannon, 2002). The systemic nature of the web of their driving forces frequently produces unexpected changes (surprises) in both the natural and the human system that are difficult to trace to specific causes lest to address through direct, single-purpose policy interventions.

Socio-economic globalization has dissolved and blurred political and institutional boundaries, has created or intensified multiple and multi-level interdependencies among people, environmental media, social and economic issues, and policies and has changed the spatio-temporal scale at which problems can be effectively addressed. National and subnational governments are rarely in a position to tackle alone and provide effective and coherent policy responses to increasingly complex and interdependent problems (OECD, 1996b; Robert et al., 2001; Hajer, 2003). Moreover, the international dimensions of several policies and membership in supra-national organizations press governments to present a coherent policy picture to the external world (OECD, 1996; Lewanski, 2002).

In the EU, the process of European integration and the Europeanization of several policy areas have increased the number of formal and informal actors involved and their interrelationships,[74] have changed the territorial reach of policy interventions and have spurred administrative restructuring (decentralization) and the devolution of decision making power to lower levels of government (Liefferink and Jordan, 2002). The state-centric, hierarchical model of governance is being replaced by multi-level forms of governance (Marks et al., 1996; Hooghe and Marks, 2001). Decision making authority and influence are shared across multiple levels necessitating horizontal cooperative relationships among formal and informal actors at all levels and participatory, joint development and use of public intervention instruments (Hajer, 2003).

These developments have increased the institutional complexity of addressing contemporary policy problems. The interdependent decisions of the numerous actors involved in both the problems and their solutions produce environmental

---

[74] The institutions of the European Union with their administrative services as well as issue or policy-based coalitions of interests (Christiansen and Piattoni, 2004).

modifications that are rarely possible to redress through the action of a single agent within a *reasonable* time frame.[75] Solutions to contemporary problems are devised and implemented in the context of polycentric, highly complex and interdependent networks of formal and informal actors, procedures and instruments, increasing accordingly the need for their co-ordination[76] (Shannon, 2002).

Regional and sub-regional planning employ resources originating in numerous policies and administering organizations[77] to address the needs of specific client groups, of lagging regions, or to promote regional sustainable development more generally. This requires that policies be organized not on the basis of function, as it has been historically the case, but on the basis of client groups or of particular regions (Peters, 1998; Hakkinen, 1999; Gibbs et al., 2003; Moss, 2004). If policies are coordinated by design, there will be fewer implementation costs and inefficiencies caused by conflicts and overlaps (Peters, 1998; Robert et al., 2001).[78]

Finally, sustainable development requires consideration of all and not only of the environmental dimensions of policy problems. Even the environmental dimensions concern not a single but a suite of policies. Because a policy is not a unitary operation, the effectiveness of the narrow, vertical, intra-sectoral EPI requires that the objects and goals of sectoral and environmental policies are congruent, policy actors communicate, and policy procedures and instruments are compatible, coordinated and non-conflicting, or even shared among policies. Suitably linking sectoral and environmental policies may achieve EPI more efficiently and effectively rather than vertically incorporating environmental considerations into sectoral policies. Therefore, both the goals of sustainable development and of EPI demand horizontal synergies among policies and more comprehensive policy integration.

On the supply side, the nature of public policy making and the performance of the current institutional and administrative apparatuses in the context of multi-level governance do not appear to serve satisfactorily the demand identified above. For one thing, policies are 'moving targets' (Wittrock and de Leon, 1986). They are not discrete, disembodied events whose impacts[79] occur in isolation from other policies (Greenberg et al., 1977). On the contrary, policy decisions are interlocked making

---

[75] Although complexity theory suggests that sometimes the action of a single agent may produce the necessary beneficial change (the butterfly effect), for practical purposes, it is not certain who this agent is, and how long it will take to produce the beneficial change for those who suffer from the problem.

[76] The role of informal governance is important to note here, as it appears to be indispensable for 'solving', under qualifications, several problems (Christiansen and Piattoni, 2004).

[77] Such as regulations, economic instruments, and financial incentives.

[78] Peters (1998) argues that policy effectiveness requires both policy integration and administrative coordination (to coordinate the many smaller, specialized departments that administrative restructuring has created).

[79] Which are more visible now than in the past because the public is better informed (Peters 1998).

it increasingly difficult for one policy area to ignore and function independently of other areas.[80] In the EU, in particular, Weale (1999) argues that environmental policy is related to other policy areas *by design* as its development was influenced, among others, by the Monnet method of European integration that encourages issue linkage and spillovers from one policy sector to the other.

Public policymaking is multifarious, variegated, following diverse styles. Policy types differ from one area to the other[81] (Richardson, 1982; Vogel, 1986; Knill and Lenschow, 2000; Moss, 2003). Historically, under the influence of the Weberian model of rational and effective public administration, policy issues have been dealt with by relatively autonomous policy sectors supported by separate administrations, leading to sectoral/functional specialization and vertical organization of administration at both the EU and the national level[82] (Robert et al., 2001; Hertin and Berkhout, 2003; Zahariadis, 2003). The result is a well-documented, general lack of coherence, coordination and cooperation among policies, that is not unique to the environmental arena (OECD, 1996a; O' Riordan and Voisey, 1998; Persson, 2002; Shannon, 2002), detracting from the achievement of sustainable development, generating costs and inefficiencies and taxing limited government budgets (Peters, 1998).

The implementation of several EU policies has revealed problems of cooperation at the level of the EU and of the member states (MS) (Robert et al., 2001). These include divergent political objectives and interests, lack of collaboration among the Directorates-General of the European Commission, lack of clear position of the Commission and different political prospects among the Community and national and sub-national actors at the stage of negotiations with MS. At the MS level (Objective 1 regions) the coordination problems are attributed, among others, to the high degree of policy and administrative sectoralization and centralization, weak coordination among administrative units and low degree of consultation of sub-national authorities. These problems have delayed negotiations or programme adoption, created tensions among actors, jeopardized sometimes the smooth operation of partnerships, barred synergy[83]

---

[80] This particularly evident at the local level where problems arise and policies are implemented eventually.

[81] Some being regulatory, others are market-oriented, still others voluntary-sector oriented.

[82] "Both national governments and the Commission are organised vertically, with Ministries or Directorates General dealing with their own portfolios, applying their own logic and methods, and taking account of interest-groups and pressures without always paying sufficient attention to the wider interests of society" (Avery, 2001, p.27).

[83] Robert et al. (2001, p.115) define "co-ordination or *synergy* as a situation whereby the sectoral policy supports the aim of territorial policy... A region with a GDP below the average EU level is likely to receive above-average Structural Funds that would raise the GDP level. A synergy with sectoral policy is then ... a situation where the average sectoral funding levels for that region is also relatively high, thus raising the GDP level for that region as well. In this case, both Community policies would contribute to economic and social cohesion". *Non-co-ordination* is defined the other way around, as a situation where the expected benefits of one policy are dampened, neutralized or off-set by another.

among programmes implemented in the same territory, and diverted funds from their original purposes to other goals, not necessarily beneficial for the regions concerned. Also, policy implementation revealed another aspect of policy integration; namely, that lack of demand management policies in the sectors that caused pollution in the first place accounts for the ineffectiveness of environmental protection policies.

The preceding discussion suggests that policy integration is needed to hold the policy system together, to overcome its entropic tendencies towards disorder, to manage the numerous policy interconnections so that policy supply meets policy demand successfully supporting the effective resolution of complex, cross-cutting socio-environmental problems,[84] such as desertification, and the transition to sustainable development. However, PI should be conceived and analyzed more broadly and thoroughly along many more dimensions than it is currently the case.

Construed as 'integration of policies', PI refers to a process of sewing together and coordinating various policies, both over (horizontally) and across (vertically) levels of governance, modifying them appropriately if necessary, to create an interlocking, hierarchical, loosely-coupled, multi-level, policy system that functions harmoniously in unity. The output of such an integration process is not necessarily an integrated policy but an *integrated policy system*, supported by participatory processes aiming to achieve multiple complementarities and synergies among policies. Although a perfectly integrated policy system may be a utopian ideal, the more policies 'talk to one another' and the right hand know what the left does, the more satisfactory will be the response of policymaking to the demands of contemporary problems. Chapter 2 negotiates the conceptual and methodological details of this task.

Introducing the Book Chapters

The contributions of this book represent modest, preliminary attempts to explore aspects of the current state and future prospects of integration among selected policies of the European Union from the perspective of various policy areas and academic disciplines. They adopt a more or less common theoretical and methodological stance to the analysis of the PI as construed above and as detailed further in Chapter 2. Secondary analysis and a limited number of interviews are the main research aids used in this early exploration because adequate and suitable empirical studies to support a full-fledged analysis of all aspects of PI, as approached in this volume, were not available. The common aim of all contributions is to examine how and to what extent PI at the EU level is conceived and approached at present to address complex policy problems and to support planning at national and sub-national levels. Mediterranean desertification has

---

[84] Avoiding the compartmentalization of otherwise indivisible socio-environmental phenomena.

guided the selection of policies considered in this volume and is used as an illustrative example in the individual chapters where appropriate and within the limits of available information. Although several procedural aspects of PI are tackled, because information was more readily available, emphasis is placed on its more essential, substantive aspects where the literature is found most deficient.

Chapter 2 focuses on the conceptual and methodological aspects of PI aiming to frame more holistically its analysis. It discusses the object and dimensions of PI (what should be integrated and how), suggests a general methodology to analyze PI, presents assessment criteria for PI and offers guidelines for designing PI schemes.

Chapter 3 negotiates the question of PI in the context of the EU regional policy. Although this is supposed to be a framework policy utilizing the instruments and means of other policies to achieve regional development goals, its current degree of integration with other EU policies can be questioned. The extent of substantive and procedural integration of EU regional policy with transport, rural development and environmental policy, three EU policies that are important to both regional development and to combating desertification, is examined. The focus is primarily on major trends that reveal the character of their present integration. The relevance of PI in the context of the EU regional policy for combating desertification in the sensitive EU Mediterranean regions is discussed also.

Chapter 4 examines the rhetoric and the reality of PI in the context of the EU rural development policy (RDP), the second pillar of the CAP, which has been heralded as an integrated policy designed to address the multifunctional development needs and the complex environmental and governance problems of rural areas. Guided by the contemporary discourse on and conceptualization of rural development as a multi-level, multi-actor and multi-faceted process, the chapter first analyzes the coherence and balance of the main instrument of the RDP, namely the Rural Development Regulation. Then it examines the integration of the RDP with other Community policies, in both a theoretical/conceptual and a procedural sense. It assesses the current state and future prospects of PI in the context of the EU RDP and negotiates the reasons behind the problems that are identified.

Chapter 5 attempts a broad brush analysis of the integration of the EU social with the economic and environmental policy areas, from both a theoretical and an applied perspective. It acknowledges that to alleviate social problems and to achieve sustainable development, policies should address the interrelatedness of the social, economic and environmental determinants of social welfare, especially in Less Favoured Areas which include the desertification-sensitive EU Mediterranean regions. The level of analysis is general because the EU does not have a coherent and comprehensive social policy yet and the linkages of the social with the economic, but especially with the environmental, dimensions of sustainable development are the least researched to date. The thematic, conceptual and procedural dimensions of the current state of PI among the three policy areas are examined. The discussion leans towards the substantive aspects of PI that are more fundamental and more problematic at present.

Chapter 6 considers the integration of the EU water policy with regional and agricultural development policies given the central importance of water in sustainable socio-spatial development in general and in combating desertification in the EU Mediterranean MS in particular. In the light of past experience and recent policy changes, it seeks to identify the present state of this integration and to assess whether it can be plausibly expected to materialize in the future and bring the desired results. Four principal aspects of PI are examined, substantive integration of policy goals and objectives, integration between actors and actor networks, and procedural integration.

Chapter 7 addresses the broad question whether synergies between the EU biodiversity and other EU policies can be achieved. In particular, it examines the relationship of the EU biodiversity with the rural development policy as biodiversity protection and combating desertification in the sensitive rural areas of the Southern EU member states primarily take place within the broader rural environment. The analysis concerns the general level of the integration of the two policies and the level of the integration of a particular rural policy instrument with the EU biodiversity policy.

Chapter 8 treats an unusual case of policy integration. Although there is no common EU forest policy at present, numerous provisions in various EU policies indirectly affect the forest sector. Guided by the ultimate goal of promoting sustainable forest management (SFM) in Europe, the central question addressed is whether it is preferable and possible to synthesize and coordinate existing EU policies into an integrated policy complex or whether it is better to design an integrated EU Common Forest Policy. The chapter assesses how well forest-related concerns are integrated in the EU rural development, biodiversity, water resources, and natural disasters policies and how this contributes to SFM in Europe. The assessment criteria concern the incorporation of SFM requirements in their objects, goals and objectives, actors, structures and procedures, and instruments. The Pan-European Criteria and Indicators for SFM proposed by the Ministerial Conferences for the Protection of Forests in Europe are used to evaluate the achievement of SFM.

Chapter 9 is devoted to a separate treatment of the most important perhaps dimension of PI, that of spatial integration, a question encountered in various guises in policy discourses, but which has not been rigorously conceptualized, articulated and addressed so far. The achievement of social and economic cohesion, the supreme goal of the European integration project, is inconceivable in a spatial void. The transition to sustainable, spatially fit, balanced, and interconnected development patterns requires the development of functional spatial synergies, cross-sectoral linkages and close co-operation among authorities from all territorial levels. The chapter provides an overview of the official interest in and the empirical evidence on the spatial integration of Community policies, explores theoretical frameworks to justify and found the spatial policy integration undertaking, reviews current proposals that address the issue, outlines broad approaches to elaborating further on the subject and highlights open questions that the political process and future research have to resolve.

Chapter 10 summarizes the preliminary findings of the individual contributions to offer an overview of the present state and future prospects of PI in terms of its principal dimensions. Then, it negotiates important questions upon which the feasibility and effectiveness of PI hinge critically, comments on the implications of the present state of PI for combating desertification and suggests future research directions.

References

Abel, T. and Stepp, J.R. (2003), 'A new ecosystems ecology for Anthropology', *Conservation Ecology*, 7(3), p.12.
  [online] URL: http://www.consecol.org/ vol7/iss3/art12.
Anderson, J. E. (1984), *Public Policy-Making*, 3rd edition, Holt Reinhart and Winston, New York.
Anderson, P., Arrow K.J., and Pines D. (eds) (1988), *The Economy as an Evolving Complex System I*, Addison-Wesley, Reading, Mass.
Ardy, B. and Begg, I. (2001), 'The European Employment Strategy: Policy integration by the back door?', Paper prepared for the ECSA 7th Biennial Conference Madison, Wisconsin, 31 May-2 June 2001.
Arthur, B. (1989), 'Competing technologies, increasing returns and lock-in by historical events', *Economic Journal*, 99, pp. 116-31.
Arthur, W.B., Durlauf, S.N., and Lane, D.A. (eds) (1997), *The Economy as an Evolving Complex System II*, Addison-Wesley, Reading, Mass.
Ascher, W. (2001), 'Coping with complexity and organizational interests in natural resources management', *Ecosystems*, 4, pp. 742-57.
Avery, G. (2001), 'Policies for an Enlarged Union. Report of Governance Group 6', *White Paper on European Governance Area no. 6*, Defining the framework for the policies needed by the Union in a longer-term perspective of 10-15 years taking account of enlargement, European Commission, Brussels.
Baas, N.A. and Emmeche, C. (1997), 'On emergence and explanation', *Intellectica*, 25, pp. 67-83.
Batty, M. and Torrens, P.M. (2001), 'Modeling Complexity: The Limits to Prediction', CASA Paper 36, Centre for Advanced Spatial Analysis, University College London (http//www.casa.ucl.ac.uk/paper36.pdf).
Berkes, F. and Folke, C. (eds) (1998), *Linking Social and Ecological Systems: Management Practices and Social Mechanisms for Building Resilience*, Cambridge University Press, Cambridge, MA.
Berkhout, F. (2002) 'Technological regimes, path dependency and the environment', *Global Environmental Change*, 12, pp. 1-4.
Blaikie, P. and Brookfield H. (1987), *Land Degradation and Society*, Routledge, London.
Braat, L.C. and van Lierop, W.F.J. (1987), *Economic-Ecological Modeling*, North-Holland, Amsterdam.

Briassoulis, H. (1986), 'Integrated Economic-Environmental-Policy Modeling at the Regional and Multiregional Levels: Methodological Characteristics and Issues', *Growth and Change*, 17(3), pp. 22-34.

Briassoulis, H. (2000), 'Analysis of land use change: Theoretical and modeling approaches', in S. Loveridge, (ed.), *Web Book of Regional Science*, Regional Research Institute, West Virginia University, (http://www.rri.wvu.edu/WebBook/Briassoulis/contents.htm).

Briassoulis, H. (2004), 'The institutional complexity of environmental policy and planning problems: The example of Mediterranean Desertification', *Journal of Environmental Planning and Management*, 47, No. 1, pp. 115-35.

Brown, M. L. (2000), 'Scientific uncertainty and learning in European Union environmental policy making', *Policy Studies Journal*, 28(3), pp. 576-96.

Buller, H. (2002), 'Integrating European Union environmental and agricultural policy' in A. Lenschow (ed.), *Environmental Policy Integration: Greening sectoral Policies in Europe*, Earthscan, London, pp. 103-26.

Byrne, D. (1998), *Complexity Theory and the Social Sciences: An Introduction*, Routledge, London.

CEC (1998), *Partnership for Integration. A Strategy for Integrating Environment into EU Policies*, Cardiff, June 1998, Communication from the Commission to the European Council, COM(98) 333, CEC, Brussels.

CEC (1999a), *Report on Environment and Integration Indicators to Helsinki Summit*. Commission Working Document, SEC(1999) 1942 final, CEC, Brussels.

CEC (1999b), *From Cardiff to Helsinki and Beyond. Report to the European Council on Integrating Environmental Concerns and Sustainable Development into Community Policies*, Commission Working Document, SEC(1999) 1941 final, CEC, Brussels.

CEC (2000), *Bringing our needs and responsibilities together - Integrating environmental issues with economic policy*, Communication from the Commission to the European Council, COM (2000) 576 final, CEC, Brussels.

CEC (2004), *Integrating environmental considerations into other policy areas- a stocktaking of the Cardiff process*, Commission Working Document, COM(2004)394 final, Brussels.

Christiansen, T. and Piattoni, S. (eds) (2004), *Informal Governance in the European Union*, Edward Elgar, Cheltenham.

Collier, U. (1994), *Energy and environment in the European Union*. Avebury, Aldershot.

Daly, H.E. (ed.) (1973), *Towards a Steady-state Economy*, W.H. Freeman and Co., San Francisco.

Dendrinos, D.S. and Mullaly, H. (1985), *Urban Evolution; Studies in the Mathematical Ecology of Cities*, Oxford University Press, Cambridge.

Detombe, D.J. (2001), 'Methodology for handling complex societal problems', *European Journal of Operational Research*, 128, pp. 227-30.

Dooley, K. (1997), 'A complex adaptive systems model of organization change', *Nonlinear Dynamics, Psychology and Life Sciences*, 1(1), pp. 69-97.

Dryzek, J.S. (1987), *Rational Ecology, Environment and Political Economy*, Blackwell Publishing, Oxford.

Eggenberger, M. and Partidario, M. (2000), 'Development of a framework to assist the integration of environmental, social and economic issues in spatial planning', *Impact Assessment and Project Appraisal*, 18(3), pp. 201-7.

EC (1999), *Integration Indicators for Energy – Data 1985-97*, European Commission, Luxembourg.

EC (2000), *Indicators for the Integration on Environmental Concerns into the Agricultural Policy-Communication from the Commission*, COM(2000)20 final, European Commission, Brussels.

EEA (1998), *Europe's environment. A second assessment*, European Environment Agency, Copenhagen.

EEA (1999), *Are we moving in the right direction? Indicators on transport and environment integration in the EU*, Executive summary, European Environment Agency, Copenhagen.

EEA (2000), *Common framework for sector-environment integration indicators*, Paper for the meeting of the EPRG expert group on indicators, 13-14 April 2000, European Environment Agency, Copenhagen.

EEB (2003), *Environmental Policy Integration (EPI): Theory and Practice in the UNECE Region*, Background Paper of the European ECO Forum for the Round Table on Environmental Policy Integration at the Fifth "Environment for Europe" Ministerial Conference, Kyiv, May 21-23, 2003, European Environmental Bureau.

Fergusson, M., Coffey, C., Wilkinson, D. and Baldock, D. (2001), *The effectiveness of EU Council integration strategies and options for carrying forward the Cardiff process*, March 2001, Institute for European Environmental Policy, London.

Finnigan, J. (2003), 'Earth System Science in the Early Anthropocene', *Global Change Newsletter*, Issue 55, pp. 8-11.

Friedrich, C.J. (1963), *Man and his Government*, McGraw-Hill, New York.

Gallagher, R. and Appenzeller, T. (1999), 'Beyond reductionism', *Science*, 284, p. 79.

Geyer, R. (2001), 'Beyond the Third Way: The Science of Complexity and the Politics of Choice', Paper prepared for the Joint Sessions of the ECPR, Grenoble, April 2001.

Gibbs, D., Jonas, A. and While, A. (2003), 'Regional Sustainable Development as a Challenge for Sectoral Policy Integration', Paper presented at Workshop II of the EU Thematic Network Project REGIONET, Regional Sustainable Development - Strategies for Effective Multi-level Governance, Lillehammer, Norway, 29-31 January 2003.

Gibson, C., Ostrom E. and Ahn, T.K. (1998), *Scaling Issues in the Social Sciences*, IHDP Working Paper No. 1,
(http://www.ihdp.uni-bonn.de/html/publications/publications.html).

Glasner, E and Weiss, H. (1993), 'Sensitive dependence on initial conditions', *Nonlinearity*, 6, pp.1067-75.

Goldsmith, E. (1972), 'A blueprint for survival', *The Ecologist*, 2, pp. 1-22.

Greenberg, G.D., Miller, J.A., Mohr, L.B. and Vladeck, B.C. (1977), 'Developing public policy theory: Perspectives from empirical research', *The American Political Science Review*, 7, pp. 1532-43.

Gunderson, L.H. and Holling, C.S. (eds) (2002), *Panarchy; Understanding Transformations in Human and Natural Systems*, Island Press, Washington, DC.

Hajer, M. (2003), 'Policy without Polity; Policy analysis and the institutional void', *Policy Sciences*, 36, pp. 175-95.

Hakkinen, L. (ed.) (1999), *Regions – Cornerstone for Sustainable Development*, Edita, Helsinki.

Harvey, D. L. and Reed, M.H. (1994), 'The evolution of dissipative social systems', *Journal of Social and Evolutionary Systems*, 17(4), pp. 371-411.

Healey, P. (1997), *Collaborative Planning: Shaping Places in Fragmented Societies*, UBC Press, Vancouver.

Henderson, B. (2001), 'Path dependence, escaping sustained yield', *Conservation Ecology*, 5(1), r3, [online] URL: http://www.consecol.org/vol5/iss1/resp3.

Hertin, J., Berkhout, F., Moll, S. and Schepelmann, P. (2001), *Indicators for Monitoring Integration of Environment and Sustainable Development in Enterprise Policy*, SPRU - Science and Technology Policy Research, University of Sussex, UK.

Hertin, J. and Berkhout, F. (2003), 'Analysing Institutional Strategies for Environmental Policy Integration: The Case of EU Enterprise Policy', *Journal of Environmental Policy and Planning*, 5(1), pp. 39-56.

Holling, C.S. (1986), 'The resilience of terrestrial ecosystems: local surprise and global change', in W.C. Clark, and R.E. Munn (eds), *Sustainable Development of the Biosphere*, Cambridge University Press, Cambridge, pp. 292-317.

Holling, C.S., Berkes, F. and Folke, C. (1998), 'Science, sustainability and resource management', in F. Berkes, and C. Folke (eds), *Linking Social and Ecological Systems: Management Practices and Social Mechanisms for Building* Resilience, Cambridge University Press, Cambridge, MA, pp. 342-62.

Holling, C.S. (2001), 'Understanding the complexity of economic, ecological and social systems', *Ecosystems*, 4, pp. 390-405.

Hooghe, L. and Marks, G. (2001), *Multi-level Governance and European Integration*, Rowman & Littlefield, Oxford.

ILO (2001), 'Policy integration in the ILO', Director-General's announcement, Circular No. 580, 19/10/2001.

Isard, W. (1972), *Ecologic- Economic Analysis for Regional Development*, Free Press, New York.

ISCID (2004), 'Dissipative Structures', ISCID Encyclopedia of Science and Philosophy, International Society for Complexity, Information and Design. (http://www.iscid.org/encyclopedia/Dissipative_Structures)

Jacob, K. and Volkery, A. (2003), 'Potentials and Limits for Policy Change Through Governmental Self-Regulation – The Case of Environmental Policy Integration', Paper presented at the 2nd ECPR-Conference, Marburg, 18-21 September 2003.

Jansen, R. (1991), *Multiobjective Decision Support for Environmental Problems*, Free University, Amsterdam.

Janssen, M., Walker, B.H., Langridge, J. and Abel, N. (2000), 'An adaptive agent model for analyzing co-evolution of management and policies in a complex rangeland system', *Ecological Modeling*, 131, pp. 249-68.

Jordan, A. (ed.) (2002a), *Environmental Policy in the European Union: Actors, Institutions and Processes*, Earthscan, London.

Jordan, A. (2002b), 'Efficient hardware and light green software: Environmental policy integration in the UK', in A. Lenschow (ed.), *Environmental Policy Integration: Greening sectoral Policies in Europe*, Earthscan, London, pp. 35-56.

Knill, C. and Lenschow, A. (eds) (2000), *Implementing EU environmental policy. New directions and old problems*, Manchester University Press, Manchester/New York.

Kosmas, C., Danalatos, N., Moustakas, N., Tsatiris, B., Kallianou, Ch. and Yassoglou, N. (1993), 'The impacts of parent material and landscape position on drought and biomass production of wheat under semi-arid conditions', *Soil Technology*, 6, pp. 337-49.

Lafferty, W.M. and Hovden, E. (2002), *Environmental Policy Integration: Towards An Analytical Framework?* PROSUS, Centre for Development and the Environment, University of Oslo, Oslo, Report 7/02.

Langton, C.G. (1990), 'Computation at the Edge of Chaos: phase transitions and emergent computation', *Physica D*, 42, 1-3, pp. 12-37.

La Spina, A. and Sciortino, G. (1993), 'Common agenda, Southern rules: European integration and environmental change in the Mediterranean State', in J.D. Liefferink, P.D. Lowe and A.P.J. Mol (eds), *European Integration and Environmental Policy*, Belhaven Press, London, pp. 217-36.

Lee, K.N. (1993), *Compass and gyroscope: Integrating science and politics for the environment*, Island Press, Washington, DC.

Lenschow, A. (ed.) (2002a), *Environmental Policy Integration: Greening sectoral Policies in Europe*, Earthscan, London.

Lenschow, A. (2002b), 'Greening the European Union: An introduction', in A. Lenschow, (ed.) *Environmental Policy Integration: Greening Sectoral Policies in Europe*, Earthscan, London, pp. 1-21.

Lenschow, A. (2002c) 'Conclusion: What are the bottlenecks and where are the opportunities for greening the European Union?', in A. Lenschow, (ed.), *Environmental Policy Integration: Greening Sectoral Policies in Europe*, Earthscan, London, pp. 219-33.

Levin, S. (1999), *Fragile Dominion: Complexity and the Commons*, Perseus Publishing, Cambridge, MA.

Lewanski, R. (2002), 'Environmental policy integration in Italy: I a green government enough? Evidence from the Italian case.', in A. Lenschow, (ed.), *Environmental Policy Integration: Greening Sectoral Policies in Europe*, Earthscan, London, pp. 78-100.

Liberatore, A. (1997), 'The integration of sustainable development objectives into EU policy-making: Barriers and prospects', in S. Baker, M. Kousis, D. Richardson and S. Young (eds), *The politics of sustainable development: Theory, policy and practice within the European Union*, Routledge, London.

Liebowitz, S. and Margolis, S.E. (1995), 'Path Dependence, Lock-In, and History', *Journal of Law, Economics, and Organization*, 11(1), pp. 205-26.

Liefferink, D. and Jordan A. (2002), 'The Europeanization of National Environmental Policy; A Comparative Analysis', University of Nijmegen, GAP, Working Paper Series 2002/14.

Limburg, K.E., O'Neill, R.V., Constanza, R. and Farber, S. (2002), 'Complex systems and valuation', *Ecological Economics*, 41, pp. 409-20.

Lowi, T.J. (1964), 'American business, public policy, case studies, and political theory', *World Politics*, 16, pp. 677-715.

Manson, S.M. (2001), 'Simplifying complexity: A review of Complexity Theory', *Geoforum*, 32, pp. 405-14.

Margaris, N., Koutsidou E., Giourga Ch., Loumou A., Theodorakis M., and Hatzitheodoridis, P. (1995), 'Managing desertification', in MEDALUS II, Project 3, Managing desertification, Commission of the European Communities, Contract Number EV5V-CT92-0165, pp. 83-110.

Marion, R. (1999), *The Edge of Organization: Chaos and Complexity Theory of Formal Social Systems*, Sage, Thousand Oaks, CA.

Marks, G., Hooghe, L. and Blank, K. (1996), 'European Integration from the 1980s: State-Centric v. Multi-level Governance', *Journal of Common Market Studies*, 34(3), pp. 341-78.

Meadows, D. (1972), *The limits to growth: A report for the Club of Rome Project on the predicament of mankind*, Universe Books, New York.

MEDACTION (2004a), *Module 4: Design of a Desertification Policy Support Framework*, Deliverables 33&34, European Commission, DG-XII, Contract No. ENVK2-CT-2000-00085, (www.icis.nl/medaction).

MEDACTION (2004b), *Module 4: Design of a Desertification Policy Support Framework*, Deliverables 36, European Commission, DG-XII, Contract No. ENVK2-CT-2000-00085, (www.icis.nl/medaction).

Metcalfe, L. (1994), 'International policy coordination and public management reform', *International Review of Administrative Sciences*, 60, pp. 271-90.

Meyer, W.B. and Turner, B.L. II (eds) (1994), *Changes in Land Use and Land Cover: A Global Perspective*, Cambridge University Press, Cambridge.

Moss, T. (2003), 'Regional Governance and the EU Water Framework Directive: a study of institutional fit, scale and interplay', Paper presented at the workshop of the EU Thematic Network project REGIONET "Regional Sustainable Development – Strategies for Effective Multi-Level Governance", 29-31 January 2003, Lillehammer, Norway.

Moss, T. (2004), 'Regional Sustainable development as a Cross Sectoral Task', Discussion Paper presented at the REGIONET workshop "Cross fertilization and integration of results of REGIONET". Brussels, 14-16 January 2004.

Nelissen, N., van der Straaten J. and Klinkers, L. (eds) (1997), *Classics in Environmental Studies*, International Books, Utrecht.

Nijkamp, P. (ed.) (1986), *Handbook of Regional and Urban Economics*, Vol. 1, Regional Economics, North-Holland, Amsterdam.

OECD (1994), *Environmental Indicators: OECD Core Set*, Organization for Economic Cooperation and Development, Paris.

OECD (1996a), *Building policy coherence: Tools and tensions*, Public Management Occasional Papers, No. 12, Organization for Economic Cooperation and Development, Paris.

OECD (1996b), *Globalization: What challenges and what opportunities for government?*, Organization for Economic Cooperation and Development, Paris.

OECD (1997), *Environmental indicators for agriculture*, OECD, Paris.

OECD (1999), *Indicators for the integration of environmental concerns into transport policies*, ENV/EPOC/SE(98)1/final, OECD, Paris.

OECD (2001), *Sustainable development: Critical issues*, OECD, Paris.

OECD (2002), *Improving policy coherence and integration for sustainable development: A checklist*, OECD, Paris.

O'Riordan, T. and Voisey, H. (1998), 'The Politics of Agenda 21', Chapter 2, in T. O' Riordan and H. Voisey (eds), *The Transition to Sustainability, The Politics of Agenda 21 in Europe*, EarthScan, London, pp. 31-56.

Ostrom, E. (1990), *Governing the Commons: The Evolution of Institutions for Collective Action*, Cambridge University Press, Cambridge.

Ostrom, E. (1998), 'Scales, Polycentricity, and Incentives: Designing Complexity to Govern Complexity', in D.G. Lakshman and J.A. McNeely (eds), *Protection of Global Biodiversity: Converging Strategies*, Duke University Press, Durham, NC., pp. 149-67.

Oxford Combined Dictionary of Current English & Modern English Usage (1982), Octopus Books Limited (First Edited by F.G and H.W.Fowler), Sixth Edition Edited by J.B. Sykes, London.

Patton, C.V. and Sawicki, D. (1986), *Basic methods of Policy Analysis and Planning*, Prentice Hall, Englewood Cliffs.

Perez-Trejo, F. (1994), *Desertification and Land Degradation in the European Mediterranean*, Commission of the European Communities.

Persson, A. (2002), *Environmental Policy Integration: An Introduction*, PINTS - Policy Integration for Sustainability, Background Paper (Draft), Stockholm Environment Institute.

Peters, G.B. (1998), 'Managing Horizontal Government: The Politics of Coordination', Canadian Centre for Management Development, Research Paper No. 21, Catalogue Number SC94-61/21-1998, ISBN 0-662-62990-6.

Poesen, J. and Lavee, H. (1994), 'Rock fragments in top soils: significance and processes', *Catena*, 23, pp. 1-28.

Poston, T. and Stewart, I. (1981), *Catastrophe Theory and its Applications*, Pitman, Boston.

Potier, M. (1997), 'Integrating environment and economy', *The OECD Observer*, June 1997, pp. 5-7.

Pressman, J.L. and Wildavsky, A. (1992), *Implementation: How Great Expectations in Washington Are Dashed in Oakland*, University of California Press, Berkeley, CA.

Prince, S.D. (2003), 'Spatial and Temporal scales for detection of desertification, in J.F. Reynolds and M. Stafford-Smith (eds), *Global Desertification: Do Humans Cause Deserts?*, Dahlem University Press, Berlin, pp. 23-40.

Puffert, D. (2003), 'Path Dependence', *EH.Net Encyclopedia*, edited by Robert Whaples, URL http://www.eh.net/encyclopedia/contents/puffert.path.dependence.php

Reynolds, J.F. and Stafford-Smith, M. (2002a), *Global Desertification: Do Humans Cause Deserts?*, Dahlem University Press, Berlin.

Reynolds, J.F. and Stafford-Smith, M. (2002b), 'Do Humans cause deserts?' in J.F. Reynolds and M. Stafford-Smith (eds), *Global Desertification: Do Humans Cause Deserts?*, Dahlem University Press, Berlin, pp. 1-22.

Richardson, J.J. (1982), *Policy styles in Western Europe*, Allen and Unwin, Hemel Hempstead.

Robert, J., Stumm, T., de Vet, J.M., Reincke, C.J., Hollanders, M. and Figueiredo, M.A. (2001), *Spatial Impacts of Community Policies and the Costs of Non-Coordination*, EC, DG-Regional Policy, ERDF Contract 99.00.27.156, Brussels.

Russell, R. and Faulkner, B. (1999), 'Movers and shakers: Chaos makers in tourism development', *Tourism Management*, 20, pp. 411-23.

Sabatier, P.A. and Jenkins-Smith, H.C. (1999), 'The Advocacy Coalition framework: An assessment', in P.A. Sabatier (ed.), *Theories of the Policy Process*, Westview Press, Boulder, CO., pp. 117-68.

Sanderson, I. (2000), 'Evaluation in complex policy systems', *Evaluation*, 6(4), pp. 433-54.

Schumpter, J.A. (1975), *Capitalism, Socialism and Democracy*. Harper, (orig. pub. 1942), New York.

Science (1999), *Complex Systems*, Vol. 284, pp. 79-109 (April 2, 1999).

Shannon, M.A. (2002), 'Theoretical Approaches to Understanding Intersectoral Policy Integration', Paper presented at the Finland COST Action meeting (European Forest Institute).

Simon, H. (1969), *The Sciences of the Artificial*, MIT Press, Cambridge, Mass.

Thomas, D.S.G. (1997), 'Science and the desertification debate', *Journal of Arid Environments*, 377, pp. 599-608.

Thomas, E. (2003), 'Sustainable development, market paradigms and policy integration', *Journal of Environmental Policy and Planning*, 5(2), pp. 201-16.

Thornes, J.B. (1988), 'Erosional equilibria under grazing', in J. Bintliff, D. Davidson and E. Grant (eds), *Conceptual Issues in Environmental Archaeology*, Edinburgh University Press, pp. 193-210.

True, J.L., Jones, B.D. and Baumgartner, F.R. (1999), 'Punctuated-equilibrium theory: Explaining stability and change in American policymaking', in P.A. Sabatier (ed.), *Theories of the Policy Process*, Westview Press, Boulder, CO., pp. 97-116.

Turner, B.L. II, Clark, C., Kates, R.W., Richards, J.F., Mathews, J.T. and Meyer, W.B. (eds) (1990), *The Earth As Transformed by Human Action: Global and Regional Changes in the Biosphere Over the Past 300 Years*, Cambridge University Press, Cambridge.

UNCCD (1994), *United Nations Convention to Combat Desertification and Drought* (www.unccd.int).

Underdal, A. (1980), 'Integrated Marine Policy: What? Why? How?', *Marine Policy*, 4(3), pp. 159-69.

UNECE (1999), *Integrating Environmental Considerations into Sectoral Policies*, United Nations Economic and Social Council, Economic Commission for Europe Committee on Environmental Policy, CEP/1999/3.

UNESCO (1995), *Poverty Eradication and Sustainable Development*, Commission on Sustainable Development, Third Session, April 11-28, 1995.

UNSO (1994), 'Poverty alleviation and land degradation in the drylands: Issues and action areas for the international convention on desertification', Paper produced by UNSO in collaboration with Roger Hay and Paul Steele (EFTEC), and Omar Noman – Food Studies Group, Queen Elisabeth House, University of Oxford.

Vogel, (1986), *National Styles of Regulation: Environmental Policy in Great Britain and the United States*, Cornell University Press, Ithaca, NY.

Waldrop, M.M. (1992), *Complexity; The Emerging Science at the Edge of Order and Chaos*, Penguin Books, London.

WCED (1987), *Our Common Future*, World Commission on Environment and Development, Oxford University Press, Oxford.

Weale, A. (1999), 'European environmental policy by stealth: The disfunctionality of functionalism?', *Environment and Planning C*, 17(1), pp. 37-51.

White, L. (2001), ' "Effective Governance" through complexity thinking and Management Science', *Systems Research and Behavioral Science*, 18, pp. 241-57.

Williams, M. (2000), 'Desertification: General debates explored through local studies', *Progress in Environmental Science*, 2(3), pp. 229-251.

Wilson, J. (2002), 'Scientific uncertainty, complex systems and the design of common-pool institutions', in E. Ostrom, T. Dietz, N. Dolsak, P. Stern, S. Stonich and E. Weber (eds),

The Drama of the Commons, Committee on the Human Dimensions of Global Change, National Academy Press, Washington, DC, pp. 327-60.

Wiman, B.L.B. (1991), 'Implications of environmental complexity for science and policy', *Global Environmental Change*, June 1991, pp. 235-47.

Wittrock, B. and de Leon, P. (1986), 'Policy as moving target: A call for conceptual realism', *Policy Studies Review*, 6(1), pp. 44-60.

Wollin, A. (1999), 'Punctuated Equilibrium: Reconciling Theory of Evolutionary and Incremental Change', *Systems Research and Behavior Science*, 16, pp. 359-67.

Yassoglou, N. (2000), *History and Development of Desertification in the Mediterranean and its Contemporary Reality*, Greek National Committee to Combat Desertification, Athens (mimeo).

Yassoglou, N. and Kosmas, C. (2000), 'Desertification in the Mediterranean Europe: A case in Greece', in *Desertification in the Mediterranean Europe*, RALA Report, No. 200.

Zahariadis, N. (1999), 'Ambiguity, time and multiple streams' in P.A. Sabatier, (ed.), *Theories of the Policy Process*, Westview Press, Boulder, CO., pp. 73-95.

Zahariadis, N. (2003), 'Complexity, coupling, and the future of European integration', *Review of Policy Research*, 20(2), pp. 285-310.

Chapter 2

Analysis of Policy Integration: Conceptual and Methodological Considerations

Helen Briassoulis

Introduction

Policy integration aims at achieving one or more goals, such as efficient and effective environmental protection, socio-economic cohesion, smooth and equitable service delivery, or sustainable development more generally. It is construed broadly either (a) as a process of incorporating certain concerns (e.g. environmental, social, economic) into an extant policy to produce an integrated policy (in a given sense) or (b) as a process of uniting and harmonizing separate policies to produce an integrated and coherent policy system. The former is a narrow and partial interpretation while the latter is broader and more comprehensive. The interpretation chosen determines critically what answers are given to the question 'integration of what, by whom, why, when and how'; i.e. the analysis of policy integration and the prescriptions offered to achieve it.

Chapter 1 argued for the second interpretation of policy integration for three main reasons. First, given the inevitably multifarious and departmentalized nature of policy making, integration of the multitudinous current policies better supports the transition to sustainable development. Second, the resolution of contemporary complex, multidimensional and cross-cutting socio-environmental problems is rarely possible through single-purpose, uni-dimensional, and uncoordinated policies or of a super policy that integrates all relevant problem dimensions. Third, in such a state of affairs, even narrower goals, such as say, environmental protection, cannot be achieved by simply incorporating environmental concerns into sectoral policies to create environmentally integrated policies because this necessarily requires the coordination of sectoral with policies associated with particular environmental issues (e.g. water, air, wastes, etc.).

The contemporary study and practice of policy integration lean heavily towards the procedural and instrumental aspects of the undertaking while rarely touching on deeper and more essential prerequisites for its success. The purpose of this chapter is to make a modest contribution towards the study of these latter prerequisites in order to frame more holistically and completely the analysis of policy integration, construed as integration of policies (or, inter-policy integration). The principal questions to be addressed are: (a) what should be integrated; i.e. the object of policy integration; (b) when policy integration should take place; i.e. at what stage of the policy process; (c) how to analyze policy integration; i.e. along which dimensions, (d) what criteria to use to assess the achievement of policy integration. Answers to the three first questions are provided in the second section of this chapter while the third section presents assessment criteria for policy integration. The last section offers suggestions for designing policy integration schemes and for future research on the subject. The rest of this section is devoted to conceptual and theoretical clarifications underlying the approach pursued.

The analysis of inter-policy integration (henceforth, PI for brevity) is approached from an institutionalist and actor-centered perspective, which is more fit and responsive to the policy and planning needs of complex socio-environmental problems (Healey, 1997; Opschoor and van der Straaten, 1993; Ostrom, 1990; Pressman and Wildavsky, 1992; Weale, 1996) and congruent with the prevailing definitions of policy that are discussed below. This perspective places emphasis on actors, their values, goals, position, resources, information-processing capabilities, the stakes they have in particular action situations, and the diverse ways through which they pursue their interests within a 'shared power world' (Healey, 1997; Ostrom, 1999). Actors belong to diverse relational webs or networks, operate within opportunity spaces structured by higher-level forces and power relations, and interact among them and with the environment, developing relational bonds of various strengths and reach. As reflective beings, human agents are not passively shaped by but they actively shape their social situation, too.

Institutions are "rules, norms and strategies adopted by individuals operating within or across organizations" (Ostrom, 1999, p.37) that mediate the relationships among actors on the same and across spatial/organizational levels. Ostrom (1990, pp.48-55) distinguishes among three hierarchical levels of rules: constitutional choice (formulation, governance), collective choice (policy-making, management) and operational choice rules (appropriation, provision, enforcement). Public policies, as collective choice institutions, have been defined as purposeful courses of action, which governments pursue to reach goals and objectives related to a problem or matter of concern and to produce certain results (Anderson, 1984; Friedrich, 1963; Lowi, 1964; Rose, 1969). They comprise not one but a long series of more-or-less related decisions and activities at each of the main stages of the policy process, agenda setting, policy formulation and legitimation, and implementation. This definition of policy focuses on what is actually done (or not done) as opposed to what is proposed or intended (Anderson, 1984; Dye, 1975; Greenberg et al., 1977). Numerous and diverse actors are involved, especially during implementation,[1] whose complex interactions introduce considerable

---

[1] Nakamura and Smallwood (1980) distinguish diverse types of policy implementers such as: policy makers, formal implementers, intermediaries, lobbies and other constituency

uncertainty in the process (Dryzek, 1987) and whose joint decisions determine the final policy outcome.

Sabatier and Jenkins-Smith (1999, pp.119-20) conceptualize public policies as belief systems because they:

> ...incorporate implicit theories about how to achieve their objectives (Pressman and Wildavsky, 1973; Majone, 1980) ... They involve value priorities, perceptions of important causal relationships, perceptions of world states (including the magnitude of the problem), and perceptions/assumptions concerning the efficacy of various policy instruments. This ability to map beliefs and policies on the same 'canvas' provides a vehicle for assessing the influence of various actors over time.

It is evident then that policy problems and policies are not things that just happen and policy instruments are not applied by themselves. The role of actors is central from the definition of the policy problem up to the implementation of particular policy measures. Consequently, the essential constituent elements of a policy are the policy object and the theories about it, the relevant actors and their goals, and the structures, procedures and instruments chosen to achieve the policy goals. This conceptualization of 'policy' is pivotal in discussing the object of PI, the dimensions along which it can be studied and the criteria for assessing the existence of integration, and the degree of its 'strength', in the next sections. For this reason, the constituent elements of a policy are first explained next.

The *policy object* concerns the facets and characteristics of a problem on which the policy focuses and the theory about the problem that the policy explicitly or implicitly endorses. Its discussion cannot be dissociated from that of policy actors who perceive the problem and participate in, or influence, its definition and resolution. This is inevitably done here for purposes of analysis.

The facets and characteristics of a problem refer to:

- its scope (who and what is involved in the problem);
- its spatial and geographical characteristics (spatial scale on which the problem manifests itself, its spatial boundaries, the spatial unit most relevant to the problem, particular geographic areas where the problem is most intense);
- its temporal characteristics (time scale over which the problem occurs, its duration, the temporal unit most relevant to the problem) and
- its social, economic, environmental, cultural and other features.

The theory about the problem provides the causal scheme that links together the chosen facets and characteristics. It comprises likely causes of the problem, its impacts and effects, and the relationships between them and with the external

---

groups (administrative lobbies, powerful individuals), policy recipients or 'consumers', the mass media and evaluators.

environment (what is not included in the problem). Obviously, depending on the viewpoints and interests of the actors involved, various theories are possible, reflecting the emphasis on selected characteristics of the policy object. Multidimensional problem definitions are associated with interdisciplinary and holistic theories.

Policy actors include individuals, firms and public, private and voluntary organizations involved, directly or indirectly, formally or informally, in various roles at various stages of the policy process. Their value systems, preferences, socio-economic status, psychological and cultural traits, position, resources and means they command, and the way they relate to particular aspects of the policy object critically determine this object. The relationships among policy actors influence when and how a problem reaches the political agenda and how it moves through the stages of the policy process. The policy networks associated with a problem are pivotal influences on the ultimate shape and fate of a policy (Börzel, 1997; König and Bräuninger, 1998).

Policy goals and objectives refer to states of the facets and characteristics of the policy object that policy actors deem desirable, partly reflecting their theory about the policy problem. The policy implementation literature suggests that rarely do the original goals and objectives of a policy materialize given the dynamic, evolving nature of the policy process (Mazmanian and Sabatier, 1983; Pressman and Wildavsky, 1992).

Policy structures and procedures refer to the particular organizational, administrative and institutional apparatuses, arrangements, and mechanisms provided for policy implementation. It is noted that formal state bodies, constituting part of policy structures, belong to the formal policy actors mentioned before.

Policy instruments include various types of means that, if properly utilized by the policy actors, help achieve the policy goals. Important among them are legal, institutional, financial, economic, technical, communication/educational, and infrastructural (physical and social) instruments. Policy actors, and their theory about the policy problem, directly influence their choice.[2] It can be argued that an indication of an internally consistent policy (as formulated by particular policy actors) is the congruence between the goals set, the policy approach adopted, and the policy structures, procedures and instruments chosen.[3]

---

[2] The particular types of policy instruments employed reflect the preferred policy style (command-and-control, direct state support, market-orientation, voluntary action, persuasion, etc.).

[3] According to Underdal (1980) consistency is one of the three criteria for a policy to qualify as integrated.

Object and Dimensions of Policy Integration

Introductory Considerations

In analyzing the integration among two or more policies, the first question that arises is "what should be integrated", i.e. the *object of policy integration*, which depends on the "when should policy integration take place".[4] A related question concerns the *dimensions* along which the object of PI can be analyzed. Policy integration can be studied in a horizontal direction, on the same spatial/organizational level, and in a vertical direction, across levels. The present discussion focuses on the horizontal direction although a complete integration scheme should consider both directions. The rest of this subsection offers a brief review of the literature while the next two subsections negotiate the object and the dimensions of PI.

The 'object of policy integration' is a summary term used here to denote the multi-part content of integration among policies that varies with the stage of the policy process (the 'when') and influences the 'how', the particular procedures and means prescribed to promote PI. The literature reviewed in Chapter 1 showed a heavy emphasis on vertical, intrasectoral integration whose aim is to incorporate environmental concerns into sectoral policies. However, it is not always clear exactly what, where and how should be integrated. The procedural aspects of vertical integration have received greater attention over the substantive aspects of the task (Lenshow, 2002; Persson, 2002). Horizontal, intersectoral integration, although increasingly recognized as the most essential form of PI, is more difficult to approach conceptually and analytically because its object is fuzzy,[5] multidimensional, multifaceted, and complex.

To obtain additional insights into the object and dimensions of the integration of policies two more studies are presented next. Among the early contributions is Liberatore's (1997) discussion of the integration of sustainable development objectives in EU policymaking that involves six dimensions shown and explained in Table 2.1.

Eggenberger and Partidario (2000) were concerned with the integration of environmental, social and economic issues in spatial planning focusing on the use of Strategic Environmental Assessment. Among the questions they raised were (a) how to enhance integration of environmental/sustainability considerations in the decision making process, (b) which sectors should be included in SEA, (c) how and when should public and private stakeholders be involved in the decision-making process (from policy to programme to plan to project) and how can

---

[4] This is a very fine point to which the literature does not pay proper attention, the frequently unspecified use of the term 'integration' being a source of confusion.

[5] Starting from the definition of what constitutes a policy, the conceptualization of the policy process and ending up with asking whether what is finally implemented on the ground bears any resemblance to the policy as formulated and legislated (Sabatier, 1999a).

(governmental) decision making be coordinated at various levels. Moreover, they identified several forms of integration shown in Table 2.2.

Table 2.1   Dimensions of the integration of sustainable development objectives in EU policy making

Dimension	Explanation
Sectors	The integration of environmental considerations into other policy sectors
Issues	Some issues transcend sectoral divisions and need to be dealt with by several sectors together (e.g. climate change and ozone layer depletion)
Space and time	Environmental integration requires paying attention to the interactions between different spaces: the *geographical space* of the affected environment, the *economic space* of the activities that have an impact on the environment, the *institutional space* of the relevant authorities and policy instruments, and the *cultural space* of values. *Time dimensions* also need to be matched to achieve sustainable development; for example, the timing of ecosystem deterioration and rehabilitation, political decision-making, scientific research, economic investments and changes in societal attitudes
Organization	The territorial competences of authorities in charge of environmental protection do not always match with the affected environment
Instruments	There is a need to broaden the range of policy instruments
Distributive elements	Burden-sharing and benefit-sharing are crucial for the operationalization and integration of sustainable development objectives into policy-making

Source: Adapted from Liberatore (1997)

Although the two studies were not concerned specifically with PI, adopted different conceptual and theoretical frames, and did not clarify several issues (Persson, 2002), they provide useful insights into the object and dimensions of PI, in both a vertical and a horizontal direction. The object of integration includes: cross-cutting issues, different features of policy problems, activities-environment interactions, policy goals, value systems, sector strategies, spatial and temporal dimensions of policy problems (global/local, different time frames), policy instruments (sector regulations), processes (sector approval/licensing, spatial planning and EIA), and information.

The object of integration can be studied along various dimensions, such as vertical (integrating environmental concerns into sectors), horizontal (issue-based and intersectoral integration), thematic/substantive (integration of social, environmental and economic dimensions, of sector strategies), spatial (integration on the same and across spatial levels), temporal, procedural (administrative jurisdictions, instruments, coordination, cooperation and subsidiarity, capacity for integration), methodological (harmonization/synthesis of different assessment methods, tools, terminologies, information sharing), and conceptual (terminologies).

Table 2.2 Forms of integration

1. Substantive	• The integration of physical or biophysical issues with social and economic issues • The integration of emerging issues such as health, risks, biodiversity, climate change and so on • The (appropriate) integration of global and local issues
2. Methodological	• The integration of environmental, economic and social (impact) assessment approaches such as cumulative assessment, risk assessment, technological assessment, cost/benefit analysis, multi-criteria analysis • The integration of the different applications, and experiences with the use of particular tools such as GIS (geographical information system) • The integration and clarification of (sector) terminologies (including the element of 'strategic')
3. Procedural	• The integration of environmental, social, economic planning/assessment, spatial planning and EIA • The integration of sector approval/licensing processes, spatial planning and EIA • The adoption of coordination, cooperation and subsidiarity as guiding principles for (governmental) planning at different levels of decision-making • The integration of affected stakeholders (public, private, NGO (non-governmental organization) in the decision-making process • The integration of professionals in a truly interdisciplinary team
4. Institutional	• The provision of capacities to cope with the emerging issues and duties • The definition of a governmental organization to ensure integration • The exchange of information and possibilities of interventions between different sectors • The definition of leading and participating agencies and their respective duties and responsibilities
5. Policy	• The integration of 'sustainable development' as overall guiding principle in planning and EIA • The integration of sector regulations • The integration of sector strategies • The timing and provisions for political interventions • Accountability of government

Source: Adapted from Eggenberger and Partidario (2000)

The Object of Policy Integration

Synthesizing the current thinking on policy integration and the conceptualization of policy presented before, it is proposed that the object of policy integration concerns *simple and cross relationships* among the objects, goals, actors, procedures and instruments of two or more policies (Figure 2.1). If these relationships reveal commonalities, shared perceptions and resources, communication and collaboration among actors, joint decision making, and coordination and/or complementarity of procedures and instruments, then a certain degree of PI can be reasonably expected to exist (see, also, Shannon, 2002). This may suggest that the policy system functions in a coordinated fashion that secures minimization of conflicts and overlaps, cost savings and greater effectiveness in producing desired policy outputs.

The main components of the object of PI are discussed below. It should be noted that the discussion concerns relationships among policies on the same spatial/organizational level, which are, however, mediated by policies from higher and lower levels as well as by non-policy factors.

Relationships among policy objects Two policies are integrated, or have chances of being integrated, if their objects have several aspects in common, are complementary, or are related in some way with respect to an issue of interest (e.g. rural development, desertification, and so on). This translates into the (two or more) policies having common scope, adopting compatible problem definitions, treating common or complementary facets of a problem situation (environmental, spatial, economic, social, institutional) in congruent or unified manner, framing environmental and other issues positively, and accommodating or respecting, to variable degrees, one another's concerns about the social, economic, environmental, cultural and other features of reality,[6] without, however, overlapping and duplicating one another. Moreover, they will most probably have similar or compatible spatial and temporal systems of reference and will consider cross-scale integration (of global and local issues).

If the above hold true, to some extent, the two policies are likely to draw on common or compatible and non-conflicting theories and epistemological frameworks reflecting common perceptions of the problem situation and common outlooks of the actors involved. This implies also that concepts are defined and operationalized similarly in all policies, providing a common language and communication code to those adopting and implementing them. The congruence of the theoretical and conceptual framings of two or more policies is perhaps the necessary precondition and the *sine qua non* for their substantive, and not only instrumental, and sustainable integration.

---

[6] For example, in the case of desertification, regional policy, rural development policy, transport policy and water resources policy should all relate to one another in this way.

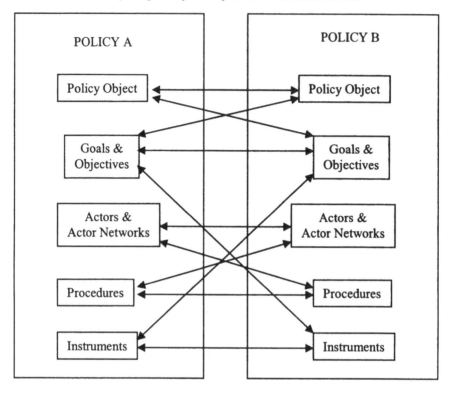

Figure 2.1   Studying the object of policy integration

Relationships among policy actors Two or more policies are, or have chances of being, integrated if the relationships among their actors take particular forms. Some actors may be common in all policies for reasons that have nothing to do with intentions to facilitate PI (such as state agencies, NGOs or other social groups routinely, formally or informally, participating in several policy fora), or they are common by design such as the integration correspondents,[7] interministerial committees, special task forces, etc. In other words, their policy networks intersect to various degrees.

The mere existence of common actors does not imply necessarily policy integration. This depends on the quality and level of communication and relationships among them. Satisfactory integration among policies can be expected if these relationships are cooperative, collaborative, non-conflicting, non-rival and non-adversarial in general, and if actors espouse similar values, share common visions, seek common goals and abide by the same rules even when these are not

---

[7] These are placed in each DG of the European Commission to liaise with DG-Environment and ensure that environmental concerns are given proper account (Lenschow, 2002).

within their organizational mandate (Shannon, 2002). This implies that their policy networks are somehow coordinated either spontaneously or by design. Note that formal rules of interaction among policy actors may be provided in the legislation even when integration is not a policy goal. These are discussed in the context of procedures below.

Besides formal, institutionalized relationships among policy actors,[8] several important relationships among formal and informal actors may be informal, providing for inconspicuous routes towards PI even when this is not required formally. These may have to do, among others, with prevailing political and policy cultures and traditional relationships among particular policy sectors. The growing literature on informal governance provides several examples of such relationships (Christiansen and Piattoni, 2004).

Several relationships among actors, contributing or detracting from PI, emerge during implementation where the stakes are clearer, many more actors are activated and winners and losers are identifiable (Pressman and Wildavsky, 1992). This is an important point because even if PI is institutionalized at higher policy making levels, it may break down at lower levels where implementation takes place and formal and informal actors may not agree with the integration idea and the related implementation arrangements.

Because actors are the critical agents of the policy system as it was discussed before, it is plausible to assume that when the objects of two or more policies exhibit commonalities this is not random or incidental. The chances are very high that the policies have common actors, with common interests and outlooks, established tradition and lines of communication and collaboration, and a genuine interest in some form of PI (not necessarily generally approved and supported). The other side of the coin is when actors are closely tied to 'favourite' policies, for ideological reasons or for the more mundane reason of reaping benefits. In this case it may be difficult to pursue efforts to integrate policies across sectors (Shannon, 2002). This may explain why the literature concludes the discussion of PI with the call for political will, cutting across all interests, or for government strategies providing a common ground for actors to comply with PI arrangements.

Relationships among policy goals   Congruent, compatible, consistent, common or complementary goals among two or more policies are favourable, necessary but not sufficient, pre-conditions for their integration. Because policy actors define both the policy object and the goals associated with it, policies may have congruent or common goals if there share common actors and/or the relationships among their actors are cooperative. When the goals and objectives of one policy consider its impacts on objects of other policies[9] or one policy is considered as a tool for the

---

[8] Usually they concern formal actors although relationships for certain informal actors have been institutionalised also.

[9] For example, the goal of an economic or of rural development policy may be avoidance of environmental degradation and/or avoidance or reduction of income inequalities; the goal of

achievement of the goals of the other, the two may exhibit some degree of integration. However, congruence at the level of goals is rather easy to achieve; note, for example, how frequently the goal of sustainable development appears in policies, because it is easier to achieve agreement on abstract statements rather than on concrete measures (Lenschow, 2002; Persson, 2002). It is conjectured also that the more multidimensional policy goals are, the more probable it may be that these are common or congruent among policies as they provide for more points of contact among them.

Relationships among policy structures and procedures Common, congruent, non-conflicting and coordinated policy structures and procedures among policies are indispensable for properly carrying out joint, cooperative and integrated solutions to a problem (or, to interdependent problems). This means that some form of horizontal linkages should exist among the organizational and administrative apparatuses of individual policies;[10] i.e. among the state agencies (formal policy actors) involved in policy formulation and implementation. Horizontal linkages are sometimes already in place at certain spatial/organizational levels which is an indication that there is a need and an interest in coordinating specific policies, for practical purposes, at least. For example, regional and local planning authorities, formulating and implementing regional and local development plans, are horizontal structures that necessarily get involved in some form of PI in the planning process. They have to make sure that observing the prescriptions of one policy does not counteract the requirements of some other(s). The same is true for agencies serving particular client groups that have to exploit and combine diverse policy resources to deliver their services (Peters, 1998). Horizontal organizational structures may be less common at higher levels where they should be created to provide joint decision making fora without which the formal, at least, conduct of joint decision making and the implementation of joint solutions may not be feasible.

Similarly, administrative procedures and other organizational arrangements and requirements for communication, joint decision making, collaboration and conflict resolution among state agencies as well as between them and non-state actors, both during policy formulation and during policy implementation, are necessary to activate the process of PI. It does not suffice, for example, to establish an interministerial committee; it should be equipped with rules and vested with power and resources to implement its decisions. Sometimes, the requirement for cooperation and collaboration among agencies on common or related issues may be already instituted, i.e. it may be included in the mandate of public agencies,

---

environmental policy may be 'achievement of environmental protection at the minimum cost'; the goal of the water policy may be the improvement in regional economic welfare and the strengthening of the rural economy; the goal of regional policy may be the protection of sensitive environmental resources; the goal of social policy may be 'avoidance of environmental disruption', and so on.

[10] It is posited here that if an individual policy is not internally consistent the chances are high that it cannot be well linked to other policies horizontally.

which means that there is interest in promoting integration among the policies concerned.[11] The same is true for informal relationships that may exist among agencies or between them and non-state actors (Christiansen and Piattoni, 2004).

The point is that by simply talking to and informing one another, actors may develop the common thinking and outlook necessary for acting jointly and coordinating their activities, thus producing integrated solutions to policy problems. This may be already happening at the local level where local actors (policy administrators and recipients) combine variously available formal with informal procedures and resources to promote particular goals. Sometimes, cooperative structures and procedures may achieve PI even when formal policy objects and goals are not integrated in the sense discussed before. This is common in 'incrementalist' politics where actors with different ideological preoccupations and goals agree on a proposal (Lindblom, 1959). This discussion points again to the pivotal role of actors in making the policy system move towards specific goals, integration in the present case.

Finally, it should be noted that both horizontal and vertical linkages among structures and procedures are necessary for effective PI. It does not suffice to integrate policies horizontally at higher spatial/organizational levels, say the European Union, if this scheme cannot be properly implemented at the lower, national and sub-national, levels because of lack of proper cross-scale linkages between policy structures and procedures. The principle of subsidiarity that applies to EU policymaking may pose difficulties because needs felt at higher levels may not be equally appreciated at lower levels where other priorities may prevail. The reverse is also true; the lack of integration felt at lower levels, during policy implementation, may not be perceived easily and directly at higher levels especially when communication is perplexed and 'noisy'. It is true, however, that if policies are not integrated when they are formulated, achieving specific forms of integration during implementation is rendered even more difficult (Peters, 1998). Policy integration may take place at lower levels as mentioned before but perhaps not for the same reasons or in the same form as contemplated and designed at higher levels.

Relationships among policy instruments The relationships among policy instruments that indicate some kind PI take various forms depending on their type, which is a function of the particular policy making style adopted. Three cases can be distinguished (a) relationships among policy instruments of the same type, (b) relationships among policy instruments of different types and (c) use of integrative instruments.[12]

---

[11] For example, the EU Water Framework Directive requires the participation and cooperation of representatives from all state agencies responsible for the various uses of water in water resources policy making and in drafting River Basin Management Plans.

[12] It is reminded that the relationships concern policies on the same and not across levels.

Policy instruments of the same type, such as legal, institutional, financial, etc., provided by different policies, which are compatible, non-conflicting, coordinated and/or complementary and mutually reinforcing, promote the goals set for the territory where they are applied eventually. For this to happen, it is reasonable to expect that the respective policy objects and goals are compatible and non-conflicting too or that one policy is used to achieve the goals of another.[13] The presence of conflicts testifies to lack of policy coordination (Robert et al., 2001). Although the testing ground of policy instrument compatibility is the level at which they are used,[14] several conflicts can be avoided at the policy formulation stage if the design of a policy's instruments takes into account the instruments of other existing policies. This may not be always possible, however, depending on how policies are designed and implemented. For example, most of EU funding for regional development purposes is provided on the basis of the Community Support Frameworks prepared by each country. It is not, therefore, straightforward to know *a priori* whether this funding will be congruent, coordinated with or complementary to funding deriving from the CAP or the Enterprise Policy or from some other policy to promote sustainable regional development. This example points to the important impact of decision and policymaking procedures on the effective coordination of policy instruments at a given level (the national and sub-national in the above example).

Policy instruments of different types, originating in different policies, which are coordinated, non-conflicting and complement each other, may promote PI and the goals this serves. Non-coordination entails costs and inefficient use of resources, leading at times to inaction. For example, legal provisions pertaining to environmental protection complemented by adequate institutional arrangements, backed by necessary funding and supported by communication and educational activities to raise awareness may have higher chances of being successfully implemented than would be the case otherwise. Financial support provided through the Structural Funds or the Cohesion Fund has frequently contradicted the legal provisions of environmental policies, indicating problematic integration between regional and environmental policy (Lenschow, 2002b). However, because the scheme for combining instruments originating in different policies, i.e. the policy instrument mix, as well as the success of its application, are context-dependent,[15] providing satisfactorily coordinated policy instruments by design at higher levels appears to be a very demanding, if not infeasible, task as actual integration materializes at lower levels. The important role of appropriate procedures to guide the combination of instruments becomes evident in this case too.

---

[13] For example, biodiversity protection regulations may serve well water- or soil-related goals and vice versa; or, financing transportation projects may promote regional development goals, etc.

[14] Which is not necessarily the local level because different instruments apply to different spatial/organizational levels.

[15] Depending, among others, on the particular purpose served, the level of implementation and the actors involved.

Use of integrative instruments is not necessarily an indication of PI. For example, the requirement for EIA or SEA, the more widely utilized horizontal environmental policy instrument, in transport or in regional policy complies more with the notion of vertical, intrasectoral environmental policy integration (EPI) rather than with the notion of horizontal policy integration. The same may apply to environmental policies (e.g. water resources) that adopt economic instruments whose purpose is to incorporate economic efficiency considerations in non-economic policies and to create markets for environmental goods and services rather than to integrate environmental with economic policies. At best, integrative instruments assist the integration between policies *indirectly* by (a) modifying their object to include concerns of other policies, (b) inducing their harmonization on theoretical grounds; e.g. environmental policies adopt the neoclassical economics theoretical paradigm and the goal of efficient resource use, (c) promoting the development and use of integrated assessment methods and of related data sets.

The limited contribution of integrative instruments to integration among policies derives also from the possibility that their use may be asymmetric. For example, environmental policies may adopt economic instruments while economic policies usually do not account for the economic value of environmental goods and services used in production and consumption. It follows that integrative instruments can promote PI only when their use is reciprocal. Moreover, the fact remains that reliance on integrative instruments is not a long-term solution to the integration of policies.

In sum, policy instruments are not neutral objects that can be objectively combined to produce automatically policy cooperation and coordination. On the contrary, prescribed instruments and the rules of their use (implementation procedures) reflect the definition of the policy object and the associated goals of the actors who have participated in policy formulation. If the latter are dissociated, then the relationships among policy instruments will suffer. The provision of integrative instruments may be a technical solution that does not remove the fundamental requirement of encouraging, facilitating and promoting proper relationships among policy objects, goals and actors primarily. Moreover, during policy implementation, users combine instruments to help them achieve their ends (i.e. solve the problem *they* perceive) in ways that are context-dependent and can neither be anticipated not specified precisely and unambiguously during policy formulation. It can be argued that policy integration *emerges* during implementation and as such it cannot be prescribed *a priori*.

Lastly, the variability of the object of PI with the stage of the policy process is briefly tackled. Although the 'stages heuristic' framework for understanding the policy process has been strongly questioned (Sabatier, 1999b), a broad distinction between agenda setting, policy formulation and policy implementation is still meaningful for EU policies, at least. All components of the object of PI are relevant at all stages but integration among policy objects and goals is more relevant and essential at the first two stages while integration among policy instruments is critical at the implementation stage. Integration among policy actors is relevant throughout the process, something well understood in EU policy circles that take steps to encourage it (see, e.g., EC, 2004). Nevertheless, all constituents of a policy change in the course of its development, necessitating, first, a constant occupation with integration among policy objects, goals, actors, procedures and instruments and, second, an unending two-way communication between formulation and implementation to facilitate as comprehensive PI as possible and feasible.

Dimensions of Policy Integration

The preceding discussion suggests that the object of policy integration has several facets that should be analyzed not only in functional and procedural terms, as it is mostly the case at present,[16] but more deeply, in substantive terms. Four broad interrelated and interdependent clusters of dimensions of PI – substantive, analytical, procedural, and practical – are discussed in the following to expand on the content and critical determinants of PI and to inform its operationalization. Table 2.3 presents the contents of each cluster.

Substantive dimensions of policy integration The substantive cluster encompasses the thematic, conceptual, and value dimensions that concern the constitution of the policy objects to be integrated. The thematic dimension negotiates the relationships among the objects of separate policies, each embodying selected characteristics of an issue – environmental, social, economic, cultural, political, and other. The term 'thematic' may be equivalent to the term 'sectoral' encountered in the literature, where the latter refers to policy sectors. The thematic integration requirement is embedded partly in the vertical environmental policy integration concept (VEPI) (Lafferty and Hovden, 2002). However, VEPI has been framed in narrow (only environmental), vague and procedural rather than broad, concrete and substantive terms thus far, leaving untouched the issue of the exact form of relationships that should be satisfied.

The need for thematic integration draws directly from the sustainable development imperative that emphasizes the interrelatedness and interdependence among all aspects of a problem because reality is unitary and integral disintegrated only during analysis and sectoralized policy making and administration. Sustainable development requires that a (sectoral) policy accounts for the relationships of the characteristics of a problem it concerns with those covered by other policies.[17] In this sense, thematic integration is not limited to environmental

---

[16] The vertical and horizontal dimensions discussed in the literature, in addition to be being unclear, overlapping and variously interpreted, are rather gross simplifications of the policy integration theme. They may be better considered as two directions along which PI can be analyzed and conducted.

[17] This is the spirit of the 'multi-functional use of space' and the promotion of 'multi-functional landscape' approaches in planning (ESDP, 1999). Any analysis of a policy

concerns but extends to relationships between economic sectors also; for example, agriculture and transport, transport and industry, etc. The development of harmonious relationships among policies associated with specific economic sectors and activities will produce fewer or no negative externalities among sectors and multiple benefits in the form of economies of scale, agglomeration, etc.

Table 2.3   Proposed clusters of dimensions of policy integration

Cluster	Components
Substantive	Thematic
	Conceptual
	Value
Analytical	Spatial
	Temporal
	Methodological
Procedural	Structural
	Procedural
Practical	Practical

It is conjectured that multidimensional policy objects and comprehensive policy problem definitions facilitate the thematic integration among policies especially if these draw on common theories about a problem in question. In fact, integrated, interdisciplinary theories, which overcome the fragmentation of reality, the compartmentalization of space and spatial development, and the separate treatment of interlinked activities, which is associated with uni-disciplinary theories, should indicate which relationships among the characteristics of a problem make sense and should be addressed by respective policies. These should constitute the substantive basis of their integration. If this is the case, the integration among policy goals and policy actors will follow naturally.

Intricately related to the thematic, but acting on deeper levels, are the conceptual and value dimensions of PI. The need for conceptual integration arises as different policies frequently define similar terms and concepts differently owing to their different origins and organizational contexts. Terms and nomenclature used may be overtly similar but are construed differently from one policy context to the other. Examples abound including definitions of 'rural',[18] 'social impacts',[19] 'spatial impact',[20] 'region', 'integrated', 'forest', 'pasture', and so on. Frequently, also, the terms and nomenclature used are not defined in the policy or they are

---

problem can demonstrate the close ties between the cultural and institutional context within which a problem arises and its socio-economic and environmental characteristics.

[18] Usually, interpreted narrowly as agricultural and in an economic sense.

[19] Usually, identified narrowly as employment and income impacts.

[20] Being vaguely defined and varying widely among policies and analyses.

defined loosely, a fact that opens the way for a multiplicity of interpretations (and mis-interpretations).

The lack of a common vocabulary, common understanding and shared meanings creates communication problems among policies that result in conceptual confusion, carrying over to differences in the operational expressions and analysis of similar concepts and, consequently, in the policy decisions that are based on them (policy instruments, implementation structures and mechanisms, etc.).

The conceptual integration problem is rooted in socio-culturally determined differences in the value systems of actors in the same or different policy contexts. When values clash, value integration is impossible, and disciplinary, sectoral, organizational, and other biases inevitably influence the ways in which policy problems are defined and policies are designed and implemented. Essential PI presupposes a *common frame of mind*, common interest in and sense of collective responsibility both for the causes and the solution of a policy problem among all those involved or, at least, a positive predisposition towards cooperation in finding solutions for the 'common good'. From the point of view of its analysis, common value systems among the actors involved are manifestations or indications of some form of, sometimes subtle and inconspicuous, policy integration or of its likelihood. The PI issue essentially boils down to a collective action problem that can be resolved only if certain conditions are met!

The combat against desertification is an excellent example of a complex policy problem demanding the coordination of several policies at various spatial levels (see Chapter 1 this volume). The substantive integration of the policies implicated is indispensable for effective solutions to the problem. However, such prospects are dim because at present a basis for thematic, conceptual and value integration does not seem to exist. An integrated conception of the phenomenon is missing as it has been approached and analyzed from different, narrow disciplinary camps and traditions, with natural sciences-based approaches dominating social sciences-based or integrated approaches. Not surprisingly, policy prescriptions are similarly colored and biased, with most of them concerning the environmental dimensions of desertification despite the recent emphasis on its socio-economic and cultural dimensions. The connections between all dimensions are still thin and not well-elaborated, especially on a spatial basis, exposing the lack of a common substantive basis for PI.

The substantive dimensions of PI bear direct relationships to its analytical, procedural and practical dimensions as the following discussion demonstrates.

Analytical dimensions of policy integration The analysis of the object of PI entails spatial, temporal and methodological considerations whose operational expressions depend on the substantive framing of PI. In a horizontal sense, the spatial dimension of PI concerns the congruence among the spatial systems of reference of different policies on the same level; i.e. the spatial classification schemes, spatial units of reference and criteria for the delineation of the spatial areas they adopt.

Liberatore (1997) suggested several different 'spaces'[21] noting also that "the territorial competences of authorities in charge of environmental protection do not always match with the affected environment" (Liberatore, 1997, p.117). This is the commonly recognized problem of incongruent spatial references between environmental and socio-economic policies. Environmental policies make sense for ecological units while socio-economic policies apply to administrative units which, however, are not necessarily suitable for addressing cross-cutting socioeconomic problems.[22]

In a vertical sense, the spatial dimension concerns the cross-level relationships among the administrative agencies involved in the various stages of policy making, among non-state actors implicated in the process and between state and non-state actors. Lack of communication, cooperation and coordination among those who define the policy problem and formulate the relevant policy at higher levels, those charged with implementation and the ultimate policy recipients (e.g. land owners, farmers, etc.) at lower levels contribute to policy failures. Another facet of vertical integration concerns the influence of policies operating at different spatial/organizational levels on the determinants of local level decision making. For example, macro-economic policies influence agricultural input and product prices, water policies regulate water use, etc. all of which influence importantly the land use decisions of individual owners.

In the same vein, the temporal dimension of PI concerns the congruence among the implicit or explicit temporal systems of reference, i.e. temporal units, time intervals, time horizons, timing of actions that policies adopt (cf. Liberatore, 1997). Frequently, the real time frame of reference of policies coincides with electoral cycles, lasting circa four years (OECD, 1996), which is not compatible with the temporal frame of reference of most social and environmental phenomena and problems. Lack of temporal integration often leads to ineffective and wasteful policy interventions as rarely do policies deliver results in isolation.

The methodological dimensions of PI concern the relationships of the methods and techniques for policy problem analysis associated with different policies. Different and incompatible methods, owing to epistemological and theoretical differences among policies, lead to incompatible or even conflicting, definitions and proposed solutions of the policy problem. Analytical techniques used in one policy context (e.g. cost-benefit analysis), or appropriate for a particular spatial level, fail when transferred to other policy contexts and spatial levels.[23] Related issues concern congruence or commonalities of methods and techniques in terms of spatial and temporal specification, terminology, and data and information collection procedures (Briassoulis, 2001). Multidimensional, comprehensive and

---

[21] The *geographical space* of the affected environment, the *economic space of the activities* that have an impact on the environment, the *institutional space of the relevant authorities and policy instruments*, and the *cultural space of values* (Liberatore, 1997).

[22] Environmentally integrated units, such as watersheds, have the potential, however, to become proper administrative units and facilitate the resolution of this problem.

[23] For example, macro-models applied to micro-level issues and vice-versa.

integrated policy problem definitions and theories, in support of substantive integration, dictate the use of integrated, multi-dimensional (in contrast to uni-dimensional) methods and techniques at all stages of policy making.

Procedural dimensions of policy integration   These concern the structural and procedural aspects of the relationships among policies that constitute the means through which PI materializes.[24] These have been covered in previous sections of this Chapter and in Chapter 1. Organizational arrangements and procedures that facilitate, accommodate and secure communication and cooperation among policy actors, participation in joint decision making as well as congruent and coordinated use of policy instruments on the same and across levels are indications of, at least, *intent* for PI. Important, if not fundamental, among them is the existence, compatibility and congruence of resource rights for all resources (e.g. land, water, labour, etc.) in all policies that are preconditions for rational resource management (Bromley, 1991). Similarly, rights to participate in decision making within and between policy areas for all types of actors are essential procedural requirements for policy integration.[25]

Two points are important in studying the procedural dimensions of PI; first, their relationships with the substantive and analytical dimensions and, second, the internal consistency and effectiveness of the procedural arrangements provided. The fact that structures and procedures exist does not automatically imply their suitability for all cases of PI, their adoption and implementation, and their effectiveness with respect to the goals of PI. Ideally, the procedural arrangements for PI should result from substantive and analytical considerations and should be such that they help realize common goals and objectives (cf. Lenschow, 2002). However, because of the context-specificity of policy problems, this ideal linkage between substance and process is difficult to achieve in most cases. Even if a perfect policy integration scheme is designed at higher spatial/organizational levels, during implementation at lower levels substantive and institutional misfits will arise that are difficult to presage at the formulation stage because policy problems are complex (see Chapter 1).

Therefore, if the procedural arrangements provided do not fit the substantive and analytical aspects of policy problems and are not accompanied[26] by integration in terms of policy objects, goals, values and actors, PI will hardly materialize. The reason is simple: who will utilize the structures and who will implement the procedures if none perceives the need, nobody is interested in, or, more realistically, no one wants policy integration? It follows that documenting PI in procedural terms does not suffice; evidence of its effectiveness is needed, in other

---

[24] Most of the literature focuses on procedural integration issues (although for the narrow case of vertical environmental policy integration).

[25] Of course, participation can take several forms and informal actor networks frequently are more effective in linking policies than formal actor networks.

[26] Or, better, preceded.

words, whether the policy system functions coherently, coordination occurs, and produces benefits while it minimizes costs. Of course, the danger here is that the functioning of the policy system may be spurious, owing not to the integration arrangements provided but to external, non-policy factors. Only a longitudinal study, based on reliable information and multiple methodologies, may perhaps elucidate the question of the effectiveness of procedural provisions for PI.

Table 2.4   Relationships between object and dimensions of policy integration

		Policy Object	Policy Goals	Policy Actors	Policy Structures & Procedures	Policy Instruments
Substantive	Thematic	X	X			
	Conceptual	X	X			
	Value	X	X			
Analytical	Spatial	X		X	X	X
	Temporal	X		X	X	X
	Methodological	X		X		
Procedural	Structural	X		X	X	X
	Procedural	X			X	X
Practical	Practical	X		X	X	X

Practical dimensions of policy integration   These concern the plethora of practical issues related to availability, compatibility, consistency and congruence of data and information needed to analyze properly the object of PI. They draw directly from the associated substantive and analytical considerations. Without spatially, temporally and conceptually integrated data and information systems integrated analysis of policy problems is not feasible and, consequently, limited essential analytical support is provided for the design of policy integration schemes (Briassoulis, 2001).

In closing, it is conjectured that there is a certain correspondence between the object and the dimensions of PI as Table 2.4 indicates. Nevertheless, because all the components of the object of policy integration are interrelated, a complete analysis requires examination of all of them along the appropriate dimensions.

Criteria for Analyzing Policy Integration

Based on the conceptualization of the object and dimensions of PI and drawing on the pertinent literature,[27] this section suggests criteria for assessing whether integration among two or more policies already exists and for proposing how it can

---

[27] Although the literature concerns environmental policy integration mostly, the criteria can be generalized to policy integration because of their predominantly procedural orientation.

be achieved or improved. The proposed criteria are organized according to the components of the object of PI and they reflect the respective dimensions. A category of general, cross-cutting criteria is included also that can be considered broadly as enabling conditions for the realization of PI.

The elusive and variable character of policies and of their integration, together with practical issues of data and information availability, preclude any definite determination of criteria to assess the degree of PI achieved in a given situation. The simple proposal made here is that the more criteria are satisfied the higher will be the achievement of policy integration. Ideally, if all criteria are satisfied, then policies are perfectly integrated. Depending on the particular criteria being satisfied, the kind of PI achieved can be gauged; i.e. whether it is substantive, analytical, procedural or practical. Lastly, those criteria that are not satisfied offer a basis for proposing what should be done to promote integration between the policies examined.

General Criteria

- Political commitment and leadership for PI in general;[28]
- Shared core belief systems across policy sectors;[29]
- Need for compliance with international and EU commitments that require PI;[30]
- Existence of an official long term Sustainable Development Strategy (or a relevant Report or Forum);
- Existence of a formal policy framework for PI in general;
- The environmental, social, economic agendas of different sectors form a consistent overall strategy (perhaps guided and coordinated by a Sustainable Development Strategy);
- Existence of favourable government architecture, fostering *horizontal cooperation* and integration through various mechanisms and instruments; a mix of centralized and decentralized decision making;[31]
- Favourable policy tradition and administrative culture (open, participatory, flexible);
- Favourable budgetary process (e.g. for 'greening' budgets);
- Environmental and/or Social Fiscal Reform;[32]

---

[28] This is the ultimate, and less easy to meet, criterion according to the literature.

[29] I.e. absence of narrow sectoral mentality and value orientation, awareness of cross-cutting issues, etc.

[30] For example, adherence to Agenda 21, EU Sustainable Development Strategy, Article 6 of the Amsterdam Treaty, etc.

[31] Relevant questions here are:

- Who has the right to elaborate a proposal and to set it on the interministerial agenda for negotiation and coordination?
- Who is entitled to have control regarding the procedures for coordination and the scope and timing of the negotiating process?
- At what level is a decision taken to solve controversial issues?

- Flexible general taxation;[33]
- Concrete implementation timetables for the integration among specific policies;
- Concrete measures, practical advice, timetables and commitments to substantive results for the integration among specific policies.

Criteria Related to the Policy Object

- Congruent, compatible, consistent and multidimensional policy objects and related or common integrated/interdisciplinary theories;
- Common and consistent concepts and terminologies.

Criteria Related to Policy Goals and Objectives

- Political commitment and leadership for PI in the case of the policies analyzed;[34]
- Common, shared, congruent, compatible and/or complementary policy goals and objectives;
- Stipulation of quantitative, measurable, indicator-based targets and timetables for PI.[35]

Criteria Related to Policy Actors

- Common formal actors at various spatial levels;[36]
- Common informal actors at various spatial levels.

Criteria Related to Policy Structures and Procedures

- Administrative capacity for PI; it concerns, among others:
  - Organization in charge of PI; such as, existence of a central unit entrusted with supervision, coordination and implementation of the integration process, or assigning existing institutions a new mandate, responsibility and accountability for PI;
  - Special unit for PI in the competent organization;

---

[32] I.e. systematic use of economic and fiscal instruments in various sectors to achieve environmental and social objectives.

[33] To allow, for example, tax differentiation schemes in favour of eco-products, development control in targeted, protected locations, such as desertification-prone areas, etc.

[34] I.e. adoption of PI as a policy objective (in either one or both policies), requirement for coordination with other policies, integration of sectoral strategies, etc.

[35] Included, for example, in their sectoral strategies.

[36] This means, among others, that the legislation prescribes that actors of one policy participate in decision making in another policy. 'Integration correspondents' is a particular case of such actors. In other words, the two policy actor networks intersect.

- Officials charged with integration tasks;
- Administrative reform (restructuring) in favour of PI;[37]
- Presence of *horizontal administrative structures* as opposed to vertical and departmentalized structures; e.g. inter-ministerial committees and task forces, issue-specific *joint* working groups, networking schemes, regular circulation of staff between sectoral departments.

- Formal/institutionalized interaction[38] among policy actors and actor networks;
- Informal interaction among policy actors and actor networks;[39]
- Interaction among formal and informal[40] policy actors;[41]
- Consistent, compatible and coordinated procedures and rules of decision-making in competent administrative bodies;[42]
- Strengthening existing administrative units with regard to procedural rights and rules relevant for coordination and *joint* problem-solving;
- *Joint decision making and joint responsibilities* of the policy sectors considered;[43]
- Provisions for implementing PI requirements;[44]
- Absence of sectoral compartmentalization, vertical alliances,[45] departmental pluralism and competition between competent organizations;[46]
- Intra-departmental power relations hindering PI.

Criteria Related to Policy Instruments

- Instruments used by different policies are compatible and consistent;

---

[37] For example, merging ministries, integrating departments and functions, creating new organizations, etc.

[38] This means (a) communication, consultation, routine early consultation on sector policies and projects, cooperation, coordination and collaboration in implementation, etc., (b) policy formulation actors interact formally with policy implementation actors and vice versa, (c) some form of interaction among actors at all spatial levels (that may correspond to the stages in the policy process).

[39] E.g. *ad hoc* meetings and informal discussions and consultations, environment (or other issue) is a regular agenda item in high-level meetings.

[40] Informal actors include: stakeholders, professionals, bureaucrats, academics, NGOs.

[41] For example, goal-related consultation and participation processes among formal and informal (strategic?) policy actors (from agenda setting to policy implementation); partnerships between government and business on cross-cutting issues.

[42] It includes the right to set formal agendas and develop policy integration proposals, early participation by departments or agencies in decision-making of other policy sectors, coordinated authorization procedures, coordination/integration of sector approval/licensing processes, spatial planning, EIA, regulatory review procedures, evaluation procedures.

[43] Includes relatively clear designations as to responsibility for policy integration goals, common provisions, shared resources, etc.

[44] They include compliance, enforcement and accountability mechanisms for PI among competent agencies.

[45] Hindering horizontal networking.

[46] I.e. turf mentality.

- Use of one policy as an instrument to achieve the goals of another policy;
- Use of integrative instruments; such as, legal, economic, financial;
- Existence of a legal framework for PI among the policies analyzed;[47]
- Common legal and institutional instruments;[48]
- Compatible, consistent and coordinated legal and institutional instruments;[49]
- Market-based integration between the two policies;[50]
- Use of financial mechanisms, such as, subsidies for PI;[51]
- Use of fiscal incentives for PI;[52]
- Balanced budgeting;[53]
- Common or coordinated/compatible sector Action Plans[54] (e.g. forest, biodiversity, desertification, transport, spatial development);
- Use of planning and management instruments for PI;[55]
- Common, shared research resources;[56]
- Common, or compatible and consistent, data and information bases;
- Common assessment and evaluation methodologies, and tools (e.g. policy integration indicators);
- Common monitoring programmes and infrastructure;
- Use of communication instruments for PI;[57]

---

[47] I.e. institutionalizing PI among policies. The specific form depends on state political philosophy, ranging from regulating PI to using persuasion and voluntary measures. It includes legal/institutional reforms in favour of PI (filling gaps in legislation), administrative rights (rights and standing to intervene in other administrative bodies), etc.

[48] Such as EIA, SEA and other provisions common in both policies; clear *and common* systems of resource and other rights shared by all agencies involved (to facilitate rational decision making and the use of market-based instruments).

[49] For example, (a) use of legal provisions of one policy as an instrument to achieve goals of the other policy – e.g. Good Agricultural Practice codes used in rural policy to achieve water protection goals of the EWFD, (b) adequacy of existing legal instruments as regards PI, (c) elimination of inconsistencies, duplications, conflicts among policies, (d) use of voluntary measures among sectors (e.g. negotiated agreements, pacts, etc.).

[50] This means (a) use of economic instruments to internalize the environmental costs of economic activities (e.g. resource pricing, environmental taxation, charges, water user fees to internalize environmental costs of agricultural activity or of regional development projects funded through the SFs), (b) economic instruments for behavior change (use of water in agriculture) and *not* for revenue raising.

[51] For example, (a) subsidy reform, i.e. removing perverse subsidies to reduce threats to water resources, (b) new subsidies (e.g. from DF-Agri or DG-Regio) to induce protection of water resources; or, use of, e.g., green public procurement, environmental liability, etc.

[52] E.g. tax relief in agriculture for the adoption of water resources protection measures, income tax deductions for environmental investments, investment tax credit, accelerated depreciation, green loans, green funds, debt-for-nature swaps, etc.

[53] Balanced distribution of financial resources among sectors.

[54] Comprehensive sectoral plan accounting for relationships with other policy sectors.

[55] Such as rural development plans, water resources management measures within sector (e.g. agriculture, SF funding), demand management measures.

[56] E.g. personnel, research centers, budgets, etc.

- Education and training services for civil servants, bureaucrats, etc. on PI issues.

Concluding Remarks – Designing Policy Integration Schemes

The need for policy integration does not appear out of the blue and does not occur in the abstract or in a vacuum. There is always an impetus for asking policies to 'talk to one another', take into account one another and act in coordination and harmony (OECD, 1996a). The sustainable development rhetoric has provided the broad backdrop of the contemporary discourse and analysis of the subject although several other interrelated issues emerged over time corroborating for policy integration. Given the documented quest for PI, this chapter focused on conceptual and methodological considerations to frame the analysis of the integration among policies on the same spatial/organizational level, an aspect of the broader theme of PI that the pertinent literature has not covered adequately yet. This closing section offers some preliminary thoughts regarding the design of policy integration schemes (PIS for brevity) and summarizes important questions that future research should address to advance the analysis of policy integration.

The following general considerations are pertinent to and influence the choice of appropriate approaches, factors of interest, and instruments for the design of effective PIS:

- Is a general, all-purpose and all-encompassing PIS possible and desirable or case- or issue-specific PIS should be developed?
- Is inter-policy (horizontal) integration sufficient to tackle crosscutting issues or intra-policy (vertical) integration is necessary too, or both?
- Is PI at a given level sufficient or is cross-level PI necessary, too, or even a grand scheme of full-blown integration on and across levels?

The ideal situation might be to design a grand PIS of horizontally and vertically integrated policies, on the same and across levels, which can accommodate all possible cases of crosscutting issues. However, the complex nature of these issues renders this ideal option utopian and infeasible (OECD, 1996a) and requires more flexible approaches that can be adapted to the particularities of each case. Before proceeding to examine one such approach, two points are in order.

First, the connectedness of human-environment systems implies that policies or PIS inevitably have positive or negative spill-over effects. In other words, a policy or a PIS designed to address a given issue provides for arrangements that may also

---

[57] E.g. encourage the use of consensus-finding procedures; use of eco-labels, Environmental Management Schemes (EMAS), etc., regular reporting of the state of water resources within the sector, e.g. agriculture (use of specific indicators), periodic reporting of progress with respect to targets at both central and sectoral levels (within and between policies).

address other issues. Of course, it is possible that a scheme suitable for one problem may prove unsuitable for another although only empirical research can support this claim. In addition, theoretically at least, some policies may be used as instruments for the achievement of the goals of other policies, as it is the case of transport policy potentially facilitating the promotion of regional development goals.

In any event, it is plausible to assume that not all possible cases of crosscutting issues need to (or, can) be considered but only a few strategic ones. This is the second point supported by Lindblom's statement:

> Clearly, everything is connected. But because everything *is* connected, it is beyond our capacity to manipulate variables comprehensively. Because everything is interconnected, the whole environmental problem is beyond our capacity to control in one unified policy. We have to find ... tactically defensible or strategically defensible points of intervention (Lindblom, 1973, pp. 11-34).

Therefore, the task is to find those strategic policies that should be integrated at each level so as to provide an enabling environment for the integration of policies on the same and on other levels.

The proposed approach builds on the adaptive management paradigm that has been developed to deal exactly with complex systems, the unpredictable interactions between people and ecosystems as they co-evolve, and to integrate *uncertainty* into the decision-making process. Initially developed to study the dynamics of ecosystems (Holling, 1978), it has also been applied to study the dynamics of linked social and natural systems (Berkes and Folke, 1998). Adaptive management can be viewed as an approach to managing risks associated with uncertainty based on learning-by-doing and experimentation. Resource management policies are considered as hypotheses, and management as experiments from which managers learn from their successes as well as from their failures. It differs from conventional resource management because it stresses the importance of two-way feedback between management and the state of the resource in shaping policy, followed by further systematic experimentation to shape subsequent policy, and so on. Its flexible, iterative, co-evolutionary and science-based character allows for *institutional learning*; i.e. changing management institutions to fit the nature of the system being managed (Berkes and Folke, 1998).

The evolving, dynamic character of policy objects and of policy implementation and the resulting uncertainty of policy outcomes justify the adoption of this paradigm for designing PIS. Transferring its main ideas to the present case, the basic tenet of the proposed approach is that any PIS is a hypothesis to be scientifically tested on the ground and revised, through participatory approaches, by incorporating systematically collected information obtained from implementation. The policies to be integrated at a given level should ideally relate to critical, influential, strategic factors associated with important crosscutting issues and the sustainability of human-environment systems on and across spatial levels.[58] In addition, given the overt and covert interconnectedness among policies, PI may start from the most instrumental and pivotal policy, orchestrating all others around it (assuming that the associated policy interests are willing to cooperate!), some of which inevitably will originate in higher or lower levels. The main steps of the approach include:

- Design a PIS following a systematic and participatory approach. The suggested lines of analysis of the object and dimensions of PI as well as the criteria presented in this chapter may be used to elaborate proper linkages among policy objects, goals, actors, procedures and instruments on and across relevant spatial levels;
- Design a monitoring and evaluation scheme to gather data to address key uncertainties;
- Implementation of the PIS;
- Systematic monitoring of all aspects of implementation and recording problems such as overlaps, conflicts, inconsistencies, etc.;
- Revision of the PIS (policy objects, hypotheses, linkages) to fit better the particular situation to which it applies based on feedback from implementation;
- Implementation of the revised PIS and repeating the monitoring and revision cycle.

The adoption, implementation and success of this approach require that certain conditions be satisfied, which are usually a function of the spatial/organizational level concerned and the scope of the PIS. The most critical requirement is interest in and commitment to policy integration. Policies driven and supported by narrow interests may be difficult to integrate because the associated actors may not see it profitable to adopt a holistic view that promotes the common good (Olson, 1971; Shannon, 2002). Absence of commitment and a 'commons' mentality[59] reduce the chances of satisfying the accountability condition; namely, 'who will be responsible for the PIS?' and 'who will be charged with coordinating the overall PI effort?' (Peters, 1998). Last, but not least, the issue of whether it is ethically acceptable to conduct social experiments, as the adaptive management paradigm somehow implies, may not be easy to resolve. Context-specific political expediencies may preclude the adoption of an adaptive management approach to PI, irrespective of its plausibility and suitability.

Despite the reservations expressed above, the proposed approach could serve as a conceptual guide for designing and testing alternative PIS to explore feasible approaches and schemes in diverse environmental and socio-economic contexts

---

[58] As, for example, economic policies and environmental, mainly water resources, policies.
[59] I.e. a concern for the wise management of Common Pool Resources (CPRs) that include a wide range of natural, manmade and human resources (Ostrom, 1990).

and problem situations. An initial application of these ideas in the case of combating desertification can be found in Briassoulis (2004). The PIS may be limited to simply incorporating environmental, social or economic concerns in sectoral policies[60] or may move to more complete integration among policies.[61] The integration process should ideally start from the integration of policy objects and move to the integration of policy instruments, ensuring the consistency of the overall process. However, as this order will most probably be difficult to preserve in reality, the integration process may start from whichever component of the PI object is handy, convenient, and easy to manipulate while trying to build reasonable linkages with the other components.

The study of policy integration is still in its early stages as one can glimpse from the literature. Future theoretical and empirical research faces a long list of open questions waiting to be explored. The question of the appropriate scale for PI and the scope of the task should be explored together with the question of intra-policy (vertical) versus inter-policy (horizontal) integration; especially the effects of integrated policies (internal, vertical integration) on inter-policy, horizontal integration. The adaptive management approach proposed may be appropriate at certain scales and for PI of a given scope which only empirical research can identify. Important are also questions of the cross-scale relationships among the objects of PI, the range of possible PIS, varying from general to case-specific schemes, and the influence of non-policy factors on the feasibility of particular types of PIS. Finally, the design of various types of integrative instruments, adapted to the broader approach to PI presented here is a practical issue deserving further research especially in the context of emerging new forms of governance (such as multi-level governance), new kinds of market-based and voluntary instruments and assessment tools (such as sustainability appraisal). More specifically, instruments should be sought to effect or improve the integration of policy objects, goals and actors and the consistency of the overall process.

References

Anderson, J. E. (1984), *Public Policy-Making*, 3rd edition, Holt, Reinhart and Winston, New York.
Berkes, F. and C. Folke, (eds) (1998), *Linking Social and Ecological Systems: Management Practices and Social Mechanisms for Building Resilience*, Cambridge University Press, Cambridge, MA.

---

[60] This is an extension of the EPI idea.

[61] It is conjectured that internally (i.e. vertically) consistent and coherent policies are more likely to be well integrated with other policies despite the risk of becoming more autonomous and 'isolated' from other less coherent policies.

Börzel, T. (1997), 'What's so special about policy networks? An exploration of the concept and its usefulness in studying European governance', *European Integration On-Line Papers (EIoP)*, Vol. 1, No. 16, http://eiop.or.at/eiop/texte/1997-016a.htm.

Briassoulis, H. (2001), 'Policy-oriented integrated analysis of land use change: An analysis of data needs', *Environmental Management*, 26(2), pp. 1-11.

Briassoulis, H. (2004), 'Design of a Desertification Policy Support Framework (DPSF)', Deliverable 36, Module 4, MEDACTION. European Commission, DG-XII, Contract No. ENVK2-CT-2000-00085 (www.icis.nl/medaction).

Bromley, D.W. (1991), *Environment and Economy*, Blackwell Publishers, Cambridge, MA.

Christiansen, T. and Piattoni, S. (eds), (2004), *Informal Governance in the European Union*, Edward Elgar, Cheltenham.

Dryzek, J.S. (1987), *Rational Ecology, Environment and Political Economy*, Blackwell Publishing, Oxford.

Dye, T.R. (1975), *Understanding Public Policy*, 2nd edition, Prentice-Hall, Englewood Cliffs, N.J.

Eggenberger, M. and Partidario, M. (2000), 'Development of a framework to assist the integration of environmental, social and economic issues in spatial planning', *Impact Assessment and Project Appraisal*, 18(3), pp. 201-7.

European Commission, (1998), *Partnership for integration: A strategy for integrating environment in EU policies*, COM (98)0333 final.

European Commission, (1999), *Report on Environment and Integration Indicators to Helsinki Summit*, Commission Working Document, SEC (1999) 1942 final, Brussels.

European Commission, (2000), *Bringing our needs and responsibilities together-Integrating environmental issues with economic policy*. Communication from the Commission to the Council and the European Parliament. COM(2000) 576 final, Brussels.

European Commission, (2001), *Policies for an Enlarged Union*, Report of Governance Group 6, June 2001, Brussels.

EC (2004), *Third Report on Economic and Social Cohesion*, COM(2004) 107, Office for Official Publications of the European Communities, Luxembourg.

ESDP (1999), *European Spatial Development Perspective: Towards Balanced and Sustainable Development of the Territory of the EU*, European Commission, Office for Official Publications of the European Communities, Luxembourg.

European Environmental Bureau (2003), *Environmental Policy Integration (EPI): Theory and Practice in the UNECE Region*, Background Paper of the European ECO Forum for the Round Table on Environmental Policy Integration at the Fifth "Environment for Europe" Ministerial Conference, Kyiv, May 21-23, 2003.

Friedrich, C.J. (1963), *Man and his Government*, MacGraw-Hill, New York.

Greenberg, G.D., Miller J.A., Mohr, L.B., and. Vladeck, B.C (1977), 'Developing public policy theory: Perspectives from empirical research', *The American Political Science Review*, 7, pp. 1532-43.

Healey, P. (1997), *Collaborative Planning: Shaping Places in Fragmented Societies*, UBC Press, Vancouver.

Holling, C.S. (1978), *Adaptive Environmental Assessment and Management*, John Wiley, New York.

König, T. and Bräuninger, T. (1998), 'The formation of policy networks; Preferences, institutions and actors' choice of information and exchange relations', *Journal of Theoretical Politics*, 10(4), pp. 445–71.

Lafferty, W.M. and Hovden, E. (2002), *Environmental Policy Integration: Towards an Analytical Framework?*, PROSUS, Centre for Development and the Environment, University of Oslo, Report 7/02, Oslo.

Lenschow, A. and Zito, A. (1998), 'Blurring or shifting of policy frames? Institutionalization of the economic-environmental policy linkage in the European Community', *Governance*, 11(4), pp. 415-42.

Lenschow, A. (ed.) (2002a), *Environmental Policy Integration: Greening Sectoral Policies in Europe*, Earthscan, London.

Lenschow, A. (2002b), 'Dynamics in a multilevel polity: Greening the EU regional and Cohesion Funds', in A. Lenschow (ed.), *Environmental Policy Integration: Greening sectoral Policies in Europe*, Earthscan, London, pp. 193-215.

Liberatore, A. (1997), 'The integration of sustainable development objectives into EU policy-making: Barriers and prospects', in S. Baker, M. Kousis, D. Richardson and S. Young (eds.), *The politics of sustainable development: Theory, policy and practice within the European Union*, Routledge, London.

Lindblom, C.E. (1959), 'The science of muddling through', *Public Administration Review*, 19(2), pp. 79-88.

Lindblom, C.E. (1973), 'Incrementalism and environmentalism', in *Managing the Environment*, Final Conference Report, Environmental Research Center, US EPA, Washington, DC.

Lowi, T.J. (1964), 'American business, public policy, case studies, and political theory' *World Politics*, 16, pp. 677-715.

Mazmanian, D. and Sabatier, P. (1989), *Implementation and Public Policy*, Revised edition, University Press of America, Lanham, MD.

Metcalfe, L. (1994), 'International policy coordination and public management reform' *International Review of Administrative Sciences*, 60, pp. 271-90.

Nakamura, R.T. and Smallwood, F. (1980), *The Politics of Policy Implementation*, St. Martin's Press, New York.

OECD (1996a), *Building policy coherence: Tools and tensions*, Public management occasional papers, No. 12, Organization for Economic Cooperation and Development, Paris.

OECD (1996b), *Globalization: What challenges and what opportunities for government?* Organization for Economic Cooperation and Development, Paris.

Olson, M. (1971), *The Logic of Collective Action: Public Goods and the Theory of Groups*, Harvard University Press, Cambridge, MA.

Opschoor, H. and J. van der Straaten (1993), 'Sustainable Development: An Institutionalist Approach', *Ecological Economics* 7 (1993), pp. 203-22.

Ostrom, E. (1990), *Governing the Commons: The Evolution of Institutions of Collective Action*, Cambridge University Press, Cambridge, MA.

Ostrom, E. (1999), 'Institutional rational choice: An assessment of the Institutional Analysis and Development framework', in P. Sabatier (ed.), *Theories of the Policy Process*, Westview Press, Boulder, Co, pp. 35-72.

Persson, A. (2002), *Environmental Policy Integration: An Introduction*, PINTS - Policy Integration for Sustainability, Background Paper (Draft), Stockholm Environment Institute, Stockholm.

Peters, G.B. (1998), *Managing Horizontal Government: The Politics of Coordination*, Canadian Centre for Management Development, Research Paper No. 21, Catalogue Number SC94-61/21-1998, ISBN 0-662-62990-6.

Pressman, J.L. and Wildavsky, A. (1992), *Implementation: How Great Expectations in Washington Are Dashed in Oakland*, University of California Press, Berkeley, CA.

Robert, J., T. Stumm, J.M. de Vet, C.J. Reincke, M. Hollanders and Figueiredo, M.A. (2001), *Spatial Impacts of Community Policies and the Costs of Non-Coordination*, ERDF Contract 99.00.27.156, EC, DG-Regional Policy, Brussels.

Rose, R. (ed.) (1969), *Policy Making in Great Britain*, Macmillan, London.

Sabatier, P. (ed.) (1999a), *Theories of the Policy Process*, Westview Press, Boulder, Co.

Sabatier, P. (1999b), 'The need for better theories', in P. Sabatier (ed.), *Theories of the Policy Process*, Westview Press, Boulder, Co, pp. 3-17.

Sabatier, P. and Jenkins-Smith H.C. (1999), 'The advocacy coalition framework: An assessment', in P. Sabatier (ed.), *Theories of the Policy Process*, Westview Press, Boulder, Co, pp. 117-66.

Shannon, M.A. (2002), *Theoretical Approaches to Understanding Intersectoral Policy Integration*, Paper presented at the Finland COST Action meeting (European Forest Institute).

Underdal, A. (1980), 'Integrated Marine Policy: What? Why? How?', *Marine Policy* 4(3), pp. 159-69.

Weale, A. (1996), 'Environmental rules and rule-making in the European Union', *Journal of European Public Policy*, 3(4), pp. 149-67.

Chapter 3

Policy Integration in the Framework of EU Regional Policy

Constantinos Liarikos

Introduction

The manifestation and evolution of socio-economic development disparities among different regions owes to the complex interactions among a variety of economic, social, environmental and historical factors which determine the distribution of population and economic activities in space. Regional policy, drawing on theories from diverse academic disciplines, seeks to manipulate this wide array of factors in order to promote a balanced spatial distribution of development benefits; or, more generally, sustainable regional development.

Regional development has always been a subject of primary importance in the European Union, insofar as the reduction of development differentials across its territory is an important prerequisite for the socio-economic cohesion, harmonious development, and, ultimately, political acceptance of the Union. The quest for regional policy has been present since the 1957 EEC Treaty, but it acquired particular momentum after the accession of Greece, Spain and Portugal in the 1980s where most EU developmentally lagging regions are found. Several of them face serious problems of land degradation and desertification; in promoting sustainable development, EU regional policy should be expected to deliver against their alleviation.

After the 1989 Dellors reform, EU regional policy adopted a multi-annual programming approach, with development targets set for a variety of policy fields, through which it seeks to reduce regional disparities. In pursuing the goal of sustainable regional development, the policy is increasingly employing new approaches including decentralized and bottom-up policy making, investment in social capital, promotion of soft interventions, and, lately, very important for the present purposes, policy integration.

Although, EU regional policy has not fared impressively in reducing regional development differentials, it undoubtedly presents a progressive approach to policy making, offering a structure within which conflicting approaches and interests are negotiated. The current attention to the issue of policy integration reveals that the

EU policy-makers recognize the importance of the integrative function of the policy and are preoccupied with its reinforcement.

The current degree of integration of EU regional policy with other Community policies, however, can be questioned posing thus an unusual conundrum: the very policy that purports to integrate different development concerns, is itself not integrated with the associated policies, especially if one considers that regional policy should be conceived as a framework policy utilizing the instruments and means of other policies to achieve its goals. A great deal of the EU regional policy's ineffectiveness can be attributed exactly to this contradiction.

The purpose of this chapter is to examine the extent of integration of EU regional policy with three major sectoral policies: transport, rural development and environmental policy. Transport policy is chosen because of its prominent role in regional development, rural development policy because it constitutes a rather novel component of EU cohesion policy and environmental policy for its obvious importance for sustainable development. All three policies bear significantly on desertification: rural development policy targets the development of rural areas that include most desertification-sensitive regions; transport policy represents an important determinant of long-term spatial development and land use patterns that may intensify the desertification risk of these regions and environmental policy offers important tools for the management of environmental degradation.[1]

Of the various components of EU regional policy, the analysis is confined to the Structural Funds (SFs) that are implemented through Community Support Frameworks (CSFs) or Single Programming Documents (SPDs). The Cohesion Fund and the various Community Initiatives are not considered, despite their importance as regional policy tools, because they are project-based meaning that they cannot be examined within the same context as the programmatically implemented SFs.

Following Chapter 2 of this volume, policy integration (PI) can be conceptualized as a process either of coordinating and blending policies into a unified whole, i.e. as integration of policies, or of incorporating concerns of one policy into another. Adopting the first conceptualization, PI is defined as a process of interrelating and coordinating various policies, both across (vertically) and over (horizontally) levels of governance, modifying them appropriately if necessary, to create an interlocking, loosely-coupled, multi-level, policy system that functions harmoniously in unity. Due to the largely decentralized nature of EU regional policy, the present examination cannot cover the issue of PI comprehensively; it focuses primarily on major trends seeking to shed light on the character of substantive and procedural integration[2] of EU regional policy with the selected policies. Desertification is used as an illustrative example of a complex, socio-

---

[1] Several other policies influence desertification; the three policies have been selected as they offer significant illustrative cases of policy integration in the context of EU regional policy.
[2] See Chapter 2, this volume, for a typology of the dimensions of policy integration.

environmental problem,[3] directly relevant to sustainable regional development whose control calls for integrated policy approaches.

The chapter comprises five sections. The second section negotiates the regional development problem, the evolution of regional policy and the relevance of PI for regional development. The third section briefly presents the content and evolution of EU regional policy while the fourth is devoted to the analysis of its integration with the selected policies. The fifth section discusses the implications of PI for combating desertification while the last section offers policy recommendations and suggestions for future research.

Regional Development, Regional Policy and the Quest for Policy Integration

The Regional Development Problem in Perspective

Numerous, interlinked factors acting on various spatial levels and over variable time spans shape the spatial distribution of population and economic activity and determine the productive structure and potential for socio-economic development and welfare of a region. Included among them are natural resource endowments and geographical features, socio-economic organization, cultural, political, and institutional capital and historic events and circumstances. The spatio-temporal variability, relative importance and mode of interaction among these factors explain differences in the characteristics and level of socio-economic development among regions, over space and time.

Traditionally, natural resource endowments and geographical characteristics have featured prominently within the regional development discussion as major determinants of development patterns and dynamics. With the rapid transformation of the global economy and society, their prominence has given away to concerns with the qualitative aspects of the development process and issues such as human resources, governance and institutions.[4] This transformation has been the product of major alterations in the patterns and factors of economic production (improved communication and transportation, new production modes and products, new information technologies, increasing importance of the tertiary sector) and of a novel understanding of the development process, as represented by the 'post-fordism' and the 'new geography' schools that are discussed below.

In both traditional and contemporary regional development paradigms, historical events and circumstances are important considerations as they determine whether, when and how the socio-economic development potential of a region will be achieved and they mould its socio-economic trajectory. The occurrence of wars, conquests, revolutions, large-scale natural disasters, technological accidents, disease outbreaks, and terrorist attacks, in certain historic moments and periods,

---

[3] See Chapter 1, this volume, for more details.
[4] See, for example, the REGIONET site: http://www.iccr-international.org/regionet.

reinforces or weakens the effects of environmental and socio-economic factors. Such events shock the socio-economic system and create stereotypes and predispositions about certain localities, influencing thus human decision making and ultimately the local development process. Similarly, human intellect and cognition, through filtering socio-economic circumstances, contribute significantly to the development process, adding to its complex nature and defying direct operationalization and treatment in the related analyses (Martens, 2004).

The complex interplay among the factors affecting the development process produces the uneven spatial distribution of economic activities, and the resultant regional development disparities. Such disparities, traditionally measured by differences in GDP per capita, are usually characterized by strong persistence which gives rise to path dependency in regional development. Path dependency could simply mean that 'history matters' and that 'where we are today is a result of what has happened in the past' (Liebowitz and Margolis, 1995); i.e. past decisions and choices affect the current situation and future development paths (Puffert, 2003). Such decisions may refer to most trivial economic acts or historical events which, in a complex socio-economic world, may fuel important changes and future path dependencies (Ormerod, 1998).

The vicious circle associated with the persistence and circular causation inherent in the complex nature of regional disparities often gives rise to lock-in situations, whereby certain areas are stuck at low economic performance levels. Liebowitz and Margolis (1995) distinguish three distinct forms of path dependence with different implications for lock-in; the first regards situations where initial decisions and conditions set a lock-in situation, which though might be optimum; the second regards lock-in situations which are identifiably sub-optimal, but where the costs of overcoming them are prohibitive; and the third regards sub-optimal but feasible to overcome lock-in situations, where the necessary changes are not realized however.

Lock-in situations of the first type are evidently irrelevant to the problem of regional disparities; the other two are highly relevant, however. The second type is akin to the new economic geography outlook, whereby, while strong inefficiencies in the current spatial distribution of activities and the benefits of a more balanced spatial distribution for the economy as a whole[5] are recognized, the gains arising from agglomeration economies and the forbidding costs of any spatial restructuring dictate the preservation of the current spatial layout. The third type of lock-in cases is even more interesting; feasible solutions to overcome low regional economic performance are not realized, an assertion going against the standard neoclassical outlook of economic rationality, although not necessarily for the case of regional development. National economic development policies that reinforce spatial polarization and agglomeration, and vice versa, despite the possibility of certain trade-offs between spatial equity and efficiency, illustrate this case (Martin, 1999 and 2003). The acceptance of a high regional disparities lock-in situation, when

---

[5] By unlocking its development potential.

there are evident ways out, could well be a rational decision favouring national efficiency over spatial equity. Numerous studies demonstrate how actions undertaken under EU's regional policy seem to favour national competitiveness while policy measures that could spur regional integration remain neglected (Martin, 1999; Rodriguez-Pose and Fratesi, 2004).

The possibility of lock-ins at unfavourable regional development levels and the persistence of the associated disparities, together with the complexity of regional development problems, demonstrate the need for public policies that support the necessary restructurings and foster the transition to more favourable development paths as well as the need for their integration to promote the goal of socioeconomic cohesion.

Regional Development Policy: Aims, Tools and Evolution

Regional development policy (henceforth, regional policy for brevity) seeks to counter the forces behind persistent regional disparities and even out the development levels of different regions. In this respect, it is essentially differentiated from macroeconomic development policy, which seeks to improve the overall performance of an economy, and from local economic development policy, which seeks to spur the development of a local economy *vis-à-vis* other local economies. Although these distinctions are not always clear, this clarification is useful for the present purposes.

Begg (1998) distinguishes three axes of concern that make regional policy relevant. The first regards purely economic reasons related to the stabilization of the economy against region-specific asymmetric shocks and the promotion of higher aggregate demand, without risking inflation pressures. The second axis concerns equity issues, as regional inequalities are socially acceptable only to some finite degree. The third axis refers to issues of economic convergence and the use of interregional transfers to foster economic restructuring, improve the competitiveness of lagging regions and spur long-term growth. Begg et al. (1995) note also that, while the first axis was particularly relevant prior to the 1973 crisis, attention has shifted to the third axis in the post-70s.

Three different regional policy schools of thought can be distinguished that address these concerns. The first school prevailed during the 1960s and adhered to the Keynesian legacy, considering regional policy as an essentially redistributive policy. It prescribed direct transfers and other mechanisms of income support, often including public sector employment. The second school prevailed in the late 1970s and the 1980s, stemming from a neo-liberal legacy and seeing to development through supply-side policies. It favoured measures for market deregulation, entrepreneurship support and investments in transport and communication infrastructures that assumed a prominent role in opening-up regional economies (Amin, 1999).

Although drastically different in their prescriptions, these first two schools share a common approach to the regional development problem, seeing a need for external interventions to catalyse and spur development in lagging regions. Moreover, both support central government-administered policies (top-down), universally applicable for countering regional problems. These characteristics differentiate them from the third school, collectively termed 'new regionalism'. This is essentially an umbrella term for a collection of theories stressing the particularities of regional or local economies and attributing importance to their non-transferable and non-traded characteristics. This approach is fuelled by various theoretical legacies, including institutionalism, which stresses the importance of social relations and network rationalities, and the school of flexible production, which advocates economic restructuring towards 'post-fordist' modes of production and seeks to draw policy implications by examining specific examples.[6] The 'new economic geography' school is also a very important contributor to this discussion, providing the technical/economic analysis which backs many of the claims put forward, and which is often lacking from the more qualitative analysis of the other schools.[7]

Although the subject of much criticism, on both conceptual and analytical grounds,[8] the 'new regionalism' school has greatly influenced the conception of the regional development problem as well as many applied aspects of regional development policy. Its policy prescriptions emphasize the utilization of endogenous regional potential and the importance of strengthening networks and economic associations. They take a territorial approach to policy, underscoring path dependencies, location-specific historical, social and economic attributes and regional governance issues. Consequently, they stress the need for endogenously derived, context-specific, decentralized, bottom-up, and locally managed policies tailored to the particular socio-economic circumstances of each area.[9] These new prescriptions co-exist side-by-side with more traditional approaches, which still dominate regional development funding.

In general, regional policy is practiced through the application of positive incentives, i.e. pecuniary benefits for undertaking certain positive investments and kicking-off certain economic activities. These include investments in public capital, most importantly transport, telecommunication and energy networks, subsidies for the adoption of new technologies and the development of private enterprises, especially SMEs, fostering of vocational training, financing of actions countering social exclusion, etc. In practice, a variety of sectoral policy measures

---

[6] For various manifestations of these theoretical approaches, see Streeck (1990), Sengenberger and Pyke (1992), Trigilia (1992), Cooke and Morgan (1994, 1998), Rodriguez-Pose (1994), Amin (1999).

[7] Krugman (1998) offers a comprehensive account of the development of the New Geography School and its links to more traditional paradigms.

[8] See, for example, Amin and Robins (1990), Harrison (1991), Markusen (1996).

[9] Seeds of these prescriptions are to be found also in the literature mentioned in note 7. Charles et al. (2004) offer a comprehensive report on the development of these new paradigms and the policy prescriptions included thereafter.

that contribute to development are used in addition to pure regional policy measures.

Therefore, regional policy is better conceived as an umbrella policy that finances and spurs sectoral activities to promote regional development and cohesion. In this perspective, regional policy fulfils two important roles: first, it finances or co-finances investments and activities in various sectors and, second, it coordinates indirectly those sectors. This second role manifests itself subtly, but is of extreme importance in conceptualizing the breadth and potential of a coherent regional policy programme. While regional policy does not regulate activities and investments, the mere fact that it constitutes a major financier for the implementation of related projects allows it, through the project selection procedure, to catalyse the investment pattern of many sectors in target areas, thus affecting their spatial structure.

Besides using investment and activity incentives, regional policy also utilizes fiscal measures and economic sanctions. The first usually include tax breaks to enterprises willing to settle or relocate in targeted areas, thus, being relevant to the attraction of inward investment. Economic sanctions are rarer, usually imposed on economic entities deciding to establish their activities in overdeveloped regions. They are indirect regional policy measures, as they target mainly the internalization of external costs due to congestion, pollution, etc.

The Relevance of Policy Integration for Regional Development

Regional policy, treated and articulated as a separate policy field, is essentially a spatial policy, seeking to engage, coordinate and support[10] policies from various spatial levels in promoting favourable spatial distributions of economic activities to achieve regional development goals. Therefore, successful implementation depends on its proper integration with sectoral policies. This view is most akin to the 'new regionalism' school that considers regional policy as a multi-level practice were synergies between a very wide array of disciplines and policies is sought (Charles et al., 2004), making the relevance of PI and the analysis of its various dimensions imperative. Adopting the methodological framework proposed in Chapter 2 of this volume, the following discussion negotiates the object and dimensions of PI in the case of regional policy.

For regional policy to act as a coordinating mechanism of a large number of sector-specific policy tools, both horizontal and vertical integration is needed. Horizontal integration refers both to the integration among separate sectoral policies and to the integration of those policies with regional policy on the same spatial/organizational level. Vertical integration concerns the coordination of policies across different spatial/organizational levels. In both cases, the analysis of the integration of regional policy with other policies should be articulated along three basic dimensions. The first is the *substantive dimension* regarding whether

---

[10] Mostly through financing.

regional policy and sectoral policies share the same understanding of the regional development problem and, thus, their goals and objectives are congruent, compatible, complementary, or even common. The second is the *procedural dimension* concerning whether administrative and institutional arrangements exist for coordinating and harmonizing different sectoral policies to effectively join forces towards regional development goals. Intertwined with the other two is the *spatial dimension* referring to the coordination of policy formulation and implementation on the same and across different spatial levels.

EU Regional Policy: Past and Present

The 1957 Treaty establishing the European Economic Community (EEC) included the notions of cohesion and harmonious development as the aim of its establishment was to spread the benefits of the Community throughout its territory. Because at that time the EEC consisted of a quite homogeneous group of member states (MS), the structures provided initially purported to strengthen trade links and assist the formation of the Common Agricultural Policy. The European Social Fund (ESF), the European Agricultural Guarantee and Guidance Fund (EAGGF) and the European Investment Bank (EIB) were set-up for that purpose.

The first signs of a regional policy proper appeared after the 1973 enlargement to the UK, Ireland and Denmark, when the European Regional Development Fund (ERDF) was set up. This is attributed to the pressures the UK exerted during its accession negotiations, but has also been underpinned by the effects of the first oil crises, which demonstrated the importance of Community-wide solidarity policies while signalling a step-out of national governments from regional support policies. The ERDF was designed as a redistribution instrument, targeting lagging regions and providing assistance to productive investments, infrastructures and SMEs.

The 1981 enlargement to Greece signalled the broadening of regional disparities, while the 1986 enlargement to Spain and Portugal deepened this effect. In compensation for the 1986 enlargement, the Integrated Mediterranean Programmes (IMPs) were set up to assist earlier Mediterranean MS, while, in view of the continuing regional disparities, the 1987 Single European Act set out to assist peripheral countries, called for a rationalization of relevant funding and combined the ERDF, the ESF, the Guidance Section of the EAGGF and the EIB under the common umbrella of the Structural Funds (SFs). These developments marked a gradual turn to a more coherent and comprehensive regional policy, which was to be fully capitalized under the 1988 first reform of the SFs and the so-called Delors Package (after the then Commissioner Jacques Delors).

The Delors Package I (1989-1993) established the doubling of the available resources and resolved that finances would be determined by indicative ranges, made available through five-year programme financing. It stressed that the policy interest should be with the regions and not the states, it described the five kinds of regions that were eligible for assistance (Objectives 1-5) and determined that financing should only come additionally to national policies and expenses (the additionality principle). Finally, it recognized the role of sub-national administration tiers and non-governmental organizations and facilitated their participation in decision making and implementation processes. In addition, the 1988 reform called for a control of the CAP expenditures and a change in direction away from infrastructure expenditures and towards more composite development considerations.

The 1988 reform was succeeded by the Maastricht Treaty and the 1993 Reform (Delors Package II), which resolved the increase of the Community's own resources and a substantial increase of the SFs. Furthermore, it provided for the establishment of the Committee of the Regions and the tri-annual issue of Cohesion Reports on the progress of regional policies in Europe.

The 1999 Reform, treading on the lines of the Agenda 2000 strategy and governing the 2000-2006 programming period, signalled the addition of the Cohesion Fund (CF) to the SFs. The CF is especially designed to assist countries with GDP lower than 90 per cent of the Union's average in meeting the Monetary Union criteria. At the same time, it resolved the reduction of Objective regions from seven to three, on grounds of better targeting Community expenses, and the inclusion of the various Commission initiatives (also reduced from 13 to 4) under the SFs umbrella. With respect to the CAP, Agenda 2000 provided for the rationalization of its expenditures and the emphasis on the regional development-oriented Guidance Section of the EAGGF.

The last foreseen reform, underway at the time of this writing, shall govern the implementation of EU's regional development policy for the period 2007-2013. The proposed Regulations (CEC, 2004c and 2004d) outline a more streamlined, flexible and decentralized operation of the SFs, as recommended by the 3rd Cohesion Report (CEC, 2004a). The proposed measures, which are meant to cater to the needs of the EU-25, prescribe, among others, the abolition of the Community Support Frameworks and the introduction of a three-tier programming process consisting of the Community Strategic Guidelines (CSG), the National Strategic Reference Framework (NSRF) and the Regional or Sectoral Operational Programmes. The proposals provide for thinner programme detailing at the initial phases of planning (more flexibility for programme managers), more open and participatory procedures, stricter provisions for management effectiveness and more attention to issues of cross-border cooperation, environmental protection, etc. They also provide for the creation of a new European Agricultural Fund for Rural Development (EAFRD) to fund rural development measures.

Thus, the EU regional policy, through successive changes and reforms, has gradually grown into a coherent policy framework for the development of the European territory. Accordingly, both its targets and its instruments have widened to include issues of territorial cohesion, environmental protection, social inclusion, etc., and to cater to increasingly complex policy considerations; albeit with a persistent focus on hard infrastructures. As shown in Tables 3.1 and 3.2, the policy has grown in importance both as a budgetary line of the EU finances and as a part of the benefiting MS budgets.

Table 3.1   Evolution of Regional Policy Funds

	1984	1988	1993	1999	2000	2006
Funds for Structural interventions	3220	6419	20478	30950	30045	29170
% of EU Budget	11,5	15,1	31	36	32.6 (36)	27.2 (45,8)

Source: Adapted from Dall' Erba (2003)
Figures in €millions – percentages in brackets those including pre-accession funds

Table 3.2   Structural Funds at the national level

	1989–93 % of GDP	1994-99 % of GDP	1994-99 € per Capita per Annum	2000-2006 € per Capita per Annum
Portugal	3.07	3.98	299	275
Greece	2.65	3.67	284	288
Ireland	2.66	2.82	346	123
Spain	0.75	1.74	181	157
Italy	0.27	0.42	63	71
UK	0.13	0.25	37	38
France	0.14	0.22	43	36
Germany	0.13	0.21	45	49
Austria		0.19	39	26
Belgium	0.11	0.18	34	26
Netherlands	0.07	0.15	28	24

Source: Adapted from Dall' Erba (2003)

Despite the widening of the policy and of the prescribed instruments, the available policy finances have been heavily targeted on operations for infrastructure investments. As shown in Table 3.3, during the 1st and 2nd programming periods, infrastructure development has absorbed well above 40 per cent of funding in the majority of benefiting MS, with the second place been taken by agriculture and rural development or business development, depending on the State. Some writers have considered this financial focus as a major cause for the ineffectiveness of regional policy (Rodriguez-Pose and Fratesi, 2004), while it is also an inherent part of what Martin (2003) identifies as a conflict between equity vs. efficiency concerns.

Within this framework, EU regional policy can be hardly considered as having effectively contributed to the reduction of development disparities between European regions.

Table 3.3   Percentage Appropriations of SF Funding

	1989-1993				1994-1999			
	Agri & Rural	Business & Tourism	Human Capital	Infra	Agri & Rural	Business & Tourism	Human Capital	Infra
Austria					15,0	68,7	16,3	0,0
Belgium					0,0	66,2	17,2	16,6
France	28,6	15,9	10,1	45,4	9,6	32,8	18,7	39,0
Greece	11,2	18,4	16,6	53,8	18,7	13,4	13,6	54,3
Ireland	14,7	33,7	26,4	25,2	0,0	54,7	3,8	41,4
Italy	14,4	35,0	1,9	48,8	21,0	21,3	27,0	30,7
Netherlands					22.2	20,4	21,0	36,4
Portugal	11,5	6,1	35,3	47,2	0,0	15,2	8,6	76,1
Spain	26,7	13,2	8,8	51,4	0,6	14,3	7,5	77,6
UK	10,5	38,1	20,9	30,4	12.2	25,0	33,1	29,7

Source: Adapted from Rodriguez-Pose and Fratesi (2004)

These disparities had been narrowing until the mid-70s oil crises but started widening after that. After the 1980s, and especially during the 1990s, a slow growth in regional disparities has run parallel to a convergence trend among national economies. This fact, owing to a large extent to the fiscal austerity imposed by the Maastricht Treaty, has steamed many researchers to identify a trade-off between equity and efficiency in the European context and an internal contradiction in EU policies trying to simultaneously promote both (Martin, 2003). Boldrin and Canova (2003), criticizing the EU policy regional development outlook, show that regional development policies have been inefficient in spurring the aggregate growth of the EU and in reducing regional disparities. They claim that the facts do not support the empirical predictions of the European Commission. Other writers identify the underpinnings of regional policy's failure with the instrument mixes utilized, and especially with the unwarranted emphasis on large transport infrastructures (Rodriguez-Pose and Fratesi, 2004).

Despite these criticisms, the argument that disparities might have been a lot more acute in the absence of those active regional policies may still be valid. Undeniably the policy implementation framework set by the SFs has contributed to the 'Europeanization' of the regional policies of the MS exemplified, among others, by improvements in the programme management capacities of benefiting regions, the ability of regional policy to reinforce and facilitate the mainstreaming of important legal and political functions of the European Union and the reinforcement of public participation and transparency procedures. Similarly, the role of EU regional policy in developing and maintaining political support for the European integration project is undeniable.

EU Regional Policy: Current State of Policy Integration

This section presents a systematic analysis of the current state and future prospects of integration of EU regional policy with the selected EU policies along the thematic and procedural/spatial dimensions. Evaluation criteria pertinent to each dimension of policy integration (PI) are used selectively for each pair of policies analysed (Tables 3.4 and 3.5).

Table 3.4   Criteria for analysing the substantive dimension of PI

- Political commitment for PI
- Existence of a PI framework
- Favourable administrative culture for PI
- Congruent and consistent theories/views about the policy problem (regional development and regional disparities) and, thus,
- Compatible and congruent policy problem definitions on the same and at different spatial levels
- Adoption of PI as a goal by either one or both policies
- Congruent, compatible and/or complementary goals and objectives among policies

Table 3.5   Criteria for analysing the procedural/spatial dimension of PI

- Administrative capacity for and/or reform in favour of PI
- Organization in charge of PI
- Common actors between the policies
- Formal interaction between policy actors on the same and on different spatial levels
- Procedures for the coordination of formal (and informal) actors on the same and on different spatial levels
- Existence of a legal framework for PI
- Use of EIA and SEA (integrative instruments)
- Common or coordinated action plans
- Common or harmonized assessment (result quantification) and evaluation procedures
- Inclusion of quantitative targets for PI
- Subsidy reform in favour of PI
- Good practice codes[11]

---

[11] Specifying actions in one policy with respect of the object of the other.

EU Regional Policy and Transport Policy

Transport policy constitutes a major instrument of EU regional policy, absorbing the bulk of relevant funding from the SFs. Its role in producing positive development effects remains heavily questioned, however. Many analysts have noted that transport infrastructures offer only short-term benefits to demand (for local products and services) and employment, with negligible contribution to long-term regional development (Martin, 1999; Rodriguez-Pose and Fratesi, 2004; Begg et al., 1995). These authors contend that the criteria that make transport investments preferred policy instruments are irrelevant to the issue of regional development, relating more to political considerations such as visibility and policy implementation ease. Some consider that transport investments are negative for the long term development of lagging regions. For example, Vickerman (1991, 1995), Vickerman et al. (1999) and Button (1998) identify aspects of large transport infrastructures that are more relevant to the development of the economic core than to the development of lagging regions. Exemplifying this issue for Greece, MEPPW (1997) presents the significant spatial inequalities that arise from such large infrastructures, while Ioannides and Petrakos (2000) go further to suggest that the better performance of some Greek regions might be attributed to their isolation from the economic core.

The study of the integration of transport with regional policy is interesting, as practice demonstrates that the former utilizes the financial support of the latter, but is articulated towards independent goals, namely connectivity between economic cores, which often go counter to the aims of regional policy. The case becomes even more relevant as it is accepted that transport infrastructures cannot spur or support regional development in the absence of factors that are important to the structure and performance of local economies.

The study of the transport-regional PI is also very relevant to the nature of regional policy making. As Martin (1999) points out, regional policy is self-contradictory in seeking to promote simultaneously regional convergence, national convergence and efficiency. He notes that policy makers cannot determine the purpose of the implemented policy and that the market failures underpinning regional disparities are yet to be clearly identified. Other authors have noted the absence of a common policy evaluation culture and framework among the regions involved (Bachter and Michie, 1994; GMF, 2003), something that obviously hinders the identification of the problem and the designation of suitable policies. These observations underscore the importance of the lack of substantive integration between the two policies, and will be touched upon in this analysis.

Finally, the issue is very relevant to desertification control as the majority of desertification-prone areas are economically lagging and are included in regions targeted by the EU regional policy. Transport policy decisions, and especially the creation of relevant infrastructures, are the main determinants of land uses through their potential to affect location considerations, alter land values and transform logistic systems. With land use change been among the main contributor to desertification, the relevance of the transport-regional PI becomes obvious. Furthermore, transport infrastructures cause important impacts on soil and generate various other environmental concerns indirectly linked to soil conservation (vegetative cover loss, pollution, forest fires, landscape fragmentation, etc.).

The Trans-European Transport Networks (TEN-T): Brief description The TEN-Ts constitute an important instrument of the EU Common Transport Policy (CTP), focusing essentially on the removal of internal borders, the improvement of the connectivity of remote regions and, ultimately, the enhancement of the functioning of the common market and the improvement of territorial cohesion. The latter goals draw directly on Article 2 of the EU Treaty (sustainable and balanced development) on which the CTP is rooted. According to the European Spatial Development Perspective (ESDP, 1999), the notion of territorial cohesion translates the content of economic and social cohesion into territorial terms, including issues of fair access of citizens and business to Services of General Economic Interest, as referred to in Article 16 of the Treaty (CEC, 2004b).

Within this framework, the 1992 White Paper on Transport set the goal of promoting the development of the TEN-Ts, while the 1997 Amsterdam Treaty (Article 154) restated the content of Article 16 of the Treaty and defined a specific goal for future developments in eliminating bottlenecks and filling-in missing network links. Furthermore, the Treaty defined the goal for the harmonization of transport systems and the establishment of common rules for international transport, outlining specific operations to be undertaken by the Commission and options for a CTP action plan. In 1998, the Commission document, "Sustainable mobility: Perspectives for the future" was adopted, setting out the priorities of the CTP and the initiatives to be taken until 2004 (CEC, 1998a). A prominent goal of the actions proposed was the completion of major missing links of the TEN-Ts. Finally, the 2001 White paper "European Transport Policy for 2010: Time to decide" (CEC, 2001a) redefined the role and goals of the EU CTP, restated the goals of the TEN-Ts and re-established the original focus of the policy (see above).

Based on the above, the CTP, and more specifically the TEN-Ts, emerge as both an autonomous EU sectoral policy, but also, very importantly, as a policy seen to promote regional development and cohesion. Its relation with the EU regional policy is further reinforced by the fact that the latter finances the majority of transport infrastructure investments, especially those targeting the completion of missing network links. Therefore, transport policy can be considered simultaneously as a sectoral policy financed under the regional policy umbrella and as a regional development instrument. Given the questionable contribution of transport infrastructures to regional development and spatial equity, it is important to recognize this dual nature of transport policy and seek to always evaluate it both as an independent policy and as a development instrument.

The substantive integration of the EU regional policy with the CTP In principle, both policies have complementary and highly compatible goals. The aim of the

CTP to provide the missing network links and enhance the connectivity of remote areas through the TEN-T concerns directly the issue of territorial cohesion, i.e. the balanced and equitable distribution of socio-economic activity across the Community space (CEC, 2004b). As such, the CTP goals and targets complement and are compatible with those of the regional policy in spurring development in lagging regions.

Moving on from the rhetoric of the CTP, though, a great deal of studies demonstrates that the final effect on regional development of enhancements in transport infrastructures and improved inter-regional nodes is uncertain. The socio-economic integration of lagging regions amidst a highly competitive economic environment, the induced changes in the spatial pattern of activities, the inequalities of access brought about by new large transport corridors, and several other possible effects demonstrate that the deal is not so straightforward. Well-put in a relevant report:

> ...this is because transport investments are a 'two-way street', so that for example linking a core to a peripheral region might be more to the benefit of the former rather than the latter (Goodbody Consultants, 2003, p.14).

Yet, this straightforward fact is not well represented in the planning of the TEN-T instrument which remains a predominantly investment tool for the creation of new transport infrastructures. A policy document co-signed by the major EU environmental NGOs, "Trans-European Transport Networks: Options for a Sustainable Future" (T&E, 2002), notes that while the TEN-T has "no explicit requirements for additional infrastructures", the Commission and the MS "have consistently placed the emphasis on implementation of the TEN-T via infrastructure construction" and have not given "sufficient consideration to the link between the TEN-T and other parts of Community law" (p.3). It emphasizes also that what is "persistently missing from the Commission's evaluations is an assessment of whether the TEN-T are in themselves desirable", while the "zero-option is constantly dismissed with economic growth arguments" (p.3). In making these arguments, the document draws on academic and policy research to suggest that the TEN-Ts are given a much higher profile and priority status than is warranted, given the economic, social and environmental arguments against their realization.

It follows then that the effects of transport investments on regional development are dealt-with in an axiomatic manner that leaves little to be questioned. Transport policy states its targets well and is right in recognizing its potential for regional development, but this potential is raised to a status of an axiom and is never well investigated and evaluated, even in the face of new regional policy stances and approaches stressing the supremacy of 'soft' interventions and of more complex policy concerns. It is tempting to suggest that both policies share a biased view of their relationship; as a way to enhance funding possibilities for the TEN-T implementation and to strengthen the political content of regional policy. Three preliminary conclusions can be drawn:

1. The autonomous workings and targets of the TEN-Ts manifest themselves in the context of regional policy dominating other considerations, prioritizing transport infrastructure investments and obtaining an often unwarranted, prominent role as regional development instruments.
2. In understanding this manifestation, the potential role of transport networks for the macroeconomic development of the EU territory, via the growth of its major development poles, is important. Back to Martin's (1999) comments, the goal of regional policy as a stimulant for efficiency and EU-wide growth seems to prevail over regional cohesion concerns.
3. The above are facilitated by the role of infrastructure development as a 'political' instrument; large-scale, visible projects demonstrate government commitment for development and serve national and international political goals.

These observations lead to the question of whether there exists a truly favourable political climate for the integration between the two policies. It seems that the political will currently concerns not the integration of the two policies so as to better achieve mutual goals, but favours the 'one-way' incorporation of TEN-T within regional policy. This also results from the lack of a clear definition of a PI framework, largely owing to a lack of understanding of the real causes underpinning regional disparities and of thorough evaluations of the policy instruments utilized.

The procedural/spatial integration of the EU regional policy and the CTP The examination of the procedural/spatial integration between the two policies distinguishes between the policy formulation stage and the actual planning and implementation of policy measures.

At the policy formulation stage, the two policies neither share similar procedures nor direct links seem to exist between the procedures for formulating each policy. The two policies are shaped under the aegis of different DGs of the European Commission, while ministerial councils of differing compositions finalize their regulations. More importantly, the time frames of reference for the determination of each policy differ substantially. The TEN-Ts set long-term goals for transport organization and infrastructure, which are revised on an *ad hoc* basis whenever the need arises. Regional policy regulations, on the other hand, have a pre-determined time frame of reference, the duration of the programming period, and are revised at the end of each such period. Consequently, the time frames of reference differ substantially on matters of length and regularity. Regional policy of the current period is implemented according to a Regulation finalized in 1999 and drawing on Agenda 2000, drafted in 1997. At the time Agenda 2000 and the Regulation were drafted, the strategy for the EU CTP, and hence the TEN-T, was based on the 1992 White Paper for Transport, while little after the start of the new programming period, a new White Paper on transport was adopted.

The procedural differences between the two policies extend beyond their temporal coordination to their content. When it comes to the designation of priorities for the TEN-T, the EU CTP becomes very detailed, literally determining the exact content of target projects, without making direct links, however, to their effects on regional development. On the other hand, regional policy making remains at this phase very general, offering broad directions and provisions only, and, while it provides for the possibility of funding TEN-T projects through the SFs,[12] it makes no reference to specific integration requirements.

At the planning and implementation stage, the two policies are linked in principle only through the connections the territorial cohesion approach provides, whereby the contribution of transport infrastructure investments to development is taken for granted and the SFs are one of the major funding tools. In this context, despite the fact that administrative capacity for better integration between the two policies exists, their goal setting and policy-making procedures remain largely isolated.

The study of procedural integration becomes more fascinating at the lower level of policy content determination and implementation where the procedures of the two policies are identified and coordinated mainly within regional policy programming as expressed by the CSFs and the SPDs. This administrative procedure translates into an operational integration between TEN-T projects with other development instruments, but not into an integration of the structure and targets/goals of the two policies. This is underpinned by the following characteristics:

1. The content and objectives of the TEN-T are detailed at the policy formulation stage, where the workings of the two policies remain largely independent and isolated. Thus, while the actual design and implementation of the projects is articulated within the framework of regional policy, it is regional policy that has to coordinate itself with the content and workings of the TEN-T, which enter its operation externally.
2. The general lack of policy impact evaluations produces a general 'ignorance' of the true impacts of planned or implemented projects on regional development and cohesion. Therefore, the necessary knowledge basis for mutual planning and integration of goals, objectives and practices is lacking. The findings of numerous studies demonstrating the questionable role, or even proving the null contribution, of transport infrastructures to regional development, do not seem to affect the respective planning and implementation processes. Much of this problem owes to a lack of a 'monitoring and evaluation culture' among policy makers (ECOTEC, 2003, p.221).

---

[12] And especially though the CF which is not covered in this analysis.

Synopsis   Regional policy and the TEN-T instrument possess spatially-defined targets seeking to upgrade the functioning of the European territory. Multiple links exist between the two policies, as transport policy is closely bound to the outlay of economic activities and beholds the potential to alter their spatial distribution and the development potential of different regions. The fact that the CTP is largely implemented through the funding mechanisms of the EU regional policy creates an immense potential for the integration between the two policies towards the harmonious development of the European territory. Yet, this potential is not been realized, because their actual relationships regarding development impacts have not been rigorously mapped and they are, to a large extent, independently defined.

In terms of policy implementation, the TEN-T almost exclusively translates into new infrastructure projects, whose contribution to the harmonious development of the European space is taken as granted. Although these projects are implemented under the aegis of regional policy, their actual planning enters regional policy externally, and thus, whatever coordination and integration are materialized, concern essentially management practices and not the substance of the two policies. At the same time, the SFs do not evaluate the true impacts of transport infrastructures on regional development while the numerous extant studies do not seem to influence policy making, due to, among others, a lack of evaluation culture in policy making.

Essentially, this is a case, on the one hand, of an operational integration of activities and, on the other, of a 'one-way' integration with regional policy being adapted to the implementation needs of the TEN-T. The political attractiveness of large infrastructures acts to strengthen and reinforce this peculiar integration pattern, as strong political will for continuing 'business-as-usual' prevails.

EU Regional Policy and Rural Development Policy

Rural development policy (henceforth, rural policy for brevity) represents a rather novel EU policy field introduced with the Agenda 2000 reform. As Van Depoele (2000) notes, the cornerstone of the new rural policy rests with the European model of agriculture, which stresses the multifunctional character of agriculture, the need for sustainable agriculture and multidimensional rural development, and the importance of non-trade concerns. The new rural development rhetoric represents a whole new perception of rural space as not merely a space of agricultural production but as a space within which a multifunctional economic system is articulated and the preservation of the European environment should be targeted. This perception dictates that measures for the development of rural space should not be confined to agriculture only, but should rather support a holistic approach to rural development which has a three-fold aim: to restructure the rural economic basis, to harmonize rural activities with the environment and to support rural livelihoods. The role of farmers and the rural economy in the provision of public goods are deemed also very important (Van Depoele, 2000).

Within this new approach, the relationship between regional and rural policy becomes essential: on the one hand, regional policy provides a set of useful tools for the development of rural space, and on the other, the two policies need to be well coordinated and integrated, especially when they concern identical spatial units. This need for integration and coordination was identified by the 1999 Rural Development Regulation (RDR), which placed a large part of rural policy under the aegis of regional policy, thus constituting the former as an integral part of the latter. Therefore, the analysis of the integration of rural with regional policy is imperative to show, among others, how well the new perception of rural space has been incorporated in regional policy practice.

Rural development policy: A brief review The design and application of EU rural policy is governed by Council Regulations 1257/1999 (CEC, 1999b), the RDR, and 1650/1999, as well as by the Commission Regulation (EC) 1750/1999 (CEC, 1999a). These regulations prescribe nine core policy measures: a) investment in agricultural holdings, b) setting-up of young farmers, c) training, d) early retirement, e) Less Favoured Areas, f) Agri-environmental measures, g) improving the processing and marketing of agricultural projects, h) forestry measures and i) integrated rural development measures.

These measures are financed by the European Agricultural and Guarantee Fund (EAGGF). Measures a, b, d, e and g are financed by the Guarantee part of EAGGF while the rest are financed by the Guidance section of the fund when Objective 1 regions are concerned, and by the Guarantee section in other cases. Accordingly, where funding originates in the Guarantee section of the EAGGF, it is programmed independently, according to the provisions of Regulations 1257/99 and 1750/99, while where funding comes from the Guidance section of the EAGGF, its programming constitutes an integral part of regional policy, programmed and governed by the CSFs of each country, according to the provisions of Regulation 1650/99.

The majority of the measures included represent the re-articulation of old independent rural development measures that have been now combined under a common regulation. Bryden notes that it is essentially a "repackaging of measures some of which have been around since 1972, directed at farmers, rather than the rural population" (2000, p.15) and that the essential change characterizing the new rural policy regards issues of governance and the perception of rural development, while it remains to be seen whether this new 'rhetoric' will translate into reality.

The present analysis of the integration of rural and regional policy will focus not so much on the content of the new rural development measures, as on the issue of whether the necessary preconditions for the territorial integration of rural policy with regional policy concerns and practises are in place.

The substantive integration between EU rural and regional development policies
The new approach to rural policy[13] is characterized by the adoption of a territorial understanding of the rural economy and a shift away from a uni-dimensional agricultural towards a multi-dimensional rural development policy. In this sense, rural policy not only comes close to the working of regional policy but essentially confers a rural focus on the latter's goal of territorial cohesion.

This conception is not endorsed, however, by all policy actors involved. Side-by-side with the new understanding, there exist intense political pressures for a 're-agriculturalization' of rural policy, expressed by proponents of the traditional understanding of the function of rural space. This conflict, far from being strictly theoretical, is underpinned by important practical and political concerns related to the small voting power of rural areas and the very large capacity of farmers for raising direct action against new measures (Ward, 2000). The latter concern has to do with the huge direct costs (income reduction, land depreciation, etc.) to be borne by farmers in the case of a transition from farm support to rural development (Schrader, 2000) and a lack of enthusiasm with the new development outlook on their part (Ward, 2000). The conflict is also buttressed by many agricultural administrations, which, when and where they are in relative power *vis-a-vis* other administrative bodies, seek to internalize differing concerns into agricultural policy (Lowe, 2000)

In the case of Greece, it has been demonstrated that, while in theory rural development policy seems to be favouring a non-agricultural stance towards rural development, in practice rural development has a clear agricultural sector bias (Papadopoulos and Liarikos, 2003). In the case of Spain, Sumpsi (2000) distinguishes also between an 'agriculturalist' and a 'ruralist' vision for the future of rural Spain, with the agriculturalist vision prevailing in the implementation of rural development policies (p.6). In examining the UK case, Ward (2000) mentions the lack of a clear definition of rural development and discusses the problem of leverage between different departments and services, identifying a pro-agriculturalist view among politicians.

This divide is encountered not only within but also between different MS states. The negotiations for the Agenda 2000 reform demonstrated a clear divide between countries favouring (Denmark, Sweden, Netherlands and the UK) and countries opposing the reform (lead by France) (Swinnen, 2001). The considerable strengthening of the pro-reform MS coalition in the events preceding the 'Fischler reform' is attributed to societal pressures for food security, together with the active participation of consumer and environmental groups in the discussions (Swinnen, 2001). Lowe (2000) ascribes the differing national stances to differences in administrative structures (sectoral vs. territorial), territorial outlooks of public services and lobbying powers of environmental agencies.

The proposals for the new 2007-2013 rural development regulations (CEC, 2004d) prescribe the creation of an all-new fund for rural development (the

---

[13] See also the discussion of Chapter 4, this volume.

EAFRD). Based on what has been discussed so far, this foreseen evolution can be perceived as one more reversal in the power balance between the two opposing coalitions, or a capitalization of the pressures for a re-agriculturalization of the rural policy. It could be perceived, however, as an attempt to better integrate rural and agricultural policies, preserving the close bonds of the former with regional policy.

The procedural/spatial integration between EU rural and regional development policies With the procedures for the implementation of rural development policy essentially adhering to those of regional policy, it could be easily claimed that the procedural integration of the two is almost complete. Indeed, especially in those cohesion countries where both policies are implemented under a common CSF, the level of procedural integration is high, taking advantage of the administrative capacity provided by the CSF administration and the almost identical procedures.

Implementation experience, however, has revealed several problems that put in question the actual level of procedural integration between the two policies. Competition and conflict between the rural development agencies themselves and with the central administration (Sumpsi, 2000), administrative compartmentalization, and isolation of agricultural services frequently render coordination between different services at various spatial/organizational levels involved in policy implementation problematic (Lowe, 2000). This is the case sometimes for the same issue in the same territorial unit (Sumpsi, 2000; Papadopoulos and Liarikos, 2003a).

An important issue concerns the divide between agricultural and economic or environmental administrative services, which is often strong enough to bar the coordination of the two policies at the implementation level. Collins and Louloudis (1995), in discussing the prevalence of a strong bureaucratic prerogative against policy reform in Greece, note that the bureaucracy of agricultural interests maintains a divide between 'insiders' (agricultural services), whose views are 'legitimate', and outsiders (those expressing alternative concerns for the development of rural space), whose views are presented as 'illegitimate'. This administrative and conceptual divide may be seen as supporting the agricultural bias in the practice of rural policy (Papadopoulos and Liarikos, 2003a).

The adequacy of institutional capacity to implement these policies is another serious concern. In both Spain and Greece, rural assistance accrues mostly to intermediate or well-developed areas, as the poorer areas lack the required critical mass of institutions to design and implement project proposals (Sumpsi, 2000; Papadopoulos and Liarikos, 2003b). Sumpsi (2000) argues that the spatial level of the reference of the policy (the region) is not appropriate for the issue, while he suggests that the very principle of bottom-up policy making may have to be questioned as far as poorer regions are concerned. In his view, the only appropriate structures are the Local Action Groups created by the LEADER initiative which agrees with the view of the LEADER as the only purely territorial rural development policy measure. Ward (2000) diagnoses the same lack of institutional capacity in the case of the UK and claims that the only regions that proved to be more capable to manage rural development assistance were those formerly receiving assistance under Objective 5b of the SF.

Synopsis The preceding brief analysis suggests that the substantive integration of rural and regional policy is satisfactory, although still missing those political prerogatives that could guarantee its permanence. Presently, it is too early to conclude whether the innovative nature of the new Rural Development Regulation will indeed develop into a 'special-interest' segment of a general policy for territorial cohesion, or whether it will regress into a slightly more spatially-focused component of agricultural policy. Proposals for the RDRs of the fourth programming period seem at this point to support the second option.

The procedural integration of the two policies is adequate at the higher management tiers, as both policies are implemented under the same planning framework. However, moving down the spatial/organizational hierarchy it gradually weakens as strong colluding forces are met, fed by administrative compartmentalization and the dominance of agricultural interests that see rural policy as related to agriculture only. Inappropriate horizontal and vertical coordination between departments and services, as well as inadequate institutional capacity at sub-national levels, only act to reinforce this situation. As, in Bryden's (2000) words, the reality of rural policy implementation differs from the rhetoric supporting it, the operational integration of rural with regional policy has not materialized yet.

EU Regional Policy and Environmental Policy

Introduction The present discussion refers to the broad EU environmental policy domain that includes sectoral[14] and horizontal[15] environmental policies. The integration between regional and environmental policy is of supreme importance for three main reasons. The notion of Environmental Policy Integration (EPI), enshrined in Article 6 of the Amsterdam Treaty,[16] is an important theme in the PI discourse and still leads most relevant discussions (see, Chapter 2, this volume). Regional policy traditionally constitutes the framework through which environmental legislation has been mainstreamed and through which many policy measures (e.g. EIA) have been enforced. Lastly, environmental protection lies at the heart of combating desertification to which this chapter makes special reference.

---

[14] Such as, water resources, biodiversity, etc.

[15] I.e. policy provisions, of a procedural character mostly, concerning general environmental management issues.

[16] This Article defines EPI as: "Environmental protection requirements must be integrated into the definition and implementation of the Community policies and activities referred to in Article 3, in particular with a view to promoting sustainable development" (Lenschow, 2002).

The substantive integration between EU regional and environmental policy
Environmental protection concerns entered the design and application of EU regional policy since its very early stages. With the provision for integrating environmental concerns in Community policies present since the First Environment Action Programme (EAP) (1982-86), the first round of regional policy implementation (1989-1993) featured requirements for an elementary environmental appraisal of proposed plans. Despite these requirements, the environmental dimension did not assume an important role in plan design (Taylor et al., 2001). As the Court of Auditors (CoA) ruled in 1992, there was very little evidence of conformity with environmental policy, while the vagueness of consultation documents and the lack of co-ordination between relevant authorities diminished the opportunities for integration. As a result, the CoA required that all Community funded operations should be in conformity with Community policies "including those concerning the environment".

In response to these requirements, benefiting MS and regions, when planning for the 2nd programming period, were required to (Roberts, 2001):

- Conduct an appraisal of the state of the environment in applicant regions;
- Prepare an evaluation and assessment of the expected environmental impacts of programmes;
- Include environmental bodies and authorities in the planning of programmes;
- Demonstrate compliance with EU environmental policy and regulations.

These requirements led to a marked improvement in the inclusion of environmental concerns in regional policy design, demonstrating better overall conformity than that of the first programming period. However, considerable variations in performance between different regions still existed, while many impact assessments were inadequate, lacking consistency or detail (Roberts, 2001). From an analysis of SPDs of that period, Clement (2001) demonstrates that great variation in the extent of environmental performance existed and that, except for regulatory conformity very few cases signified an environmental integration approach.

Many of these problems were addressed in the current 3rd programming period (2000-2006), whereby the new Regional Development Regulation (CEC, 1999c) preserves the horizontal character of environmental policy and strengthens the environmental requirements of programmes. Clement (2001) considers that the most marked requirements of this regulation are:

1. A scope for differentiating rates of EU financial contribution according to regional importance attributed to the environment.
2. A provision for partnerships to include environmental organizations.
3. A scope for the ERDF to be supportive to efficient resource use and the development of renewable resources.
4. A provision for environmental concerns to form a greater part of evaluations.

As an important evolution, Roberts (2001) adds the requirement of SWOT analysis to contain an environmental element and observes that the majority of new programmes have attempted an integration of sustainable development requirements.

During the 15 years of programmatic regional policy implementation, major steps have been made towards the integration of environmental concerns in the process of planning and implementing regional policy. This process has observed the provisions of the EAPs (from the 3rd onwards) and of the Cardiff Integration Process (CEC, 1998b) recording some substantial progress. This evolution demonstrates a favourable political climate towards EPI, but does not necessarily signify that the two policies are moving towards a deeper substantive integration. This may owe to the different views the respective policy makers hold of the relationships between the two policies. Moreover, it should be taken into account that environmental policy is not one but many sectoral (and the horizontal) policies, implying a large number and diversity of policy actors and policy apparatuses involved at both the EU and the national and sub-national levels.

Regional policy recognizes the need for EPI and for observing environmental regulations, but often fails to recognize the direct relations between environment and development, as an ecological modernization approach would imply (Andersen and Massa, 2000; Hertin and Berkhout, 2001). As Werner Simon of EEB put it "Structural Funds have their own objective, economic and social cohesion, but in pursuing that goal *they have to take on board other policy requirements*" (EEB 2002, p.65, emphasis added). On its part, environmental policy, adopting a more integrated approach to the environment-development relationship, views environmental protection as an integral part of development, a view inherent in the call for sustainable development policies (Amsterdam Treaty) and often expressed by environmental organizations. A report for English Nature and the Countryside Council of Wales, for example, concludes that "mainstreaming has to overcome the problem that many stakeholders still perceive the Structural Funds as an instrument for the promotion of economic development rather than sustainable development" (IEEP 2002, p.18).

These differences in definitions are rooted in, among others, a lack of political will, absence of incentives and accountability on the part of policy makers and a 'bureaucratic prerogative', which grows to safeguard the 'business-as-usual' operation of departments and officials (Glasson and Gosling, 2001). It goes without saying that these factors combine with strong pressures for growth especially from interested pressure groups in lagging regions (Berger, 2003).

Obviously, then, the substantive integration between regional and environmental policy still leaves much to be desired in terms of true integration of their goals and objectives and their underlying belief systems. Clement (2000b) divides regional programmes in three categories, 1st, 2nd and 3rd order, according to the level of environmental integration they have pursued (Table 3.6). He notes that the vast majority of programmes fall into the 1st or 2nd categories, with major improvements taking place but with most programmes not attempting a true incorporation of environmental concerns in the definitions of their objects. Although exceptions do exist, a clear EU-wide pattern cannot be identified.

Table 3.6   Orders of environmental integration in EU programmes

1st Order integration	2nd Order integration	3rd Order integration
✓ Horizontal ✓ External Consultations or individual experts or other committees ✓ Limited Influence	✓ Vertical ✓ Committee of environment experts assessing projects ✓ Environment Budget	✓ Strategic ✓ Internal environmental expertise ✓ Generating regional environmental competitiveness

Source: Clement (2000b)

The procedural/spatial integration between the EU regional and environmental policy Despite its several shortcomings, the EU regional policy remains very progressive as regards EPI. Most of the implementation problems reported owe to its particular spatial outlook and the procedural aspects of its integration with environmental policy that are discussed in the following.

The spatial system of reference of EU regional policy are NUTS II regions whose suitability for representing economic, social or environmental phenomena is debatable. The artificiality[17] and size of these regions are the two main issues of concern when dealing with the protection and management of the environment: administrative regions may be too small to contain environmentally-defined units (such as river basins), or may not coincide with natural boundaries, and too large to provide a framework for specialized, thematic approaches to particular environmental issues or problems. Although the importance of regions and of the regional level as the proper units and level for sustainable development planning is undeniable,[18] their delineation is the critical parameter that judges their responsiveness to the goals they are assumed to serve. Ideally, a combination of administrative and environmental criteria should be employed for delineating an appropriate spatial system of reference for resolving environment-development

---

[17] A region being an administrative and not a functional unit.

[18] "The regional level constitutes a proper level for the definition of sustainable development priorities, lying in-between the national and the local level and presenting a working compromise between functionality and planning needs. Regions are appropriate planning units, large enough to contain a variety of activities resources and ecosystems but relevant to individual localities, and representing a level of governance at which it may be possible to reconcile frequently competing environmental, social and economic considerations" (Roberts, 2001, p.66). As such, regions constitute a 'meso-tier' of administration capable of coordinating central and local authorities, and thus conducive to the aims of the White Paper on Governance (CEC, 2001b) and the 6th EAP (CEC, 2001c) for wider participation and multi-level governance.

issues. For these reasons, the current spatial system of reference of EU regional policy may not be suitable for the articulation of environmental policy and, consequently, may not serve optimally the goal of integration between the two policies.

The next question concerns the existence of administrative capacity for promoting the integration of regional with environmental policy. At the policy formulation stage it is not easy to challenge the potential capacity of the Commission or of national governments to design necessary PI procedures. Rather, the problems arise from a lack of political commitment towards PI and, consequently, the under-utilization of that potential. The compartmentalization of policy making authority among the different DGs of the Commission and administrative compartmentalization at the national level hinder meaningful interactions between the two policies and the adoption of integrated approaches. Different policy-makers adopt different approaches and develop different bureaucratic prerogatives. In general, however, procedures for the negotiation of interests and, at least, the discussion of development and environment interactions are in place.

The situation becomes more complicated at the policy implementation stage, where the structures and procedures associated with regional and environmental policy vary substantially among countries and regions. Two categories of issues are discussed here: systematic issues regularly encountered in most programmes and non-systematic issues that vary widely among programmes.

Systematic issues initially concern the way in which environmental concerns are introduced in regional policy-making. Clement (2000b) identifies two approaches: programmes where the environment constitutes a cross-cutting, horizontal issue in all policy measures (characteristic of Objective 2 programmes), and programmes where the environment enters as a vertical issue (characteristic of the majority of Objective 1 programmes), constituting a separate target with a vertical organization. He argues that both approaches are flawed, in that they do not succeed in effectively integrating the environmental dimension in decision making; the first does not present a coherent outlook of environmental protection as a policy target, while the second tends to isolate environmental protection from the rest of development activities. Other recurring problems concern the lack of active technical assistance to stakeholders, insufficient dissemination of best practices and lack of rigorous definition of the administrative structure and roles of policy partners. This last point is very important as when formal decision making structures are missing, "informality and, maybe worse, lost transparency can emerge and predetermine policy making" (Berger, 2003, p.225).

Non-systematic issues arise in different settings and to varying degrees in the MS, two of which deserve special mention: the outlook for integrating the environment in planning and the specific institutional provisions put in place to coordinate and oversee environmental integration. As regards the first issue, Clement (2000a), in discussing Objective 2 SPDs, notes that two different approaches emerge. In what he calls 'generalist systems', planning and management rest with economic policy-makers who periodically consult with environmental authorities to ensure compliance with environmental legislation; in 'specialist systems', programme management is conducted in such a way as to steer economic development towards 'environmentally advanced practices'. This distinction is also valid for Objective 1 programmes possessing a tendency towards generalist systems.

The structures provided to coordinate and oversee the process of integrating environmental concerns in the implementation of regional policy are even more diversified. For example, while Italy has established a mechanism, the Network of Environmental Authorities (NEA), for the integration of the environment into SF application, Greece merely assigns such responsibilities to the Ministry of Environment, Public Works and Physical Planning, which, however, has no clear mandate over integration issues. The differentiation among different programmes is even more accentuated in Germany Objective 1 regions, whereby some regions have set up councils and procedures to implement integration (ex. Brandenburg), while some others present no structures what so ever (ex. East Berlin) (WWF Adena, 2003).

The largely insufficient structures for environmental integration mirror also on evaluation procedures which in their majority prove inadequate to cover comprehensively the environmental impacts of plans and programmes. This has essentially to do with an overall problematic structure of the SFs evaluations, which do not address policy impacts and outcomes and contain very little quantification of results[19] as well as with the thin attention paid to environmental concerns. Even in cases where SEAs are attempted, as in the case of West Wales, the results are included in a general framework where the indices used are inadequate.[20] In other cases, as in Greece, no sustainability indices are used at all, and the environmental part of the evaluations remains superficial, essentially focusing on the management effectiveness of project implementation (Liarikos, 2004).

Synopsis During the 15 years of EU regional policy implementation on the basis of programming principles, major steps have been made in the direction of more complete integration of environmental concerns in regional policy-making. In general, progress in the creation of a favourable political climate and commitment to attaining the goal of sustainable regional development must be acknowledged. However, much is left to be desired to achieve meaningful and effective EPI. The meaning and operational expression of EPI should be clarified in the context of regional policy operations, considering that the environment is not a uni-dimensional concept and entity but comprises various media, processes and

---

[19] Among others, this has been highlighted as an evaluation problem in the ex-post evaluation of the SF interventions for the second programming period (1993-1999); see, ECOTEC (2003).

[20] See, SDC (2004) for an overall assessment of indices used in the UK.

relationships. It should be recognized also that a divide in perception between economic and environmental policy makers signifies a limited potential for achieving a more complete integration between regional and environmental policies. At the policy formulation stage, a flawed approach to the way the environment enters the design of regional policy, either as a horizontal or a vertical theme, keeps it largely isolated from the rest of the policy not to mention the inevitable focus on selected environmental issues and not to the totality of the environment. Overall, an integration of the two policies, as an ecological modernization approach would imply, is not achieved.

At the policy implementation stage, progress has also been made, although great variations exist among the implemented programmes of different MS. In general, a systematic scheme for integration does not seem to exist, while some generally recurring issues concern, among others, the need for stronger technical assistance to policy partners and a better definition of the role of each partner in the process. An important issue mirroring the above is the lack of consistent impact and outcome evaluations that would facilitate better informed decision making and a stronger case for EPI. Overall, although EU regional policy remains a very progressive policy field as regards EPI, major improvements are still needed as regards both its design at the EU level and its planning and implementation at the national and regional levels.

EU Regional Policy Integration for Combating Desertification: Insights and Implications

The examination of the integration of EU regional policy with the selected policies concludes with a discussion of its implications for combating desertification in Southern European MS. Toulmin (1999) identifies three groups of policy-relevant factors pertinent to desertification control: *technical methods and means*, concerning the knowledge and instruments available for the direct control of desertification; *incentives and motives*, concerning the factors that shape the management decisions of land-users; and the *political context* that determines the more general political stance towards the issue. Regional policy belongs to the second group of factors as it actively seeks to alter the incentives faced by economic entities, in an attempt to rework the spatial distribution of activities.

Desertification is a complex socio-environmental phenomenon, resulting from numerous and diverse interactions among environmental, socio-cultural, demographic, economic, technological, and political factors (see, Chapter 1, this volume). Its control does not constitute an independent policy nor falls within the ambit of a particular, single policy; rather instruments provided by sectoral policies must be steered towards actions that alleviate the problem. Therefore, regional policy interventions should properly combine sectoral policy instruments associated with the determinants of desertification, within a certain programming framework, to address the environmental and socio-economic problems of desertification-sensitive regions. The introduction of soil conservation concerns into mainstream EU environmental policy and into the design and application of regional policy is more of a political decision; the major issue lies with the capacity of regional policy to deliver against complex policy problems, such as desertification. This depends critically on its proper integration with sectoral policies.

The findings of the preceding analysis appear much diversified, owing both to the nature of the policies examined and to the general political context of EU regional policy application. Transport policy is an extremely powerful policy concern with a strongly-founded independent character, whose integration with regional policy practice can be characterized as thin. With transport concerns and policy decisions entering regional policy externally, it is highly questionable whether the implementation of transport policy measures really furthers regional policy targets, or whether the implementation of these measures benefits from their inclusion under the regional policy funding umbrella. This has major indirect implications for desertification control. On the one hand, the majority of desertification-sensitive areas belong to those least developed and as such any shortfall that undermines the effectiveness of regional policy is bound to act contrary to the aim of combating desertification. On the other hand, transport policy measures may drastically alter the uses of land and, thus, exert pressures on soil, water and ecosystem resources. Lack of integration between these measures and other sectoral or spatial policy measures reduces the potential for controlling the transport-induced pressures of land-use change.

Rural development policy could constitute an example of a policy demonstrating a huge substantive integration potential, with previous agriculturalist approaches giving away to a multifunctional understanding of and a territorial approach to rural development. Rural policy has grown within regional policy to become an inherent part of it. The current Rural Development Regulation (1257/99) could be utilized as an example demonstrating the potential of policy integration to deliver against desertification control. At the policy formulation level, the integration between rural and regional policy provides for a more coherent approach to the development of rural space and, thus, an improved potential to balance and control the pressures exerted on environmental resources. At the policy implementation level, it offers a framework for coordinating the interventions of the two policies in space, recognizing and capitalizing on their synergies. The viability of this well-conceived integration scheme is questioned, however, as a variety of policy implementation problems arise; lack of coordination between relevant administrative departments and levels, lack of institutional capacity at the lower administrative levels, vested political interests supporting the preservation of the current policy regime, firmly founded bureaucratic prerogatives and competing conceptions of the rural space and its development, all act counter to policy's capacity to deliver against its potential. As pressures for a re-agriculturalization of rural policy are very strong, it is highly probable that they will permeate the relevant regulations of the next programming period, although too early to make an assessment.

Concerning environmental policy, despite the problems discussed previously and the fact that EPI is treated in rather general terms and not in terms of specific environmental media and issues, its integration with EU regional policy is satisfactory as the latter has actively pursued the reconciliation of economic and environmental concerns. Given the wide array of sectoral policy instruments that regional policy utilizes and coordinates within its framework, it is argued that it has developed a potential towards acting as a mechanism for enforcing EPI, especially as regards horizontal environmental concerns. The preceding analysis has not touched on the issue of combating desertification, and, thus, no direct links can be drawn. It can be asserted, however, that the integration of regional with the variety of environmental policies, especially with the EU water and biodiversity policies, could deliver valuable functions in providing a well-defined framework for dealing with the issue and facilitating the sustainable development of desertification-sensitive regions.

Concluding Remarks

EU regional policy exhibits considerable potential to become integrated with a variety of sectoral policies. Experience demonstrates that it has often acted as a 'Trojan horse' for mainstreaming, for example, environmental concerns and measures within development practice: the gradual mainstreaming of EIAs is only one of several such examples. A deepening of the integration content of regional policy could not only strengthen such mainstreaming functions but could provide a framework for the coordination of variegated policy instruments originating in various policies, achieving valuable synergies among them that are necessary to address complex policy problems, such as combating desertification.

To realize such a deepening of PI, it is essential to take actions with respect, first, to the conceptual and theoretical underpinnings of EU regional policy and their relationships to those of other EU policies and, second, to procedural issues of practical policy management and institutional capacity for PI. The first category includes initiatives to involve policy makers from the various EU policy quarters to engage in (a) a rigorous redefinition of the object, goals and targets of regional policy and of their relationships to the objects and goals of other EU and national policies and (b) a thorough discussion for the establishment of an widely accepted definition of the causes of regional disparities in Europe and of the contributions of various policy and non-policy factors acting on various spatial levels. Such a redefinition would facilitate a better mapping of the relationships between regional and sectoral policies and of their separate and/or combined contribution to regional development and regional disparities. In this context, it is necessary also to undertake a thorough evaluation of the actual impacts of regional and sectoral policies, and especially of the impacts of combinations of different policy instruments on regional convergence.

The second category targets the bottlenecks hindering the coordination between different policies and owe to administrative compartmentalization at all levels, the inertia of policy views that fail to recognize the importance of policy integration and the bureaucratic prerogatives that develop to defend the status quo. To attack these problems, high level political will towards countering strong administrative resistances and bureaucratic inertia is necessary in the broader context of the policy redefinition and mapping of relationships referred to above. More tangible measures include the clear communication of legislative requirements, a vigorous, target-oriented definition of policy integration, and the enhancement of the human and technological capacity of public sector services. Avoiding the temptation to suggest the establishment of new institutions for the promotion of PI, it is essential to redefine the role of certain administrative bodies, bestowing on certain of them clear mandates and competences for policy integration. It is conjectured that these two categories of measures will favour the amelioration of the socio-environmental development problems of desertification-affected regions as the incorporation of desertification concerns into regional policy and, hence, its sectoral instruments, faces exactly the same conceptual, theoretical and procedural matters as the general case of regional PI discussed.

This chapter made a modest and limited attempt to wrestle with certain aspects of the integration of EU regional policy with selected other EU policies with which it has great affinity and which are relevant to desertification control in Southern EU regions. Much more in-depth and interdisciplinary research is needed to explore the issue further. More aspects of the integration of EU regional with transport, rural, and environmental policy, as well as with all other EU policies, should be examined following the methodological framework proposed in Chapter 2 of this volume. In particular, the horizontal and the sectoral environmental policies (water, biodiversity, etc.) should be separately investigated as they constitute distinct sub-domains within the broader environmental policy domain. A variety of qualitative and quantitative research techniques should be employed, in addition to secondary analysis that was used here. The analysis should focus on the relationships among the objects, goals and objectives, actors and actor networks, procedures and instruments of the policies considered at the EU level as well as the same relationships across levels. It is important to investigate whether policy integration crafted at higher levels is maintained at lower levels or dissolves during implementation.

Another research strand should focus on how policy integration can cope with the inherent complexity of regional development problems and contribute to reducing regional disparities. It would be interesting to investigate, for example, how an integrated policy system may help avoid perverse cases of path dependence leading to lock-in phenomena of regions stuck at low development levels. Desertification control presents a great test bed for such research that may result in practical suggestions to combat through integrated approaches under the umbrella of EU regional policy.

References

Amin, A. and Robins, K. (1990), 'Industrial Districts and Regional Development: Limits and Possibilities', in F. Pyke, G. Becattini and W. Sengenberger, *Industrial Districts and Inter-Firm Cooperation in Italy*, International Institute of Labour Studies, London.

Andersen, M.S. and Massa, I. (2000), 'Ecological Modernization – Origins Dilemmas and Future Directions', *Journal of Environmental Policy and Planning*, Vol. 2, pp. 337-45.

Amin, A. (1999), 'An Institutionalist Perspective on Regional Economic Development', *International Journal of Regional and Urban Research*, Vol. 23(2), pp. 365-78.

Bachtler, J. and Michie, R. (1994), 'Strenghtening Economic and Social Cohesion? The revision of the Structural Funds', *Regional Studies*, Vol. 28(8), pp. 789-96.

Begg, I., Gudgin, G. and Morris, D. (1995), 'The Assessment: Regional Policy in the European Union', *Oxford Review of Economic Policy*, Vol. 11(2), pp. 1-15.

Begg, I. (1998), 'Structural Fund Reform in the Light of Enlargement', Sussex European Institute, Centre on European Political Economy, Working Paper 1, Sussex.

Berger, G. (2003), 'Reflections on Governance: Power Relation and Policy Making in Regional Sustainable Development', *Journal of Environmental Policy and Planning*, Vol. 5(3), pp. 219-34.

Boldrin, M. and Canova, F. (2001), 'Inequality and Convergence in Europe's Regions: Reconsidering European Regional Policies', *Economic Policy*, Vol 16, p. 205.

Boldrin, M. and Canova, F. (2003), 'Regional Policies and EU Enlargement', in B. Funck and L. Pizzato (eds), *European Integration, Regional Policy and Growth*, The World Bank, Washington.

Brulhart, M. and Torstensson, J. (1998), 'Regional Integration, Scale Economies and Industry Location in the European Union', CERP Discussion Paper 1435, Centre for Economic Policy Research, London.

Bryden, J.M. (2000), 'Is there a "New Rural Policy"'?, International Conference: European Rural Policy at the Crossroads, 29-1 July 2000, Arkleton Centre for Rural Development Research, Kings College, University of Aberdeen, Aberdeen.

Button, K. (1998), 'Infrastructure Investment, Endogenous Growth and Economic Convergence', *Annals of Regional Science*, Vol. 32, pp. 145-62.

CEC (1996), *The Cork Declaration*, LEADER II Magazine, Winter 1997, No.13, Special Issue, Brussels.

CEC (1998a), *Sustainable Mobility: Perspectives for the Future*, Commission Communication to the Council, European Parliament, Economic and Social Committee and Committee of the Regions, COM(98) 716, Brussels.

CEC (1998b), *Cardiff European Council – presidency conclusions*, SN.150/98, Cardiff.

CEC (1999a), *Commission Regulation (EC) 1750/1999 of July 23 1999: laying down detailed rules for council regulation (EC) No 1257/1999 on support for Rural Development from the European Agricultural Guarantee and Guidance Fund (EAGGF),*

Brussels.

CEC (1999b), *Council Regulation 1257/1999 of May 17 1999: on Support for Rural Development from the European Agricultural Guarantee and Guidance Fund (EAGGF) and amending and repealing certain regulations*, Brussels.

CEC (1999c), *Council Regulation (EC) 1260/1999 of June 21 1999: laying down general provisions on the structural funds*, Brussels.

CEC (2001a), *White Paper: European Transport Policy: time to decide*, Official publication of the European Communities, Brussels.

CEC (2001b), *European Governance: A White Paper*, COM (2001)428, Commission of the European Communities, Brussels.

CEC (2001c), *Environment 2010: Our Future, Our Choice* – The 6th Environmental Action Programme, Commission of the European Communities, Brussels.

CEC (2004a), *A new Partnership for Development: 3rd Report on Economic and Social Cohesion*, Commission of the European Communities, Brussels.

CEC (2004b), *Interim Territorial Cohesion Report: Preliminary results of ESPON and EU Commission studies*, DG Regional Policy, Brussels.

CEC (2004c), *Proposal for a Council Regulation laying down general provisions on the European Regional Development Fund*, the European Social Fund and the Cohesion Fund, COM(2004) 492 final, Brussels.

CEC(2004d), *Proposal for a Council Regulation on support for rural development by the European Agricultural Fund for Rural Development (EAFRD)*, COM(2004) 490 final, Brussels.

Clement, K. (2000a), Environmental gain and Sustainable Development in the Structural Funds, Insights from reviews of Single Programming Documents across Europe and from a Study of SEA in Sweden, Finland and Austria, Workshop Proceedings, WP 2000(9), Nordregio.

Clement, K. (2000b), *Economic Development and Environmental Gain*, Earthscan, London.

Clement, K. (2001), 'Strategic Environmental Awakening: European Progress in Regional Environmental Integration', *European Environment*, Vol. 11, pp. 75-88.

Collins, N. and Louloudis, L. (1995), 'Protecting the Protected: the Greek Agricultural Policy Network', *Journal of European Public Policy*, Vol. 2(1), pp. 95-114.

Cooke, P. and Morgan, K. (1994), 'Growth Regions under Duress: Renewal Strategies in Baden Wurttemberg and Emilia-Romagna', in A. Amin and N. Thrift, *Globalization, Institutions and Regional Development in Europe*, Oxford University Press, Oxford.

Cooke, P. and Morgan K. (1998), *The Associational Economy: Firms, Regions and Innovation*, Oxford University Press, Oxford.

Charles D., Perry B. and Benneworth P. (2004), *Towards a Multi Level science policy: Regional Science Policy in a European Context*, Regional Studies Association report, January 2004.

Dall' Erba, S. (2003), 'European Regional Development Policies: History and Current Issues', Paper of the European Union Centre, University of Illinois at Urbaa-Champaign, May 2003.

ECOTEC (2003), *Ex-Post Evaluation of the Objective 1 1994-1999: A Final Report to the Directorate General for Regional Policy*, European Commission, Ecotec Research and Consulting, Birmingham.

EEB (2002), *New Chances for Better Enforcement of EU Environmental Legislation*, European Environmental Bureau, Seminar Report November 2002, Brussels.

ESDP (1999), *European Spatial Development Perspective: Towards Balanced and Sustainable Development of the Territory of the EU*, European Commission, Office for Official Publications of the European Communities, Luxembourg.

European Court of Auditors (1992), *Special Report No 3/92: Concerning the Environment*, Official Journal of the European Communities, OJ C245, Brussels.

Glasson, J. and Gosling, J. (2001), 'SEA and Regional Planning – Overcoming the Institutional Constraints: Some Lessons from the EU', *European Environment*, Vol. 11, pp. 89-102.

GMF (2003), *Interim Report on the 3rd Community Support Framework*, Greek Ministry of Finance, Management Authority for the 3rd CSF, Athens (in Greek).

Goodbody Consultants (2003), *Transport and Regional Development*: a report in association with the Department of Urban and Regional Planning UCD and Oscar Faber Transportation, Dublin.

Harrison, B. (1991), 'Industrial Districts: Old Wine in New Bottles?', *Regional Studies*, Vol. 26(5), pp. 469-83.

Hertin, J. and Berkhout, F. (2001), *Ecological Modernization and EU Environmental Policy Integration*, University of Sussex, Science and Technology Policy Research, Working Paper No 72, Sussex.

IEEP (2002), *Environmental Sustainability in UK Structural Funds Programmes, 2000-2006*, Institute for European Environmental Policy Report for English Nature and the Countryside Council for Wales, (available online at www.ieep.org.uk).

Ioannides, Y. and Petrakos, G. (2000), 'Regional Disparities in Greece: The performance of Crete, Peloponese and Thessaly', EIB Papers 5(1).

Krugman, P. (1998), *Development, Geography and Economic Theory*, MIT Press, Cambridge.

Liarikos, C. (2004), 'Regional Policy in Greece: Brief Overview, Description of the 3rd CSF and an Analysis of its Relevance to Environmental Protection ', World Wide Fund for Nature, WWF-Greece, Working Paper 3/2004, Athens (in Greek).

Lenschow, A. (2002), 'Greening the European Union: An introduction', in A. Lenschow, (ed.), *Environmental Policy Integration: Greening sectoral Policies in Europe*, Earthscan, London, pp. 1-21.

Liebowitz, S.J. and Margolis, S.E. (1995), 'Path Dependence, Lock in and History, Journal of Law', *Economics and Organization*, Vol. 11, pp. 205-26.

Lowe, P., Flynn, B., Just, F., De Lima, A.V. and Povellato, A. (2000), 'National Cultural and Institutional Factors in CAP and Environment', in F. Brouwer and P. Lowe, *CAP Regimes and the European Countryside: Prospects of Integration between Agricultural, Regional and Environmental Policies*, CABI Publishing, New York.

Markusen, A. (1996), 'Sticky Places in Slippery Space: A typology of Industrial Districts', *Economic Geography*, Vol. 72(3), pp. 293-313.

Martens, B. (2004), *Cognitive Mechanics of Economic Development and Social Change*, Routledge, London.

Martin, P. (1999), *Are European Regional Policies Delivering?*, EIB Working Papers 4(2).

Martin, P. (2003), 'Public Policies and Economic Geography', in B. Funck and L. Pizzato (eds), *European Integration, Regional Policy and Growth*, The World Bank, Washington.

MEPPW (1997), 'Spatial effects of Community Policies: Conclusions from the research project', Greek Ministry of Environment, Planning and Public Works in cooperation with the Aristotelian University of Thesaloniki, Spatial Development Research Unit, Thesaloniki, project coordinator Kafkalas (in Greek).

Ormerod, P. (1998), *Butterfly Economics*, Faber and Faber Publishers, Suffolk.

Papadopoulos, A. and Liarikos, C. (2003a), 'The Rural Development Policy Network in Greece: Issues of Policy Implementation and Policy Integration', XXth European Society for Rural Sociology Congress: "Work, Leisure and Development in Rural Europe Today", Sligo, Ireland, 18-22 August 2003.

Papadopoulos, A. and Liarikos, C. (2003b), 'Towards What kind of Development for the Less Favoured Areas of Greece?', Scientific Conference: Less Favoured Areas and Development Strategies: Economic, Social and Environmental Dimensions and Support Mechanisms, 21-22 November 2003, University of Aegean, Mytilene: (in Greek).

Puffert, D. (2003), 'Path Dependence', EH.Net Encyclopaedia, R. Whaples (ed.), URL: http://www.eh.net/encyclopedia/contents/puffert.path.dependence.php.

Roberts, P. (2001), 'Incorporating the environment into Structural Funds Regional Programmes: Evolution, Current Development and Future Prospects', *European Environment*, 11, pp. 64-74.

Rodriguez-Pose, A. (1994), 'Socioeconomic Restructuring and Regional Change: Rethinking Growth in the European Community', *Economic Geography*, Vol. 70, pp. 325-43.

Rodriguez-Pose, A. and Fratesi, U. (2004), 'Between Development and Social Policies: The Impact of European Structural Funds in Objective 1 Regions', *Regional Studies*, Vol. 38(1), pp. 97-114.

Schrader, J.V. (2000), 'From a Common Agricultural Policy to a Common Rural Policy in the EU?', International Conference: European Rural Policy at the Crossroads, 29-1 July 2000, Arkleton Centre for Rural Development Research, Kings College, University of Aberdeen, Aberdeen.

SDC (2004), *Shows Promise But Must Try Harder: An Assessment by the Sustainable Development Commission of the Governments Reported Progress on Sustainable Development Over the Past Five Years*, Sustainable Development Committee, London.

Sengenberger, W. and Pyke, F. (1992), 'Industrial Districts and Local Economic Regeneration: Research and Policy Issues', in F. Pyke and W. Sengenberger (eds), *Industrial Districts and Local Economic Regeneration*, ILO Publicatons, Geneva.

Sumpsi, J.M. (2000), *Actors, Institutions and Attitudes to Rural Development: The Spanish National Report*, Report to the World-Wide Fund for Nature and the Statuory

Countryside Agencies of Great Britain, Department of Agricultural Economics, University of Madrid, Madrid.

Steeck, W. (1991), 'On the institutional Conditions of Diversified Quality Production', in E. Matzner, and W. Streeck (eds), *Beyond Keynesianism: The socio Economics of Production and Full Employment*, Edwar Elgar, Vermont.

Swinnen, Johan S.M. (2001), 'A Fischler Reform of the Common Agricultural Policy?', Centre for European Policy Studies (CEPS), Working Document No. 173, Brussels.

T&E (2002), 'Trans European Transport Networks: Options for a Sustainable Future', Common NGO policy document, co-signed by T&E, WWF, Birdlife International, CEE Bankwatch and Friends of the Earth, Transport and Environment (T&E) 03/02, Brussels.

Taylor, S., Polverari, L. and Raines, P. (2001), 'Mainstreaming the horizontal themes into structural funds programming', IQ-NET Thematic Paper 10, Glasgow.

Toulmin, C. (1999), 'International Experience in Desertification Policy: the Global Picture', in P. Balabanis, D. Peter, A. Ghazi, and M. Tsogas (eds), *Mediterranean Desertification: Research Results and Policy Implication*, European Commission DG Research, Brussels.

Trigilia, C. (1992), 'Italian Industrial Districts: Neither Myth nor Interlude', in F. Pyke and W. Sengenberger (eds), *Industrial Districts and Local Economic Regeneration*, ILO Publicatons, Geneva.

Tsoukalis, L. (1997), *The New European Policy Revisited*, Oxford, London.

Van Depoele, L. (2000), 'The European Model of Agriculture (EMA): Multifunctional Agriculture and Multisectoral Rural Development', International Conference: European Rural Policy at the Crossroads, 29-1 July 2000, Arkleton Centre for Rural Development Research, Kings College, University of Aberdeen, Aberdeen.

Vickerman, R.W. (ed.) (1991), *Infrastructure and Regional Development*, Pion Ltd, London.

Vickerman, R.W. (1995), 'The Regional Impacts of Trans-European Networks', *Annals of Regional Science*, Vol. 29.

Vickerman, R.W., Spiekerman, K. and Weneger, K. (1999), 'Accessibility and Economic Development in Europe', *Regional Studies*, Vol. 33(1), pp.1-15.

Ward, N. (2000), *Actors, Institutions and Attitudes to Rural Development: The UK National Report*, Report to the World-Wide Fund for Nature and the Statuory Countryside Agencies of Great Britain, Department of Geography, University of Newcastle upon Tyne, Newcastle.

WWF Adena (2003), Comparative study for the integration of the environment in the 2000-2006 Structural Funds Programmes in Various Member States of the EU, Report to the Spanish Ministry of Environment by WWF Adena, Project Coordinator: Guy Beafuy (in Spanish, appendices in English).

Chapter 4

EU Rural Development Policy: The Drive for Policy Integration Within the Second Pillar of CAP

Apostolos G. Papadopoulos

Introduction

The EU rural development policy (RDP), commonly known as the second pillar of the CAP, has been heralded as an integrated policy designed to address the multifunctional development needs and the complex environmental and governance problems of rural areas. This chapter examines the rhetoric and the reality regarding policy integration (PI) within the second pillar of the CAP with the aim to assess its current extent and form and future prospects and to inquire the reasons for the problems identified. More specifically, the examination concerns the coherence and balance of the main instrument of the RDP, namely the Rural Development Regulation (RDR), and the integration of the RDP with other Community policies, in both a theoretical/conceptual and a procedural sense.

This section briefly discusses the new conceptualization of rurality and of rural development and its influence on the theoretical underpinnings and conceptualization of the RDP. The next section offers a selective description of the main characteristics of the CAP, the mother policy of the RDP, the emergence and the milestones of the RDP domain, and the RDR. The analysis of the PI in the context of the RDR is undertaken then, focusing on the vertical integration within the RDR and on the horizontal integration between the RDR and two important Community policies, the environmental and the regional policy. Lastly, the main findings of the analysis are summarized and reconsidered in view of the new conceptualization of rural development.

Since the late 1980s, rural development has gradually gained a prominent position in the policy agendas of the European Union (EU) and the member states (MS) among several other interdependent issues such as continuing urban growth, urbanization of the countryside, regional inequalities and the crisis of the agricultural sector. The introduction of the concept of rural development into public policy needs further exploration to examine how important are the terms 'rural' and 'rurality' in the new context, and if they are still relevant for conceiving and analyzing policy actions.

The evolution of modernity has led to the demise of the self-evident connotations of the 'rural', necessitating at the same time the reconsideration of its meaning. Two principal conceptualizations of the terms 'rural' and 'rurality' are found in several recent studies: the *rural as locality* and the *rural as social representation* (Mormont, 1990; Halfacree, 1993; Gray, 2000; Terluin, 2001). The first refers to empirical traits of rural areas such as physical and population characteristics, social and economic organization, social relations and relations with urban and other areas. On the basis of this 'positivist' conceptualization, it is suggested that rurality belongs to the past and that the concept of 'rural' has eroded and needs to be replaced by other more encompassing concepts (e.g. regional, local, etc.). The second, 'constructivist' conceptualization involves a de-spatialized cultural image of rurality which may reflect landscape images, remote areas and/or idealized reflections of society. The rural does not constitute a unidimensional reality, but it is something experienced and socio-culturally constructed.

The current dilemma is that the 'rural' ought to encompass the socioeconomic, cultural and spatial diversity of the countryside or it will perish. 'Rurality' can be viewed as "the ongoing co-production of man and nature" (van der Ploeg, 1997, p.42), meaning that both society and nature are contingent and heterogeneous and that the way they interact does not imply any determination. The encounter between society and nature constitutes a rurality which is not exclusively agricultural; it rather possesses various economic, social, political and cultural facets. The contemporary 'expansion of rurality' (van der Ploeg, 1997) crosses the boundaries of several conventional distinctions, such as between production and consumption.

Drawing on the vast pertinent literature, four important points inform the notion of rural development:

1. The distanciation between rurality and agriculture has led to a conceptual discomfort; this is further intensified as it is also becoming difficult to define agriculture due to the introduction of industrial, (bio)technological and post-industrial processes which challenge the very essence of primary production (Friedland, 2002).
2. The contested, diverse and imbued with numerous meanings notion of rurality can be considered as a 'hybrid' which conflates initially separate features on a common spatial platform (Whatmore, 2002).[1]
3. The re-conceptualization of rurality, resulting from complex interactions and conflicts between globalizing systems and rural particularities, owes to the contemporary importance ascribed to agency; the 'rural' is interpreted and

---

[1] However, the hybrid geographies are "partial, provisional and incomplete" (Whatmore, 2002, p.7).

experienced by various social actors and is embodied in their practices (van der Ploeg, 1997).

4. The reconstructed rurality, which bridges the gap between geographical localities and their social representation, is the result of re-invention in the context of the EU policy serving different purposes (Gray, 2000).

These theoretical considerations do not point towards a coherent view of rurality or of rural development, but rather stress the pluralistic and/or fragmented nature of any attempt to define these terms. Thus, there is an inherent general weakness in constructing a 'theory in rural development', a problem frequently attributed to the absence of a single, widely accepted, and formally defined model of rural development. Moreover, there is a number of factually-based notions or informal models of rural development. However, it seems more sensible to accept that a theory of rural development may emerge through the systematic analysis and consideration of national rural development practices (Ward and Hite, 1998; van der Ploeg et al., 2000).

The development process of rural areas has gradually displaced agriculture from its dominant position. Rural development has become a contemporary issue of socioeconomic practice and policy, expanding beyond the agricultural modernization project and the practices of farm families and encompassing many actors, sectors, practices, spatial contexts and innovative policy measures (van der Ploeg et al., 2000).[2] Thus, rural development should be conceived as *a multi-level, multi-actor and multi-faceted process* (van der Ploeg, 2003). It represents more than the diversification of the rural economy and the commodification of rurality and can be identified with changes in power relations, involving new contests and conflicts, new conceptions of rurality and new forms of participation in the countryside (O'Hara, 2002).

Rural development policy, taking its examples from rural development practices, has expanded its focus to incorporate wider issues than those pertaining to the agricultural sector. RDP is conceived as an integrated policy concerned with the immanent linkages between economic sectors, policy issues/themes, political and administrative domains, spatial levels, social actors and policy processes. Thus, the concept of rural development expands further, encompassing too many (often inconsistent) characteristics/objectives within the same policy domain often making the distinction between rural and regional development extremely difficult (Bryden, 2000).

---

[2] A recent review argues that a "new rural development paradigm" has been, actually, constructed through the cumulation of development practices and the formulation of policy measures and guidelines which challenges the "modernization paradigm" (van der Ploeg et al., 2000). The rural development paradigm provides a number of alternatives to modernization, although agriculture still holds a central place in rural development (van der Ploeg, 1997). However, the construction of *an empirically grounded theory* of rural development practices is still far from being achieved.

The recent emergence of the EU RDP is the result of a policy shift from the sectoral (agricultural) to the territorial (spatial) development of the countryside. Some have viewed this movement as a promising, although yet incomplete, challenge (Bryden, 2000) while others are more critical, considering the implicit dangers arising from the predominance of a spatial perspective over rurality and its agricultural particularity (Richardson, 2000; Hadjimichalis, 2003).[3]

Saraceno (2002) contends that it cannot be said whether EU rural policy has been developed in recognition of the need to deal with the specific problems of rural areas or because of the need for CAP reform. She claims that, in the context of the RDP initiated as the 'second pillar' of the CAP, there are two functions or visions of rural development. The first addresses the structural needs of the agricultural sector in a reformed CAP (sectoral function) while the second addresses the development of the countryside through the use of a multisectoral, integrated approach (territorial function). Both functions are considered legitimate and relevant for rural policy and, therefore, the way forward appears to be their compromise and articulation under the inscription of rural policy.[4] However, a number of pressing questions arise in relation to this seemingly rational suggestion: How this compromise/ articulation will be achieved? Which actors will advocate or fight against it? What are the social forces behind the articulation and/or the disarticulation of the two functions? Who will benefit from such a development? What is the likely outcome of a potential confrontation of social actors on the issue of articulation?

The challenge of articulating and implementing a RDP leads to the essentially political issue of PI. PI is conceived variously and has been approached from many directions.[5] A normative, rationalist conception of PI sees it as the unification of different policy elements into a whole, materialized through consensual mechanisms, while a functionalist, instrumental conception sees PI as the incorporation of certain concerns into a particular policy whose realization tends to be hierarchical (Persson, 2002). Because the meaning of integration depends on who advocates or fights against it, the question of power and hierarchy is raised, a reality that cannot be easily sidestepped when advocating PI.

PI can be horizontal or vertical or both. In a horizontal sense, PI concerns more what Peters (1998) calls "policy coordination", which should be a basic priority of public sector officials. It is considered as an end-state where policies are

---

[3] From a 'neutral' spatial perspective, the central issue is something like starting from scratch and asking how to re-integrate agriculture into the countryside. The emphasis on the 'spatial' aspect actually institutionalizes the 'invisibility' of rurality and of its particularity and culture and, therefore, preludes the marginalization of agricultural interests, the dislocation of local communities and the legitimization of an 'urban bias' in RDP (see, also, Hadjimichalis, 2003).

[4] 'Rural policy' is a term somewhat wider than that of 'rural development policy' which is a more formalistic expression. However, both terms are used interchangeably in the text.

[5] For a general discussion of PI, see Chapter 1, this volume.

characterized by minimal redundancy, incoherence and lacunae.[6] "Coordination is inherently a political process" (Peters, 1998, p.300) because the politics of coordination reflect the relative powers of interest groups and, therefore, the set up of specific policy priorities and hierarchy.

However, the dominant conception of PI seems to be that of vertical integration, i.e. of incorporating social, economic and environmental concerns into an extant policy. Naturally, a complete integration of policies involves both vertical and horizontal integration.

A central question pertaining to PI is what should be integrated: policy objectives, decision-making structures, knowledge and capabilities, or policy instruments? (Hertin and Berkhout, 2003). Some approaches consider the integration or coherence of *policy outcomes* (objectives or practices) in specific policy domains while others refer to the *process* of integrating policies (Jacob and Volkery, 2003). Thus, two broad possibilities arise: (a) to study the coherence of individual policies and (b) to analyze the process of integration of different policies. Both possibilities are important and require equal attention.

Most of the emphasis of the PI literature is on the procedural rather than on the substantive aspects of PI (Lenschow, 2002). The move towards PI implies that there is a transition process towards a new type of policy-making (Hertin and Berkhout, 2003; Jacob and Volkery, 2003). PI can be argued to be part and parcel of a new type of governance in the EU context (Hooghe and Marks, 2001; Eberlein and Kerwer, 2002; Berger, 2003) which goes hand-in-hand with the process of Europeanization or European integration (Bache, 2003; Zahariadis, 2003).

In addition to the theoretical challenges, several problems in conceiving and analyzing PI exist in practice. On the one hand, particular interests have invested in the current sectoral, less-integrated policy-making system and which certainly oppose any movement towards PI in their policy domains. On the other hand, competing interest groups favour the movement towards PI in particular policy domains. Both strands have reasonable arguments and are frequently associated with particular national contexts.

PI is not a unified process proceeding with the same speed in all policy domains. Zahariadis (2003) uses two broad criteria, *complexity* and *coupling*, to answer the question of: "why is there more integration in some areas of the EU than in others." Complexity refers to the nature of interaction; systems characterized by complex interactions contain more unplanned and/or unexpected sequences, which are not immediately comprehensible. Coupling refers to the strength of connectivity among units or actors. The higher the dependence of one unit upon others for its performance, the more tightly coupled the system will be. Zahariadis (2003) suggests that low complexity and tight coupling make PI more likely. Contrary to conventional wisdom, he contends that PI is not only a matter of

---

[6] Peters (1998) makes the important point that integration within a policy may reduce the capacity to coordinate across policies.

will and choice, but also the unintended consequence of greater interaction at the EU level.

The EU Rural Development Policy

The CAP Context

The CAP is the most important policy of the EU in terms of expenditure. In 1998, the European Agricultural Guidance and Guarantee Fund (EAGGF) represented the 55 per cent of the EU budget (€ 43.3 billion). For the period 2000-2006, the total appropriation of agricultural expenditure will represent 46 per cent of the EU budget (CEC, 2001).

When it was created in the 1960s, the CAP aimed at supporting countries facing problems of food shortages; hence, its objectives were dominated by production-based considerations. Since then, agriculture has declined as an economic activity while other issues came to the fore, especially in the period following the Agenda 2000, such as food security and the protection of the environment, defining the issues of the future agricultural and rural policy.

Agenda 2000 was an initiative of the Commission to reform the CAP and the Structural Funds (SFs) for the period 2000-2006, deriving its impetus from two interconnected types of drivers:

- Global World Trade Organization (WTO) agreements and the pressure to reduce price support in farm subsidies and decouple subsidies from production in the CAP;
- The prospect of EU enlargement to include new MS in Eastern Europe that would make the cost of the current arrangements insupportable.

To meet the new goals of the CAP, Agenda 2000 proposed the formulation of two interrelated policy domains considered as the two pillars of the CAP: (a) the market-oriented agricultural policy (pillar one) and (b) the rural development policy (pillar two). This reform implied a movement from a sectoral to a spatial policy, which was thought to reconnect agriculture with the local community opening the prospect for the 'regionalization' or the 're-territorialization' of rural policy. Moreover, this meant a shift in emphasis in farm subsidies from price support to a more integrated approach which, nevertheless, treats agriculture as an essential and integral part of rural areas. The changing emphasis towards 'integrated' rural development at the EU level has been reflected in subsequent proposals to reform the CAP.

The 1999 CAP reform was a significant step forward in the agricultural policy reform process. For many writers, this reform represented a deepening and an extension of the 1992 CAP reform.[7] It included a significant reform of Common Market Organizations (CMOs) with a further shift from price support to direct payments and a consolidation of rural development policies in view of the second pillar of CAP.

A number of important changes between the old and the new objectives of the CAP can be observed in Table 4.1. There is a movement from favouring increases in productivity and yields to supporting the competitiveness of the agricultural sector, which stresses the primacy of market forces, and a qualitative change towards the support for public goods in the countryside in justifying market intervention with respect to the production of agricultural commodities. The emphasis on farmers' living standards and income expands to include the whole rural community based on the wider rural economy. The issue of food security is not treated any more in commodity terms but there is a clear emphasis on public health and quality. The model of agricultural modernization seems to be substituted by agricultural diversity. The environment and amenity are added in the agricultural policy agenda.

Table 4.1   The evolution of the objectives of CAP

Old CAP objectives art. 39 of the Treaty of Rome, 1957	*The new objectives of CAP following the Götenborg Summit of 2002*
Increasing agricultural productivity	A competitive agricultural sector
Ensuring a fair standard of living for farmers	Production methods that support environmentally friendly, quality products that the public wants
Stabilizing markets	A fair standard of living and income stability for the agricultural community
Warranting food security	Diversity in forms of agriculture, maintaining visual amenities and supporting rural communities
Ensuring reasonable prices for consumers	Simplicity in agricultural policy and the sharing of responsibilities among Commission and MS
	Justification of support through the provision of services that the public expects farmers to provide

Source: CEC, 2002.

---

[7] Some radical reform-minded writers argue that the 1999 reform was compromised at the very end, avoiding the confrontation with the established agricultural interests which favoured inaction against the current system of agricultural support (Daugbjerg, 1999; Ackrill, 2000). Others find evidence on the 'path dependency' of CAP reform, which explains the poor record of radical reform (Kay, 2000; 2003).

However, environmental goals were incorporated explicitly in the CAP only after the Agenda 2000. Finally, there are signs of renationalization of rural policy accompanied by a plea for the simplification of agricultural policy.

Table 4.2 presents the appropriations of the two pillars in terms of the budgeted expenditure of the CAP in the period 2000-2006. The RDP measures represent an expenditure of only 10 per cent of the total CAP funding, an indication that the importance of rural development measures is rather claimed than actual.

Table 4.2　　CAP expenditure 2000-2006 (billion €, 1999 prices)

	2000	2001	2002	2003	2004	2005	2006	Total
Total CAP	40.92	42.80	43.90	43.77	42.76	41.93	41.66	297.74
Markets	36.62	38.48	39.57	39.43	38.41	37.57	37.29	267.37
Rural development	4.30	4.32	4.33	4.34	4.35	4.36	4.37	30.37
(2) as % of total	*10.5*	*10.1*	*9.9*	*9.9*	*10.2*	*10.4*	*10.5*	*10.2*

Source: CEC (1999, p.5).

The Evolution of the EU RDP

It is important to stress that the EU RDP has been elaborated as an accompanying policy to the CAP rather than as an autonomous policy domain. Saraceno (2002) distinguishes three periods or 'waves' of policy making:

a) *Since the mid-1960s* a first generation of measures, intended to provide accompanying policy for market support, addressed the modernization of farm structures based on a compensatory approach. The side effect of these measures was to facilitate the exit of small subsistence farms and to compensate farms operating in marginal agricultural land, i.e. in Less Favoured Areas (LFAs).

b) *In the second half of the 1980s*, a second set of measures introduced the territorial function of the CAP. The pressure to accommodate the development needs of the newly entered southern European MS, partly expressed in the design and implementation of Integrated Mediterranean Programmes (IMPs), revealed the need to tie together different forms of agricultural intervention, adapting them to the specific needs of rural areas within a specific time period (multi-year programming). This shift is signified by the reform of the SFs and the rising problematic on *The Future of Rural Society* (CEC, 1988) both of which were initiated in 1988. The SF reform allowed more flexibility to regional authorities in designing and programming policy interventions and in implementing multisectoral programmes. It was recognized then that the modernization of farm structures and the sectoral approach were not sufficient to support the viability of rural areas. In the beginning of the 1990s, the launching of LEADER Community Initiative was an actual pilot implementation of territorial policy functions in rural areas. By 1992, the McSharry reform, which introduced a set of accompanying measures (agri-environmental, extensification and afforestation) to the existing agricultural policy support measures, significantly challenged the rationale of the CAP and brought a major policy change in the form of support to farmers by shifting emphasis from price subsidies to direct payments (Baldock and Lowe, 1996; Brouwer and Lowe, 2000; Buller, 2002). In this period the territorial function seems to develop while the sectoral function comes gradually under criticism.

c) *Since the second half of the 1990s* a third set of measures, more fully developed in Agenda 2000, aimed to incorporate both the sectoral and territorial functions into the CAP. In 1996, the Cork Declaration, at the conclusion of a European Conference on Rural Development, emphasized, among others, the particular cultural, economic and social value of Europe's rural areas and claimed that "sustainable rural development must be at the top of the rural development agenda". RDP should be multidisciplinary, multisectoral and with a clear territorial dimension (CEC, 1996).

The design and implementation of RDP cannot be seen apart from the evolution of the CAP in general. Table 4.3 outlines the main recent events, agreements and decisions taken in relation to the formulation of rural policy in the EU context. The period 1992-1996, was a critical and preparatory phase for the emergence of RDP. The second period was the phase of RDP initiation which took some years until the relevant policy matured. The third period was the implementation phase, when the first problems were recognized and set the stage for the fourth, very recent period (2002-2003). This is the agricultural and rural policy review phase, during which a further reform of the CAP was completed and a number of commitments were taken for the next programming period 2007-2013.

The Rural Development Regulation

The Agenda 2000 reform has signified the passage to a new era for the integrated development of rural areas. The new RDP was designed to meet a number of contemporary needs, aiming to put in place a consistent and lasting framework for guaranteeing the future of rural areas and promoting the maintenance and creation of employment. The basic principles of this policy, according to the Commission, are the following:

- The *multifunctionality of agriculture*, i.e. its varied role over and above the production of foodstuffs. This implies the recognition and encouragement of the range of services provided by farmers;

- A *multisectoral and integrated* approach to the rural economy in order to diversify activities, create new sources of income and employment and protect the rural heritage;
- *Flexible* aids for rural development, based on subsidiarity and promoting decentralization, consultation at regional, local and partnership level;
- *Transparency* in drawing up and managing programmes, based on simplified and more accessible legislation.

The RDP aims to be an integrated policy governed by a single instrument to ensure better association between its two pillars, the market and the rural development policy, and to promote all aspects of rural development by encouraging the participation of local actors. Its main goals include: (a) strengthening the agricultural and forestry sectors, (b) improving the competitiveness of rural areas, and (c) preserving the environment and the rural heritage. These strategic goals are translated into a significant number of objectives which are brought together under the same framework (see Table 4.1).

The main instrument of the RDP is the Rural Development Regulation (RDR)[8] which established the framework for Community support for sustainable rural development beginning January 1st 2000. An additional tool of the RDP is the LEADER+ Community Initiative, which integrates and networks the projects completed during the former programming periods.

The RDR was explicitly designed to accompany and complement other instruments of the CAP and the Community's structural policy. It is an attempt not only to organize and simplify the existing policy agenda, but also to launch a 'menu' of measures of which only the AEMs and the LFA measure are compulsory. It is an umbrella Regulation which collects together and builds on measures representing nine policy themes which were separate in the past; namely the measures funded under agricultural structural policy (Objective 5a), agricultural measures in regional policies (Objective 5b) and the older accompanying measures introduced through the 1992 CAP reform.

These measures fall into two broad groups which illustrate the convergence of objectives under the rural development agenda: (a) measures aiming at the modernization and diversification of agricultural holdings and (b) accompanying measures of the 1992 reform generally aiming at introducing environmental concerns in agriculture.

---

[8] Council Regulation No 1257/1999, of 17 May 1999 on *"Support for Rural Development from the European Agricultural Guidance and Guarantee Fund (EAGGF) and Amending and Repealing certain Regulations."*

Table 4.3   Recent history of the EU Rural Development Policy

1992-1996: From CAP Reform to Cork

1992	McSharry Reforms agreed
1995	Commissioner Fischler's Agricultural Strategy paper produced
November 1996	Cork Conference and Cork Declaration

1997-1999: From Cork to CAP Reform

July 1997	Commission publishes Agenda 2000 proposals
March 1998	Commission publishes draft legislation
11 March 1999	Agriculture Ministers agree on CAP reform package
26 March 1999	Agenda 2000 reforms agreed by Heads of Government
17 May 1999	Text of Rural Development Regulation (1257/1999) agreed
23 July 1999	Text of Rural Development Implementing Regulation (1750/1999) agreed

1999-2002: Implementing the Rural Development Regulation

8 September 1999	Commission decision on allocations of EAGGF funds
2000	Rural Development Plans drawn up and submitted for approval
End 2000	Programmes commence
26 February 2002	A new text of Rural Development Implementing Regulation (445/2002) is agreed
July 2002	Commission publishes proposals for MTR of the CAP including changes to the RDR

2002-2003: From Mid-Term Review to Salzburg

October 2002	European Council reaches an agreement on the budget ceiling
23 January 2003	Commission proposals for reforming CAP are presented
26 June 2003	The Council of Agriculture Ministers of the EU reaches an agreement
September 2003	No agreement is reached on agricultural trade at the WTO Ministerial held in Cancun
29 September 2003	The Regulations (1782/2003 – 1788/2003) implementing the agreement were adopted by the Agriculture Council meeting
November 2003	Salzburg Conference and Salzburg Declaration

Sources: Adapted from Dwyer et al. (2002). Information is also drawn from Edwards (2003) and CEC (2003).

The specific RDR measures are the following:

1. *Investment in Agricultural Holdings*, which contributes to the improvement of agricultural incomes and the betterment of living, working and production conditions.
2. *Setting up of Young Farmers*, which facilitates the establishment of young farmers as a means to revitalize the countryside and create a more favourable age structure for rural populations.
3. *Training*, which is vocational training provided with a view on improving the occupational skills and competences of farmers and other persons involved in the sector. It could also encompass measures for their occupational conversion.
4. *Early Retirement*, which aims at providing income for elderly farmers who decide to withdraw and to encourage their replacement by farmers who will be able to improve the business or alternatively to encourage the conversion of land to non-agricultural use.
5. *Less Favoured Areas and Areas with Environmental Restrictions*, which contributes to the viability of areas endangered by their natural characteristics or their environmental restrictions.
6. *Agri-Environmental Measures*, which aim at supporting agricultural practices that protect the environment and conserve the countryside.
7. *Improving Processing and Marketing of Agricultural Products*, which aims to increase the competitiveness and added value of products, through a rationalization of processing and marketing techniques.
8. *Forestry*, which supports the maintenance and development of the economic, social and ecological functions of forests, by contributing to sustainable forest management, the development of forestry, the maintenance and improvement of forest resources and the extension of woodland areas.
9. *Promoting the Adaptation and Development of Rural Areas*, which supports measures of farming activity and conversion as well as measures relating to rural activities, that do not fall in the aforementioned categories. This last RDP instrument (Article 33 of RDR) includes a large number of specific measures which aim at environmental protection, improvement of rural infrastructures, diversification of economic activities, improvement of quality of life in rural areas, etc.

Two types of financial arrangements are provided for funding RDP measures. The accompanying measures, classified under the label of AEMs, and the LFAs measures are co-financed by the EAGGF-Guarantee section in all MS. The other RDP measures are co-financed by the EAGGF-Guidance section. As for the rate of co-financing, the EU must contribute at least 25 per cent of eligible public expenditure and no more than 50 per cent of total eligible costs in areas outside Objective 1 and 2 regions. Regarding AEMs, this contribution should amount to 75 per cent in Objective 1 regions and 50 per cent in other areas. Specific provisions also exist for income-generating investments.

The RDR requires the design and implementation of Rural Development Plans (RDPLANs) at national and regional levels for the programming period 2000-2006 and foresees that the MS will set out verifiable standards entailing compliance with general obligatory environmental requirements in these plans. The Commission, and more specifically DG-Agriculture, is responsible for providing both the framework and the necessary instructions and consultation for the design and implementation of RDPLANs. The MS prepare the RDPLANs on the basis of the development objectives of their regions, at the geographic level deemed most appropriate by each MS, for a period of 7 years (2000-2006). There are two types of RDPLANs. One type involves Regional Operational Programmes (ROPs), Operational Programmes and/or Single Programming Documents (SPDs) (Objective 1 regions) and the other involves regionalized and/or horizontal programmes (Objective 2 regions), depending on the MS. The role of regions varies according to their development objectives and the administrative preferences and traditions of the MS.

The MS were obliged to submit their RDPLANs to the Commission within six months of the entry into force of the RDR. The Commission appraises and adopts the programmes proposed by the RDPLANs within six months from their submission. Responsibility for selecting actual projects falls exclusively on the competent national and regional authorities, following the best schedule they are able to devise. After the projects are launched, the MS and the Commission monitor the initiatives together; where necessary, monitoring committees can be set up. The programmes are subject to annual implementation reports; independent auditors carry out mid-term and ex-post evaluations.

Policy Integration within the Second Pillar of the CAP

Methodological Approach

This section examines the extent to which the RDR, the main instrument of the EU RDP, actually promotes PI in the RDP domain. The analysis is undertaken from two interconnected, although relatively distinct, angles. First, the internal coherence and balance of the RDR measures, i.e. the vertical integration within the RDR, is assessed. Secondly, the interconnections between the RDR and other Community policies, i.e. the horizontal integration of RDP with other Community policies, are examined. In both cases, integration is examined in terms of the articulation of policy objects, goals and objectives, actors, structures and instruments.

Because only four years have passed since the implementation of the RDR, it is extremely difficult to detail all aspects and prospects of PI due to the complicated nature of the RDR and the lack of systematic in-depth and comparative studies at the EU level. Thus, the present attempt is confined necessarily to a broad brush analysis based on the official text of the RDR, the formal requirements for the design of RDPLANs[9] and limited evidence from the implementation of the RDR in the EU-15 so far.

Vertical Integration within the RDR

The analysis of the vertical integration of the RDR aims to uncover whether the RDR rhetoric has been followed in practice, as reflected in the design of the RDPLANs. The main criteria employed concern: (a) whether the object of the RDR is clear and coherent; (b) whether the goals and objectives set in the RDR text are integrated among themselves and can function as guidelines for the design of RDPLANs by the MS; (c) whether the roles and interactions of actors involved in RDPLAN design and implementation are adequately clear, allowing for proper coordination of the design and implementation of the RDPLANs; (d) whether the basic financial, administrative, and managerial procedures foreseen by the RDR are coordinated, at what level and how they contribute to PI; (e) whether the specific policy instruments (measures) can be coordinated and the extent to which they can be integrated within a RDPLAN.

The object criterion The object of the RDR is defined pluralistically, comprising the more specific objects of the nine groups of policy measures it encompasses. Some of these objects refer to agricultural (sectoral) development, others refer to spatial (territorial) development and the rest concern the economic development of rural areas. Moreover, the various specific objects concern different socioeconomic and/or administrative-territorial entities; individuals, agricultural holdings, households, enterprises, localities, regions, ecologically-defined areas.

Each specific object of the RDR is relatively coherent as it provides incentives to specific entities to achieve particular goals. For example, it is reasonable to assume that investments in agricultural holdings are positively related to improvements of agricultural structures because support offered to certain viable agricultural holdings may increase their competitiveness and the well-being of the individual farm owners and households. The internal coherence of each specific object is normally assumed drawing on the experience from the application of the respective policy instruments. However, it is not conceivable how all these objects can be realistically tied together within a single RDPLAN as they are not necessarily related or, even worse, some may contradict each other. For example,

---

[9] The LEADER Community Initiative is not covered here because its conception and rationale concern predominantly the local level, it spans three different programming periods and it represents a relatively negligible financial intervention. The treatment of the LEADER together with the RDR would further complicate the present discussion. However, such a comprehensive treatment should be undertaken because the proposals for the revised version of the RDR of the new age, planned for the next programming period (2007-2013), incorporates LEADER as one of its axes.

how is it possible to harmonize the object of the LFAs and areas with environmental restrictions with the object of investment in agricultural holdings? The former poses constraints on the latter and the latter contradicts the former. Moreover, the different objects imply policy interventions in different types of entities, which are not necessarily compatible with each other. For example, the good ecological condition of a territorial unit does not imply improved living conditions for its inhabitants. Or, investments in agricultural holdings do not necessarily promote the adaptation and development of rural areas, but only the improvement of certain holdings or the well-being of certain individuals. Hence, the issue of compatibility and coherence among the specific objects subsumed under the RDR arises.

Nevertheless, there are groups of specific objects which are compatible with each other. For example, on a sectoral basis, there are clear links between investments in agricultural holdings, the setting up of young farmers, training, early retirement and the improvement of the processing and marketing of agricultural products. Another group of objects may be combined on a territorial basis; namely, the LFAs and areas with environmental restrictions, the agri-environment, forestry and the promotion of the adaptation and development of rural areas. Thus, it seems that certain objects, due to their conception and function, tend to be compatible with certain other objects, but not all objects are compatible with every other. The identification of inconsistencies and incompatibility among the components of the RDR poses fundamental conceptual and theoretical issues concerning the rationale behind the attempted integration of the RDR.

The rhetoric concerning the RDP claims that the RDR promotes the integration of formerly separate policy measures, while an unabashed observer realizes that the RDR is a container for both the former sectoral (market-oriented) priorities and the new territorial (cohesion and social) priorities. The transition from the former to the latter has not been completed yet, as some writers have argued already (Buckwell et al., 1997). In other words, a divide seems to exist between the sectoral and the territorial objects within the RDR. Three relatively distinct approaches are open to address this divide as discussed below:

a) The *territorial approach to rural development* supports the further territorialization of rural policy, which needs to move closer to the pattern of regional and local development (Bryden, 2003), considering that "there is a misfit between the rhetoric surrounding the Agenda 2000 and the proposed regulations" (Buckwell, 1998, p.25). The significant shift in the principles of agricultural and rural support from price support to an integrated and balanced rural policy, the considerable price cuts and the expansion, deepening and integration of environmental concerns in rural development policy are acknowledged, but "the extent to which there is a real switch in the principles underlying policy for the period until 2006 is quite limited in scope" (Buckwell, 1998, p.25). This approach favours the full decoupling of agricultural support and looks forward to the fragmentation, localization and customization of rural development policies.

b) The *multifunctional European Model of Agriculture (EMA)* argues that rural development needs to take into account the fact that agriculture produces food but it also provides environmental and public goods and it is a primary factor for the viability and maintenance of the quality of life in the countryside (Herview and Beranger, 2000). This approach builds on the concept of multifunctionality, a holistic vision of rural development referring to "a policy statement about the unity between society, landscape and agriculture which has become an important tool, rather than a normative framework, for agriculture and rural policies in the future" (Depoele, 2000). It recognizes the interconnections between the territorial organization of farming and external factors, and argues that a new contract is required between farmers and the wider society to formulate a rural development strategy (Colson and Mathurin, 2002).

c) Finally, an intermediate approach argues that "the sectoral and territorial functions should be conceived as complementary dimensions of rural policy that acknowledges that sector and space no longer coincide" (Saraceno, 2002, p.10). The *mix of the sectoral and the territorial functions* varies with respect to the relevance of agriculture in rural areas, its need for restructuring and the degree of diversification of those areas (Mantino, 2003a).

All in all, it is extremely difficult to suggest how to bridge the divide between the sectoral and territorial objects of the RDR because the EU RDP is the result of successive amendments of measures with different rationales.[10] The Commission, in order to accommodate the different development demands of rural regions, different public administrations, various social actors and collective interests has formulated a potentially disjointed RDP agenda. The growing penetration of liberalization arguments and rationales into the CAP and in RDP has intensified the tensions between the sectoral and the territorial function (Buller, 2003).[11] The multifunctional EMA appears to be the last bastion of a decoupled territorialized sectoral RDP. Morever, it is not an easy task to decide whether multifunctionality represents a legitimization of non-trade concerns or constitutes a form of disguised protectionism (Potter and Burney, 2002). Both interpretations appear equally plausible despite the problems associated with the operationalization of the concept of multifunctionality (Colson and Mathurin, 2002; Sumpsi and Buckwell, 2002).[12]

---

[10] For example, it is no longer possible to specify what kind of farming practices are recommended or what kind of structural farm adjustment is desirable in terms of common policy (Saraceno, 2002).

[11] Some writers have argued that the liberalization of agricultural policy is a more realistic prospect today than ever before since it allows for the decline of the agricultural intensification model through the decoupling of agricultural support and also for environmental, 'green re-coupling' of support which is to the benefit of rural environment (Potter and Goodwin, 1998).

[12] One of the most important obstacles which the RDP faces in its movement towards a more integrated rural policy is the higher transaction costs of the new reformed CAP. More

The discussion of the conceptual and theoretical issues surrounding the coherence of the object of the RDR brings to the fore the question of *rural development strategy* which the MS are obliged to develop in designing RDPLANs. However, the formulation of such a strategy requires a proper analysis of the goals and objectives of the RDR.

The goals and objectives criterion The broad aim of the RDR is that its measures should: (a) "be integrated into the measures promoting the development and structural adjustment of those regions lagging in development" (Objective 1 regions), and (b) "accompany the measures supporting the economic and social conversion of areas facing structural difficulties" (Objective 2 regions). The Regulation was designed to comply with the Community's different development priorities and applies to all rural areas. It stresses that the rural development measures will be applied in those regions which are in need, taking into account the specific targets of Community support as laid down both in the Treaty and in Regulation 1260/1999, which includes the general provisions relating to the operation of SFs. Therefore, the application of the RDR follows the territorial targeting of the development priorities foreseen by the operation of SFs. Notably, the different Funds will finance the various rural development measures on the basis of their own specific objectives.

Moreover, the RDR includes a large number of agricultural and non-agricultural objectives such as:

* Improvement of structures in agricultural holdings and structures for the processing and marketing of agricultural products (agricultural);
* Conversion and reorientation of agricultural production potential, the introduction of new technologies and the improvement of product quality (agricultural);
* Encouragement of non-food production (economic);
* Sustainable forest development (environmental);
* Diversification of activities with the aim of promoting complementary or alternative activities (economic);
* Maintenance and reinforcement of viable social fabric in rural areas (social);
* Development of economic activities and maintenance and creation of employment aiming to ensure a better exploitation of existing inherent potential (economic);
* Improvement of working and living conditions (social);
* Maintenance and promotion of low-input farming systems (agricultural);
* Preservation and promotion of a high nature value and a sustainable agriculture respecting environmental requirements (environmental);

---

particularly, the detailed targeting of the decoupled agri-environmental payments entails high transaction (administrative) costs (Colson and Mathurin, 2002; Sumpsi and Garcia-Azcarate, 2002).

- Removal of inequalities and promotion of equal opportunities for men and women, by supporting projects initiated and implemented by women in particular (social).

Obviously, the RDR accommodates different types of objectives, corresponding to one or more of the measures it includes. However, the question is how these objectives can be balanced within a single RDPLAN. For example, how is it possible to link the objective relating to the improvement of agricultural structures with the objective referring to the maintenance and promotion of low-input farming systems? The first objective contradicts the second. Moreover, the objective of improving agricultural holdings and structures counters, to a large extent, the social objective of maintaining and reinforcing a viable social fabric in rural areas.

At the rhetorical level, the RDP is considered a 'pluralistic' policy for rural areas reflected in the adoption of a large number of highly diverse objectives, thus accommodating various interpretations and conceptualizations of rural policy. These objectives reflect the demands of different actors, interest groups and/or public administrations as well as the established national, regional and/or local agendas. Several Community texts give rise to or reflect various ideas and approaches about the objectives and content of the RDP. The *Second Report on Social and Economic Cohesion* (CEC, 2001), for example, conceives the RDP in terms of diversification of rural areas. According to this viewpoint, the economic improvement of rural areas depends on the development of off-farm activities, which will increase the role and size of services and enterprises in these areas. Within the CAP context, the RDP is destined to provide for the necessary farm modernization, the enhancement of agricultural competitiveness and the all-round support for the development of rural areas. However, a sectoral division of roles between regional policy and the RDP underlies the rhetoric over the RDP (Mantino, 2003a).[13]

The existence of so many objectives under the RDP owes to the evolution of the CAP. The RDP is significantly influenced by the drivers of the CAP reform and the competing paradigms of agricultural support. Buller (2003) distinguishes between four such paradigms: (a) the *classic interventionist paradigm*, which has largely defined the CAP and its history for the first 20 years, (b) the *free-market* or *liberalist paradigm* which gained momentum during the late 1980s and 1990s, (c) the *multifunctionalist paradigm*[14] and (d) the *rural development paradigm*.[15] The

---

[13] The defenders of the territorial approach to RDP see regional policy as actually encompassing rural development, which means that the sectoral dimension of the RDP is completely downplayed under the pretention that the regional includes the rural. This reductionist thinking attributes sectoral objectives to rural development especially for those regions which have retained a sizeable agricultural sector.

[14] This paradigm is for the maintenance of agricultural support arguing that agriculture fulfills more than one function (production); namely, it provides for the reproduction of the European rural landscape and more generally public goods which should be conserved for

first two paradigms co-exist today but compete with each other for the upper hand. The other two are placed in-between the former two paradigms, combining elements of the classic interventionist approach with a large number of liberalist arguments.

Buller's (2003) remarks are particularly instructive in relation to the rhetoric of the 'pluralistic' RDP. The dominant paradigm of EU intervention remained largely unchallenged until the 1980s, despite the fact that other visions had been expressed in relation to the direction of policy reform. Since then, different visions and/or models of agricultural support have competed for influence on the direction of CAP reform, resulting in policy instability and a resurgence of national agenda setting. The division of agricultural funds between two pillars implies a justification of multiple paradigms for agricultural support. Pillar one seems to lean towards the liberalist paradigm, whereas pillar two is more relevant to the rural development and the multifunctionalist paradigms that suggest a more territorial, regional and localized approach. The fragmentation of the CAP leads to a growing re-nationalization and regionalization of the design and implementation of the RDP. The increasing impact of the liberalist, the multifunctionalist and the rural development paradigms has led to significant changes and diversification of the mechanisms (e.g. contracts) used for channelling funds to rural areas. Finally, the new approaches to agricultural and rural support build their strength upon the objective identification of the costs of maintaining public goods and of the benefits arising from farmers' or stewards' actions in the countryside.

Thus, in the very end, the RDP seems to be still strongly tied to a sectoral vision of rural development and the sectoral rationale of the CAP (Bryden, 2000; Dwyer et al., 2002; Saraceno, 2002; Mantino, 2003a). Although the objectives included into the RDR have shifted in response to the new socio-economic demands of rural territories, there are still important inconsistencies among them as well as with the agenda of a territorial rural policy. There is no real integration of objectives because the market competition objective, i.e. of increasing the competitiveness of farming and of rural areas, is placed at the top of the agenda dominating all others (Thomas, 2003), a fact that seriously undermines the internal balance and coherence of the RDR.

The actors criterion   The role of policy actors is instrumental in both the design and the implementation of the RDP. The main actors involved are the Commission, the Council of Agricultural Ministers and the MS but they do not participate equally in the policy process. The Commission holds a strategic position during policy design, while the MS pursue their own particular interests and apply their own interpretations during this process. The Council of Agriculture Ministers

the coming generations. Thus, farmers are the 'gatekeepers' of the countryside and should be supported thereof.
[15] The rural development paradigm aims at the multi-sectoral, multi-level and territorial development of rural areas.

has the power to reform and redirect policy objectives and instruments. The organized agricultural interests exert pressures on the MS and the Commission for minimum or no reform of the existing sectoral policy orientation. The national administrative structures,[16] with their particular institutional and organizational capacities, undertake the task of designing and implementing the specific RDR measures. Before actual implementation, negotiations over the form and content of the RDP (reflected in the RDPLANs) between the Commission and the MS take place. Thus, the coherence of the RDP is severely constrained by the imbalance of power between the actors involved in the policy process. The integration-minded Commission is far ahead, but relatively weak when compared to the large majority of sector- or budget-minded MS which are relatively stronger in influencing both the policy process and its outcomes (Daugbjerg, 1999; Ackrill, 2000). Moreover, there is not a unified interpretation of the RDP, despite the Commission's attempts; each MS reflects on it its own agricultural, regional and/or environmental policy objectives.[17]

The organizational and administrative structures of policy design and implementation of the RDR are similar to those of the CAP. Despite the territorial orientation and the concomitant territorialization of CAP measures (Shucksmith, 2003), DG-Agriculture (DG-VI) and the Ministries of Agriculture of the MS are the manifest vehicles of agricultural interests, which are very selective and hesitant in integrating other policy concerns and instruments that may endanger the integrity of their policy domain.[18] Therefore, it is reasonable to suggest that many organizational and administrative resistances will exist in the MS with respect to the exploitation of the integrative potential of the RDR.

The procedures criterion   The particular roles of institutional actors become more evident when the procedures of the RDR are examined. Three types of procedures are relevant: (a) the financial, (b) the administrative, which includes the programming process, and (c) the managerial, which encompasses the participation process.

It was mentioned earlier that different financial provisions by type of development priority region exist. The RDR measures are basically funded by the

---

[16] National administrations may be assisted by NGOs and/or other consulting agencies interested in participating in policy design and/or policy implementation.

[17] See, for example, how France integrated the environmental aspect in its agricultural policy (Montpetit, 2000). Moreover, the different national interpretations and approaches (e.g. the British and the French) to the RDP illustrate divergent patterns of the policy process (Lowe et al., 2002).

[18] There are some national actions of adaptation to the new conditions set by the reformed CAP. Among the EU-15 MS, only 7 non-integrated Ministries of Agriculture remain. The other MS either include "Environment" or "Rural Development", or omit "Agriculture" altogether (Thomson and Psaltopoulos, 2004, p.42). The recent example of Greece, which few months ago renamed its Ministry of Agriculture to Ministry of Rural Development and Food, is illustrative of the trend.

EAGGF-Guarantee section, which is traditionally related to market policies, and not by the EAGGF-Guidance section which is designed to fund measures of a structural character. Objective 1 regions are an exception, however, since the AEM and the LFAs are funded by the EAGGF-Guarantee section, whereas the other measures are funded by the EAGGF-Guidance section. Moreover, certain rural development measures may enjoy assistance from the ESF or the ERDF, as the orientation of their actions justifies such a funding diversification. Thus, the realization of rural development measures is based on the prior amalgamation of separate funding budgets which is usually limited by the different legal provisions and rules for each budget and considerably restricts the switch of funds between different measures. Consequently, the financial integration of the RDR would require a formal act of budgetary consolidation (Thomson and Psaltopoulos, 2004).

Finally, in terms of RDP implementation (in the EU-25), the financial resources allocated to the RDR represents around 15 per cent of the total CAP funds. Nearly two thirds of them come from the EAGGF-Guarantee section and the rest is contributed by the EAGGF-Guidance section. Moreover, 56 per cent of the total RDR funds are allocated to Objective 1 regions and the rest 44 per cent is left for Objective 2 and other regions (Mantino, 2003b). The RDP, despite its claims in favour of an integrative policy potential in rural territories, remains a highly localized and partial policy within the wider EU economic area in terms of financial resources.

It is reminded that the Community rural development measures aim to supplement and not replace existing national measures. For this reason, only basic support criteria are laid down at the Community level, meaning that the measures proposed and accepted are and will remain under the co-financing principle. The RDR makes explicit reference to two basic framework rules that are to be specified by the Community and which the MS have to observe in order to increase the consistency and balance between rural development measures and contribute to the simplification of the system. The first is that specific State Aid rules should be established and the second that well-defined indicators need to be agreed and established prior to programme implementation (Pezaros, 1999). These two controlling rules represent significant constraints on the discretion of the MS regarding the formulation of RDPLANs.

However, some important points have to be made regarding RDP programming. The Commission's role is to appraise the RDPLANs submitted by the MS and determine whether they are consistent with the RDR. On the basis of the procedure foreseen by Regulation 1260/1999, the Commission should approve rural development programming documents (Article 44). The whole process of plan design and negotiation exceeds 12 months. This exerts pressure on the MS to prepare their RDPLANs as quick as possible which need to include the following:

- A quantified description of the current situation showing problems and issues to be tackled, the financial resources deployed and the main results of operations undertaken in previous programming periods;

- A description of the proposed strategy, its quantified objectives and rural development priorities selected as well as the geographical area covered;
- An appraisal of the expected economic, environmental and social impact as well as the employment effects;
- An overall financial table summarizing the national and Community financial resources provided for and corresponding to each rural development priority submitted in the context of the plan;
- A description of the measures selected for implementing the plans and in particular State Aids;
- More detailed information on the specific content of demonstration projects, technical assistance, etc.;
- The designation of competent authorities and responsible bodies;
- Provisions to ensure the effective and correct implementation of the plans including monitoring and evaluation.

Moreover, the MS should include in the RDPLANs an appraisal of the compatibility and consistency of the rural development measures considered appropriate and an indication of how this compatibility and consistency will be ensured. However, there are no clear instructions how to achieve this target. The call for compatibility and consistency remains rhetorical; the strict timetable to design and implement RDPLANs severely restricts the actual process of PI.

In terms of managerial procedures, the RDP should be designed and implemented, through the RDPLANs in the MS, as decentralized as possible, using a 'bottom up' approach and emphasizing the participation of private social and economic actors.

During the implementation of the RDR a number of problems were detected: (a) the process of programme formulation by the MS is too short; (b) the process of programme approval by the Commission is too long; (c) the financial thresholds for modifications which do not require formal Commission approval are low; (d) the procedure for formal Commission approval of modifications is too slow; (e) the number of modifications allowed per year is too small; (f) all amendments to single measures require Commission approval; (g) the eligibility criteria for granting support for some measures is too detailed and problematic for the purposes of the RDR; (h) a number of actions necessary for the operation of existing rural development measures[19] are not eligible for funding (Mantino, 2003b).

The EU promotes a new system of governance by supporting and encouraging the implementation of 'bottom-up' decision making, thus, strengthening the active participation of regions and other interested parties in drafting, approving and implementing the RDP and offering flexibility and legitimization of policy actions.

---

[19] For example, technical assistance, agricultural research, promotion of agricultural products.

Although the Commission restricts itself to the role of steering the policy process,[20] leaving to the MS the task of devising an operational system, it is argued that, in reality, it strongly influences both the programming and implementation of the RDR. The Europeanization process[21] intervenes in the integration of the RDP as the latter depends upon formal and informal interactions between the Commission and the MS (Hennis, 2001; White, 2001; Bache, 2003; Mantino, 2003a).

The implementation of the RDP differs widely among the MS. Some countries have prepared only one national RDPLAN (e.g. France), while some others have chosen the region as the most appropriate level for RDPLANs (e.g. Italy, Germany). However, the number of programmes dealing with structural and rural development measures has been substantially reduced compared to the previous programming period (1994-1999). The RDPLANs of certain MS that are financed by the EAGGF-Guarantee section are designed and managed by the central administration. Most of the northern European MS have adopted a centralized model of RDP design. Some of them are small in size and, therefore, central planning is justified. Moreover, large countries like France and smaller countries like Austria and Sweden, with a tradition of decentralization, have opted for policy management at the central level. A second group of countries (Germany, Italy, United Kingdom and Spain) have adopted a more decentralized model. It should be noted, however, that the degree of decentralization is higher for RDR measures supported by the EAGGF-Guidance section (Mantino, 2003a).

The instruments criterion   The RDR recognizes that "rural policy is currently carried out through a range of complex instruments", whereas in the following paragraph it suggests a "reorganization and simplification of the existing rural development instruments" on the basis of experience gained by the long-term application of existing instruments. However, this latter statement remains a rhetorical suggestion which is only partly realized by the RDR itself.

A significant premise of the RDR is that support for several measures may be combined only if these are consistent and compatible with each other (Article 38). This statement indirectly recognizes the existence of conflicts among many measures, a situation owing to the fact that nearly all RDR measures comprise a legacy of past periods of CAP and Structural Policy reform. Their objectives and rules have not been modified and their simple incorporation into an umbrella Regulation does not imply an essential reorganization. Simplification and harmonization.

---

[20] There are many discussions of the rationale behind the Commission's role in steering the policy process. It may well be that the Commission seeks new ways to promote the Europeanization of public policy and to legitimize its policies (Sbragia, 2000; Heritier, 2002; Bache, 2003; Roedeger-Rynning, 2003), but it may also the case that the new conditions of multi-level governance are not easily controlled and/or dominated by one powerful policy actor.

[21] Which refers to the spread of particular types of institutional capacity, of administrative rules and practices to the MS.

Although the focus of the RDR is still on the structural adjustment and territorial contextualization of farming activities, it considers as equally important the role of rural development in enhancing the competitiveness of rural areas and in maintaining and creating employment. Rural development is thus considered as a form of territorial development of rural areas. The RDP measures are taken as remedies responding to the wider socioeconomic crisis of the countryside. Consequently, the RDP is considered as a redistribution policy, aiming at correcting any emerging market inefficiencies, and not as an integrated policy.

Notably, the RDR assigns a prominent role to AEMs to support the sustainable development of rural areas and respond to society's increasing demand for environmental services. However, the strong tendency of the Commission and of some MS to stress the implementation of AEMs does not lead to the integration of the RDR measures but rather echoes a re-balancing of the RDPLANs towards environmental policy integration (EPI) depending on the characteristics and needs of certain MS.[22]

The lack of integration of the RDR measures is evident in the case of the southern European MS which contain a large proportion of LFAs. In several of them, harsh natural conditions together with weak socioeconomic structures constitute the main drivers of desertification (Briassoulis, 2004). At the same time, most of the southern European territory is considered as Objective 1 development priority which needs measures for its structural adjustment and receives higher incentives for agricultural development. However, in the desertification-sensitive LFAs, where there should be fewer incentives for agricultural development, there is a greater tendency to adopt sectoral and/or production-oriented measures that favour agricultural intensification which stresses the already fragile land resources of these areas Thus, the 'menu' of the RDR measures does not serve the goal of combating desertification, which is a complex socio-environmental problem. The need to treat such problems holistically poses again the question of who (actors) designs a strategy for rural development, how and for what purpose (van der Leeuw, 2000; Meadowcroft, 2002).[23]

As it may be expected there has been considerable variation in the content of the RDPLANs, in the weighting of the measures included in them and in the use of Article 33 sub-measures. However, in the process of negotiations and agreement over RDPLANs with the Commission, the rhetoric of a broad, integrated RDP agenda seems to have been overridden by a more pragmatic view on the RDR. The

---

[22] It has been argued that the group of Northern MS are more competent in environmental policy making (Liefferick and Andersen, 1998) and in EPI, whereas the Southern MS are presented as 'laggards'. However, the so-called "Mediterranean syndrome" should be considered as a stereotype which does not concern the institutional capacity but rather the institutional performance on environmental policy issues (Börzel, 2000).

[23] Similar issues of environmental problem contextualization and institutional complexity were raised by Hardin (1968) who studied "the tragedy of the commons", the stalemate of the management of common pool resources, and argued that this pressing dilemma "has no technical solutions".

Regulation has been taken primarily as providing a structural adjustment policy for agriculture, with only limited use in providing support beyond the farm gate. In many MS, implementation narrowed down the scope of the RDR to a largely agricultural agenda, which offers a more secure ground for both policy design and implementation. The analysis of different cases of MS shows that, for example, the application of the Article 33 measure leaned towards supporting farm structural adjustment rather than anything like purchasing or leasing land for nature conservation (Dwyer et al., 2002).

An analysis of the draft programmes of some MS shows that the AEMs and the Farm Investment Scheme occupy an important position in the RDPLANs, whereas measures relating to the wider rural economy, included in Article 33, seem to be receiving very low priority, concentrating less than 10 per cent of the overall budget (Bryden, 2000). Thus, the mix of rural development measures incorporated in the specific RDPLANs is not balanced. Moreover, the problem is deeper than that since, at the level of policy priorities, the RDR remains firmly attached to the support of the farming sector. It has been argued that the Commission has 'ditched' rural policy due to pressures exerted by the agricultural lobbies and the internal disagreements between DG-Agriculture and DG-Regional Policy (Bryden, 2000; Kay 2003). The focus of the RDP remains firmly on the farming sector because the 'Second Pillar' of the CAP is part of the Commission's strategy on multifunctional agriculture for the continuing WTO talks and for limiting the impacts of EU enlargement on the CAP budget. Consequently, the significant distance between the rhetoric and the reality of the RDR can be largely attributed to the specific sectoral and financial demands of the Commission and the MS. It should be noted, however, that the lack of national RDPs in many MS (especially the southern European) is a further reason for the low level of integration within the RDR at the MS level (Dwyer et al., 2002).

Integration of the RDP with other Community Policies

The horizontal integration of the RDP and two other Community policies, the environmental and the regional policy, is examined in the following. The criteria employed concern whether: (a) the object of the RDP is related to, agrees with or is complementary to the object of the other two policies; (b) the goals and objectives of the RDP are congruent with or complementary to the goals and objectives of these policies; (c) the RDP and the other two policies have common actors or, more generally, their actors interact and act in coordination; (d) the financial, administrative and managerial procedures of the RDR and of the other policies are coordinated; and (e) the specific RDR measures which are included in the RDPLANs are related to and coordinated with instruments of the other two policies.

The object criterion The RDP makes weak references to the objects of the environmental and the regional policy which take the form of favouring the sustainable development of rural areas and of promoting and reinforcing their territorial development. However, despite this relative progress, the RDP remains attached to its CAP context. One of the premises of the RDR is that it "should comply with Community law and be consistent with other Community policies as well as with other instruments of the CAP". This general statement is accompanied, however, by more specific claims such as, for example, that "rural development should accompany and complement market policies". Thus, the RDP mostly contributes to the better coordination of the CAP object, constituting one of its components and not being an integrated policy domain in itself. Thus, the expressed orientation of the object of the RDP is to complement the CAP and not to become coordinated with other Community policies.

The goals and objectives criterion The RDP and the regional policy goals seem to be formally integrated. More specifically, the organization of the RDR measures differs according to development priority status, as defined by regional policy, of the regions where the RDPLANs are to be implemented. On the other hand, there is no explicit integration of the RDP with any environmental policy goals.

Only at the level of objectives there is a certain integration dynamics between the RDP and the other two policies. The environmental policy objectives that have been incorporated into the RDR are limited to forestry measures, the preservation and promotion of a high nature value areas and sustainable agriculture respecting environmental requirements. It seems that the environmental policy objectives operate mostly as constraints, posing barriers to and complementing the agricultural, economic and social objectives of the RDR measures, rather than being truly integrated with these objectives.

The regional policy objectives incorporated into the RDR concern the encouragement of non-food production, the diversification of activities to complement or replace existing farming activities, the development of economic activities and the maintenance and/or creation of employment. These objectives probably reflect the RDR claim that the implementation of rural development measures is to take into account and contribute to the achievement of "economic and social cohesion", which is an explicit objective of regional policy included in the EU Treaty. Many of the measures included in the Regulation comply with this objective because some of them were previously classified as regional policy measures under the SF regulations. It should be reminded that some of the sub-measures classified under Article 33 are actually infrastructures financed by the ERDF, the financial instrument of EU regional policy. Moreover, the RDR incorporates training and other measures relating to the improvement of human capital which are financed by the ESF, the instrument of EU employment policy.

Overall, a better picture is presented in terms of integration of objectives. However, the regional (and social) policy objectives are more integrated as they fit better to the sectoral (agricultural) objectives of the RDR, whereas the environmental policy objectives are less truly integrated because they contradict most of the RDR production-oriented objectives.

The actors criterion The design and implementation of RDP involves the same actors as those involved in the CAP; namely, DG-Agriculture, the Ministries of Agriculture and the regions of the MS. In most cases, no significant interaction and/or coordination of actions between the RDP and the environmental or regional policy actors seems to exist. At the EU level, despite the formal obligations of most DGs to integrate environmental objectives in their policy domains, there is not considerable and essential coordination of actions between DG-Agriculture and DG-Environment. At the MS level (especially the southern European), no significant interaction between the Ministry of Agriculture and the Ministry of Environment exists.

Due to the programming arrangements in MS with Objective 1 regions, there is some interaction and cooperation, mainly during policy design, between the Ministry of Agriculture and the Ministry of Economy (or Finance). The regional authorities, through the ROPs (in Objective 1 regions) or through the regional programmes (in Objective 2 regions), are obliged to incorporate different regional policy axes into a single programme. This formal obligation does not necessarily imply that the different RDR measures are integrated into the regional programmes. The latter contain a range of policy axes that include sets of sectoral policy measures, which are not truly integrated among them but simply formally coordinated.

The essential interaction among actors involved in different policy domains and the coordination of their actions is minimal both in the EU and the MS, although there is some formal interaction and coordination of actions at the regional level foreseen by the EU programming guidelines.

The procedures criterion At the level of procedures followed in the RDPLANs, there is a certain degree of integration of the RDP with the other two policies. In relation to financial procedures, there are close interconnections of the RDP with regional policy. The EAGGF-financed RDR measures are integrated into the ROPs (Objective 1 regions) or in the Regional Programmes (Objective 2 or other regions) where the development priorities are set at the regional level and those measures are expected to be compatible and consistent with the respective regional programming documents. There is no integration of the RDP with environmental policy in terms of financial procedures.

However, the AEMs and the LFAs measures, funded by the EAGGF-Guarantee section (that also funds pillar one interventions), may be implemented as a horizontal (in Objective 1 regions) or regional programme (in Objective 2 regions) in parallel with regional or other policy interventions.

The decentralization procedures followed in the design of the RDPLANs vary among the MS. In the more centralized MS, this design has followed a more sectoral rationale, while in decentralized MS the regionalized processes have usually favoured a less sectoral orientation. In most MS, and particularly the southern European, there is mostly limited or just formal consultation of other policy actors beyond the RDP actors during the RDPLAN design. The Ministry of

Environment, environmental NGOs, the Ministry of Finance or Economy, and regional authorities may be invited to participate, but due to the limited time of plan preparation, this consultation process reduces to a formal interaction following the Commission's guidelines and recommendations.

In terms of administrative procedures, the long delays that have been recorded in most MS in the preparation of the programming documents, which should follow the design phase and guide the implementation of the RDPLANs, indicate the administrative weaknesses of the RDP actors and their limited readiness to organize and deliver policy outcomes.

At the implementation phase, a formal procedure which secures a form of interaction between the RDP and other policy actors is the operation of the monitoring committee (MC). Different actors representing various policy domains (e.g. environment, social, regional) participate formally in the MC which convenes at least once a year. However, this committee does not function as a type of coordination mechanism between the relevant policy domains because it does not exercise real power but rather operates like a forum where different issues, problems and solutions are discussed. Theoretically, the MCs of the regional programmes may reveal more directly any cross-policy coordination problems, but interpersonal interactions and conflicts may obscure the weaknesses of structural coordination.

Some coordination between the RDP and environmental policy takes place mainly at the implementation of single projects, which require the submission of an Environmental Impact Assessment (EIA). However, again there are suspicions that this environmental condition may be observed only formally.

The bottom up approach of the design and implementation of the RDPLANs has been considered as instrumental for PI due to the significant functions attributed to the participation of different social actors in the policy process. In certain MS the participatory approach is more directly utilized since their particular socio-political organization and social regulation obliges them to take into account the demands of collective social actors. However, other MS have not exploited the formal participatory approach so much, but they have rather allowed not-so-formal routes of interest representation, involving a great deal of informal (interpersonal).

This instrumentalist thinking of pursuing PI through a participatory approach attracts criticism at least from two sides. Some ask, firstly, whether such an approach is democratic since it does not appear to meet the accountability principle, and secondly, whether it secures integration or provides a legitimation of the existing EU policy process. Thus, PI pursued through a bottom up approach needs to be reconsidered as it demands further elaboration and clarification (see Shortall, 2004).

In the context of combating desertification, the UNCCD firmly supports the bottom-up approach in preparing plans for desertification-sensitive areas. It considers it as an indispensable integrating mechanism leading to the formulation of new alliances and partnerships that bind together international institutions, states, regions, NGOs and local communities (Danish, 1995) and obviously implying increased institutional complexity. The utility of applying a bottom-up approach in designing RDPLANs for desertification-sensitive regions to address their particular development goals through actor and procedures coordination is hard to assess. First, knowledge and information has to be channeled to different spatial/organizational levels where different public and private actors (stakeholders) reside, each with their own perspective and way of acting on the problem, which will be surely affected during the policy process. Secondly, the bottom-up approach introduces informal actors and processes in RDPLAN design. Nevertheless, in both cases, the process operates based on consensus and legitimation of actions (van der Leeuw, 2000).

The instruments criterion Several measures are provided for integrating environmental considerations in the RDR (Table 4.4).

Table 4.4   Measures and actions foreseen for integrating environmental concerns within agricultural policy under the RDR

Measure	Criteria
Good Agricultural Practice	All farmers should respect basic environmental standards defined by MS in Codes of Good Agricultural Practice without receiving payments for following them
Optional Cross-Compliance	MS have the option to introduce mandatory environmental conditions (above the basic standards) to grant agricultural **support** payments to farmers
Monitoring	MS are obliged to monitor the impact of environmental measures in farming
Agri-environmental Measures	MS are obliged to draw-up agri-environmental schemes under which farmers may receive payments for maintaining or improving the environmental quality of the countryside
Modulation	MS may opt for redistributing part of farmers' direct payments under CAP commodity regimes towards schemes designed to promote or maintain environmental quality
Zoning	The territorial coverage of rural development and environmental measures is extended and are now available over much wider regions
Programming	MS are obliged to produce seven- or six-year RDPLANs (depending on the development priorities of their regions) which provide the frameworks for rural development and agri-environmental policy

Source: Buller (2002, p.119).

Buller (2002) contends that the three most important measures which address the issue of EPI in the RDR are: a) the AEMs, b) the Codes of Good Agricultural Practice and c) the optional Cross-Compliance The RDR places a major emphasis on the AEMs[24] which promote the sustainable development of rural areas and respond to society's rising demand for environmental services. One of its basic premises is that farmers should be encouraged to serve society as a whole by introducing or continuing the use of farming practices compatible with the protection and improvement of the environment, natural resources, soil and genetic diversity and the maintenance of the landscape and the countryside. In AEMs the RDR makes explicit reference to the need for applying good farming practices; in some cases farmers are obliged to undertake important commitments involving the provision of particular environmental services. Thus, farmers are given incentives in exchange for the provision of environmental goods. The design and implementation of targeted AEMs for the whole of the EU territory constitutes major evidence of environmental policy integration (EPI) in the RDR.

Many writers, however, argue that EPI is an issue of the CAP in general and not of the RDP exclusively (Lowe and Baldock, 2000; Baldock et al., 2002; Buller, 2002; Sumpsi and Buckwell, 2002) because the RDP is not separated from the CAP (Colson and Mathurin, 2002).

Table 4.5 provides a detailed description of the basic provisions and conditions of eligibility relating to environmental concerns in each of the RDR measures.

In measures promoting investments on the basis of economic efficiency[25] there is a clear reference of the Regulation to granting support to those persons/enterprises which "comply with minimum standards regarding the environment, hygiene and animal health". Of course, this relatively general statement on compliance with environmental standards is expected to be operationalized by the MS themselves.

The AEMs and the LFAs measures contain clear environmental objectives and foresee higher than basic environmental standards for beneficiaries getting Community support. The forestry measures of the RDR are adopted in the light of MS's forestry plans, taking into account the specific problems of climate change, while support is granted for activities to maintain and improve the ecological stability of forests in certain areas. The training, the early retirement scheme and the promotion of adaptation and development of rural areas measures include, directly or indirectly, environmental concerns and objectives.

Many of the measures of the RDR have included provisions which directly target environmental improvements, including biodiversity conservation. In this

---

[24] The AEMs and the LFA measures are the only compulsory elements in each and every RDPLAN.

[25] That is, investment in agricultural holdings, setting up of young farmers and improving the processing and marketing of agricultural products.

respect, the involvement and participation of environmental NGOs in the design and implementation of the RDPLANs should in any case be supported by the MS.

Overall, however, as Buller (2002) suggests, it is too early to assess the degree to which the RDP will actually facilitate EPI, although it has a significant potential. In any case, the achievement of EPI at the level of instruments, despite its importance, is rather inadequate because a truly effective EPI would also require integration at the level of objects, goals, objectives and actors.

Table 4.5 Provisions and conditions of eligibility (*) included in RDR measures which favour EPI

Chap.1, Investment in Agricultural Holdings	One of investment objectives is: "to preserve and improve natural environment, hygiene conditions and animal welfare standards". *Compliance with minimum environmental conditions.*
Chap. 2, Setting Up of Young Farmers	*Compliance with minimum environmental conditions.*
Chap. 3, Training	Among training aims is to: "to prepare farmers for the application of production practices compatible with the maintenance and enhancement of the landscape, the protection of the environment, hygiene standards and animal welfare" and "to prepare forest holders and others for the application of forest management practices to improve the economic, ecological or social functions of forests".
Chap. 4, Early Retirement	A non-farming transferee "may take over released land to use it for non-agricultural purposes, such as forestry or the creation of ecological reserves, in a manner compatible with protection or improvement of the quality of the environment in the countryside".
Chap. 5, Less Favoured Areas and Areas with Environmental Restrictions	Maintenance of extensive farming systems. Support to farming activity in Natura 2000 zones. Compliance with environmental requirements in particular through sustainable farming systems. *Application of usual good farming practices compatible with the requirements of of safeguarding environment and maintaining the countryside.*
Chap. 6, Agri-environment	Support is granted to promote: - "Ways of using agricultural land which are compatible with the protection and improvement of the environment, the landscape and its features, natural resources, the soil and genetic diversity". - "An environmentally-favourable extensification of

	farming and management of low-intensity pasture systems".
	- "The conservation of high nature-value farmed environments which are under threat".
	- "The upkeep of the landscape and historical features on agricultural land"
	- "The use of environmental planning in farming practice"
	Involving more than merely applying usual good farming practice
Chap. 7, Improving Processing and Marketing of Agricultural Products	Among the objectives are: - "To improve the presentation and preparation of products or encourage the better use or elimination of by-products or waste". - "To improve and monitor quality". - "To improve and monitor health conditions". - "To protect the environment". *Compliance with minimum environmental conditions*
Chap. 8, Forestry	The objectives are: - "Sustainable forest management and development of forestry". - "Maintenance and improvement of forest resources". - "Extension of woodland areas". Support will be based on national of subnational forest programmes or equivalent instruments existing as a result of international engagements. Application of forest protection plans. Assistance to foresters on sustainable forest management rules. Maintaining and improving the ecological stability of forests.
Chap. 9, Promoting the Adaptation and Development of Rural Areas	Among the objectives are: - "Marketing of quality agricultural products". - "Renovation and development of villages and protection and conservation of rural heritage". - "Agricultural water resources management". - "Protection of the environment in connection with agriculture, forestry and landscape conservation as well as with the improvement of animal welfare". - "Restoring agricultural production potential damaged by natural disasters and introducing appropriate prevention instruments"

Note: (*) in italics.

Conclusions

It is widely acknowledged that the EU RDP is not an independent policy, but rather an integral part of the CAP. However, since its inception, the RDP has been designed as a new policy domain which tends to be separate from the CAP and to interact with other policy domains, particularly with environmental, regional, social policy, etc. The present analysis underlined the formal character of this interaction which, however, is still far from being considered as PI. The target of PI in EU policy making is mostly implicit rather than explicit and the relevant discussion remains general and/or rhetorical. The frequent reference to the term 'integration' or 'integrated' in the discussion of the EU RDP has not led to its much needed clarification (Thomson and Psaltopoulos, 2004).

The RDP is a new EU policy domain with significant horizontal and vertical integrative potential, aspiring to address a broad range of development issues and sectors at different territorial levels in an integrated way. New forms of governance have developed characterized by decentralization, partnership, participation and new formal mechanisms of horizontal and vertical coordination. Although the EU RDP defines a new policy container for older components, it appears to be expanding and transforming into a potentially distinct policy domain. However, until today this promising domain is still overshadowed by the CAP both in terms of funding and of political decision making. The design and implementation of the RDP does not match its integration rhetoric which is still far behind the PI target.

The brief analysis of the degree of vertical integration of the RDR has provided significant evidence of policy evasiveness, in terms of containing conflicting and contradictory policy objects, goals and objectives, putting together diverse policy measures, bringing together policy actors with unequal power and involving national and regional actors with diverse institutional and organizational capacities. Formal PI at the level of procedures cannot be taken as true PI. The main problem is that the object of rural policy is too loose and, consequently, difficult to define strictly. On the other hand, the pluralism of the concept of the 'rural' does not foster any meaningful strategic alliances of social actors outside the national context.

The realization of the vertical integration of the RDP is delegated to the national and/or regional level, given the EU's emphasis on the subsidiarity principle. The implementation of the RDR has revealed numerous inconsistencies and problems as regards its application at the national level and the relationship between the Commission and the MS. Its implementation is accompanied by different degrees of decentralization and, consequently, of bottom up strategies reflecting the diverse national administrative, organizational and institutional capacities.

The horizontal integration of the RDP with the EU environmental and regional policies has proven to be even more difficult to achieve when compared to vertical integration, depending, to a large extent, on the ability and willingness of different national and regional authorities to join forces towards PI. However, the PI goal is not shared by the MS and many of their regions. For this reason, perhaps we should ask whether the quest for PI remains a normative issue for the EU while it is a political issue for MS and their regions.

The rhetoric of the RDP surpasses by far its integration dynamics. The poor level of PI within the second pillar of the CAP is determined by the high expectations built on the presumption that RDP is designed to be an integrated policy. The rhetoric, despite its apparent political and regulatory functions at the EU level, actually conceals the caveats in the design of the RDP and the lack of (political) willingness and institutional capacity of the national actors to practice PI.

The discussion of the problems related to the object of rural policy has exposed the lack of consensus over its definition. This major weakness is due to the national specificities and the social constructivism which interact in the formulation of the policy object. The pluralistic and integrative nature of rural policy needs to be interpreted by means of a rural development strategy, which is currently missing and is substituted by the bottom up approach that transfers the responsibility to the actors involved and their interactions. Thus, the object of rural policy is transformed and broadened through a policy process, which encompasses a wide range of sectors, actors, levels and institutions. Its definition becomes a matter of national and regional actors' political will, negotiations among the actors comprising the EU rural policy network and new modes of governance which allow for flexible and reflexive social and policy processes.

Despite the present difficulties, PI in the context of the RDP is an important policy issue which requires further systematic theoretical as well as empirical research to support meaningful and responsive policy choices. Future research should investigate, in greater detail and on the basis of detailed evidence (a) the internal coherence of the RDR at both the formulation and the implementation level, (b) the integration of the RDP object, goals, actors, procedures and instruments with those of several policies, including the regional and environmental that the present study has examined, that impinge on rural development (e.g. transport, water, spatial) at the EU and at the MS levels. The formal aspects of PI seem to attract most attention at the moment, but it is encouraging that a concerned group of researchers is critical of the informal and essential aspects of PI that need to be further explored as they are particularly important in the Southern EU MS. Finally, dedicated studies are needed to examine the implications of PI in the context of the RDP for combating desertification as it is a serious problem that inflicts several EU LFAs and which can be treated only by means of integrated approaches.

Acknowledgements

The author would like to thank Prof. Helen Briassoulis for her generous guidance on policy analysis and assessment and also for her helpful comments on earlier drafts of this chapter. The research on which the chapter is based was supported by the EU research project entitled: "Policies for Land Use to Combat Desertification (MEDACTION)" (EVK2-CT-2000-00085).

References

Ackrill, R.W. (2000), 'CAP Reform 1999: A Crisis in the Making?', *Journal of Common Market Studies*, Vol. 38, No 2, pp. 343-53.

Bache, I. (2003), 'Europeanization: A Governance Approach', Paper presented at the EUSA 8th International Biennial Conference, Nashville.

Baldock, D., Dwyer, J. and Sumpsi Vinas, J.M. (2002), *Environmental Integration and the CAP*, A Report to the European Commission, DG Agriculture, Brussels.

Baldock, D. and Lowe, P. (1996), 'The Development of Agri-environmental Policy', in M. Whitby (ed.), *The European Environment and CAP Reform: Policies and Prospects for Conservation*, Wallingford, CAB International, pp. 9-25.

Berger, G. (2003), 'Reflections on Governance: Power Relations and Policy Making in Regional Sustainable Development', *Journal of Environmental Policy and Planning*, Vol. 5, No 3, pp. 219-34.

Börzel, T.A. (2000), 'Why there is no 'Southern Problem'. On Environmental Leaders and Laggards in the European Union', *Journal of European Public Policy*, Vol. 7, No 1, pp. 141-62.

Briassoulis, H. (2004), 'The Institutional Complexity of Environmental Policy and Planning Problems: The Example of Mediterranean Desertification', *Journal of Environmental Planning and Management*, Vol. 47, No 1, pp. 115-35.

Brouwer, F. and Lowe, P. (2000), 'CAP and the Environment: Policy Development and the State of Research', in F. Brouwer and P. Lowe (eds), *CAP Regimes and the European Countryside: Prospects for Integration Between Agricultural, Regional and Environmental Policies*, Wallingford, CAB International, pp. 1-14.

Bryden, J. (2000), 'Is there a 'New Rural Policy'?, Paper presented in the International Conference: European Rural Policy at the Crossroads', University of Aberdeen.

Bryden, J. (2003), 'Rural Development Situation and Challenges in EU-25', Paper presented in the EU Rural Development Conference, Salzburg.

Buckwell, A. (1998), 'Agenda 2000 and Beyond: Towards a New Common Agriucultural and Rural Policy for Europe – CARPE', Paper presented in a Seminar organized by the ISAD and the SIDEA, Firenze.

Buckwell, A., Blom, J., Commins, P., Hervieu, B., Hofreither, M., von Meyer, H., Rabinowicz, E., Sotte, F. and Sumpsi Vilas, H.M. (1997), *Towards a Common Agricultural and Rural Policy for Europe*, Report of an Expert Group, DG Agriculture.

Buller, H. (2002), 'Integrating EU Environmental and Agricultural Policy', in A. Lenschow (ed.), *Environmental Policy Integration: Greening Sectoral Policies in Europe*, Earthscan, London, pp. 103-26.

Buller, H. (2003), 'Changing Needs, Opportunities and Threats – The Challenge to EU Funding of Land Use and Rural Development Policies', Paper presented in the Land Use

Policy Group Conference on: Future Policies for Rural Europe, Brussels.

CEC (1988), *The Future of Rural Society*, COM(88)501, Commission of the EC, Brussels.

CEC (1996), *The Cork Declaration – A Living Countryside*, The European Conference on Rural Development, Brussels.

CEC (1999), *CAP Reform: Rural Development*, Fact Sheet, DG Agriculture, Brussels.

CEC (2001), *Unity, Solidarity, Diversity for Europe, it People and its Territory: Second Report on Economic and Social Cohesion*, Adopted by the European Commission on 31 January 2001, Brussels.

CEC (2002), *Mid-Term Review of the CAP*, Communication from the Commission to the Council and the European Parliament, COM (2002) 394 Final, Brussels.

CEC (2003), *CAP Reform Summary*, Newsletter, Special Edition, DG Agriculture, Brussels.

Christiansen, T., Føllesdal, A. and Piattoni, S. (2003), 'Informal Governance in the European Union: An Introduction', in T. Christiansen and S. Piattoni (eds), *Informal Goverance in the European Union*, Cheltenham, Edward Elgar, pp. 1-21.

Colson, F. and Mathurin, J. (2002), 'How could the CAP Pillars be balanced for the promotion of a Multifunctional European Model?', Paper presented in the Akademie für Raumforschung und Landesplannung, Hanover.

Danish, K.W. (1995), 'International Environmental Law and the "Bottom-Up" Approach: A Review of the Desertification Convention', *International Journal of Global Legal Studies*, Vol. 3, No 1.

Daugbjerg, C. (1999), 'Reforming the CAP: Policy Networks and Broader Institutional Structures', *Journal of Common Market Studies*, Vol. 37, No 3, pp. 407-28.

Depoele, L. (2000), 'The European Model of Agriculture (EMA): Multifunctional Agriculture and Multisectoral Rural development', paper presented in the International Conference: European Rural Policy at the Crossroads, University of Aberdeen.

Dwyer, J., Baldock, D., Beaufoy, G., Bennett, H., Lowe, P. and Ward, N. (2002), *Europe's Rural Futures – The Nature of Rural Development II. Rural Development in an Enlarging European Union*, Final Report, The Land Use Policy Group and the WWF.

Eberlein, B. and Kerwer, D. (2002), 'Theorizing the New Modes of European Union Governance', *European Integration Online Papers (EIOP)*, Vol. 6, No 5 (available from http://eiop.or.at/eiop/texte/2002-005a.htm).

Edwards, T. (2003), 'CAP Reform: A New Common Agricultural Policy?', *SPICe briefing 03/86*, The Scottish Parliament.

Friedland, W.H. (2002), 'Agriculture and Rurality: Beginning the "Final Separation"?', *Rural Sociology*, Vol. 67, No 3, pp. 350-71.

Gray, J. (2000), 'The Common Agricultural Policy and the Re-Invention of the Rural in the European Community', *Sociologia Ruralis*, Vol. 40, No 1, pp. 30-52.

Hadjimichalis, C. (2003), 'Imagining Rurality in the New Europe and Dilemmas for Spatial Policy', *European Planning Studies*, Vol. 11, No 2, pp. 103-13.

Hardin, G. (1968), 'The Tragedy of the Commons', *Science*, Vol. 162, pp. 1243-8.

Hennis, M. (2001), 'Europeanization and Globalization: The Missing Link', *Journal of Common Market Studies*, Vol. 39, No 5, pp. 829-50.

Heritier, A. (2002), 'New Modes of Governance in Europe: Policy Making without Legislating?', *Political Science Series Paper 81*, Institute for Advanced Studies, Vienna.

Hertin, J. and Berkhout, F. (2003), 'Analysing Institutional Strategies for Environmental Policy Integration: The Case of EU Enterprise Policy', *Journal of Environmental Policy and Planning*, Vol. 5, No 1, pp. 39-56.

Hervieu, B. and Beranger, C. (2000), 'New Regulation of Agriculture and Rural Development in Europe particularly in France through multifunctional character of agriculture and land', Paper presented in the International Conference: European Rural Policy at the Crossroads, University of Aberdeen.

Hooghe, L. and Marks, G. (2001), 'Types of Multi-Level Governance', *European Integration Online Papers (EIOP)*, Vol. 5, No 11, (http://eiop.or.at/eiop/texte/2001-011a.htm).

Jacob, K. and Volkery, A. (2003), 'Environmental Policy Integration (EPI) – Potentials and Limits for Policy Change Through Learning', Paper presented at the Berlin Conference on the Human Dimension of Global Environmental Change.

Jokela, M. (2002), 'European Union as a Global Policy Actor: The Case of Desertification', in F. Biermann, R. Brohm and K. Dingwerth (eds), *Proceedings of the 2001 Berlin Conference on the Human Dimensions of Global Environmental Change*, Potsdam, Potsdam Institute for Climate Impact Research, pp. 308-16.

Kay, A. (2000), 'Towards a Theory of the Reform of the Common Agricultural Policy', *European Integration online Papers*, Vol. 4, No 9 (available from http://eiop.or.at/eiop/texte/2000-009a.htm).

Kay, A. (2003), 'Path Dependency and the CAP', *Journal of European Public Policy*, Vol. 10, No 3, pp. 405-20.

Lenschow, A. (2002), Greening the European Union: An Introduction, in A. Lenschow (ed.), *Environmental Policy Integration: Greening Sectoral Policies in Europe*, Earthscan, London, pp. 3-21.

Liefferick, D. and Andersen, M.S. (1998), 'Strategies of the 'Green' Member States in EU Environmental Policy-making', *Journal of European Public Policy*, Vol. 5, No 2, pp. 254-70.

Lowe, P. and Baldock, D. (2000), 'Integration of Environmental Objectives into Agricultural Policy Making', in F. Brouwer and P. Lowe (eds), *CAP Regimes and the European Countryside: Prospects for Integration between Agricultural, Regional and Environmental Policies*, CABI Publishing, Wallingford, pp. 31-52.

Lowe, P., Buller, H. and Ward, N. (2002), 'Setting the Next Agenda? British and French Approaches to the Second Pillar of the Common Agricultural Policy', *Journal of Rural Studies*, Vol. 18, pp. 1-17.

Mantino, F. (2003a), 'Rural Development Policies in the EU: The Main Progresses after Agenda 2000 and the Challenges Ahead', Paper presented in the International Conference entitled: Agricultural Policy Reform and the WTO: where are we heading?, Capri, Italy.

Mantino, F. (2003b), 'The Second Pillar: Allocation of Resources, Programming and Management of Rural Development Policy', Paper presented in the Land Use Policy Group Conference on: Future Policies for Rural Europe – 2006 and beyond, Brussels.

Meadowcroft, J. (2002), 'Politics and Scale: Some Implications for Environmental Governance', *Landscape and Urban Planning*, Vol. 61, pp. 169-79.

Montpetit, E. (2000), 'Europeanization and Domestic Politics: Europe and the Development of a French Environmental Policy for the Agricultural Sector', *Journal of European Public Policy*, Vol. 7, No 4, pp. 576-92.

Mormont, M. (1990), 'Who is Rural? or How to be Rural: Towards a Sociology of the Rural', in T. Marsden, P. Lowe and S. Whatmore (eds), *Rural Restructuring: Global Processes and Their Responses*, David Fulton Publishers, London, pp. 21-44.

O'Hara, P. (2002), 'Rural Development in the 21st Century – A View from the Middle', Paper presented in the COST Conference, Budapest, Hungary.

Persson, A. (2002), 'Environmental Policy Integration: An Introduction', Background Paper, PINTS - Policy Integration for Sustainability, Stockholm Environment Institute.

Peters, B.G. (1998), 'Managing Horizontal Government: The Politics of Coordination', *Public Administration*, Vol. 76, pp. 295-311.

Pezaros, P. (1999), 'The Agenda 2000 CAP Reform Agreement in the Light of the Future EU Enlargement', *Working Paper 99/2*, European Institute of Public Administration.

Potter, C. and Burney, J. (2002), 'Agricultural Multifunctionality in the WTO – Legitimate Non-Trade Concern or Disguised Protectionism?', *Journal of Rural Studies*, Vol. 18, pp. 35-47.

Potter, C. and Goodwin, P. (1998), 'Agricultural Liberalization in the European Union: An Analysis of the Implications for Nature Conservation', *Journal of Rural Studies*, Vol. 14, No 3, pp. 287-98.

Richardson, T. (2000), 'Discourses of Rurality in EU Spatial Policy: The European Spatial Development Perspective', *Sociologia Ruralis*, Vol. 40, No 1, pp. 53-71.

Roedeger-Rynning, C. (2003), 'Informal Governance in the Common Agricultural Policy', in T. Christiansen and S. Piattoni (eds), *Informal Goverance in the European Union*, Edward Elgar, Cheltenham, pp. 173-88.

Saraceno, E. (2002), 'Rural Development Policies and the Second Pillar of the Common Agricultural Policy', Paper presented in the Akademie für Raumforschung und Landesplannung, Hanover.

Sbragia, A. (2000), 'The European Union as a Coxwain: Governance by Steering', in J. Pierre (ed.), *Debating Goverance: Authority, Steering and Democracy*, Oxford University Press, Oxford, pp. 219-40.

Shortall, S. (2004), 'Time to Re-Think Rural Development?', *EuroChoices*, Vol. 3, No 2, pp. 34-9.

Shucksmith, M. (2003), 'Territorial Aspects of the Common Agricultural Policy', Paper presented in the 20th European Congress of Rural Sociology, Sligo, Ireland.

Sumpsi, J. and Buckwell, A. (2002), 'Greening the CAP: The Future of the First Pillar', paper presented in the Akademie für Raumforschung und Landesplannung, Hanover.

Sumpsi, J. and Garcia-Azcarate, T. (2002), Obstacles and Constaints for a New CAP, Paper presented in the Akademie für Raumforschung und Landesplannung, Hanover.

Terluin, I.J. (2001), *Rural Regions in the EU: Exploring Differences in Economic Development*, Nederlandse Geografische Studies 289, Utrecht/Groningen, Rijksuniversiteit Groningen.

Thomas, E.V. (2003), 'Sustainable Development, Market Paradigms and Policy Integration', *Journal of Environmental Policy and Planning*, Vol. 5, No 2, pp. 201-16.

Thomson, K.J. and Psaltopoulos, D. (2004), '"Integrated" Rural Development Policy in the EU: a Term Too Far?', *EuroChoices*, Vol. 3, No 2, pp. 40-5.

Van der Leeuw, S.E. (2000), 'Some Potential Problems with the Implementation of Annex IV of the Convention to Combat Desertification', in G. Enne, Ch. Zanolla and D. Peter (eds), *Desertification in Europe; Mitigation Strategies, Land-Use Planning*, EC, DG for Research Environment and Climate, pp. 249-62.

Van der Ploeg, J.D. (1997), 'On Rurality, Rural Development and Rural Sociology', in H. de Haan and N. Long (eds), *Images and Realities of Rural Life: Wageningen Perspectives on Rural Transformations*, Van Gorcum, pp. 39-73.

Van der Ploeg, J.D. (2003), 'Rural Development and the Mobilization of Local Actors', paper presented in the Second Rural Development Conference, Salzburg.

Van der Ploeg, J.D., Renting, H., Brunori, G., Knickel, K., Mannion, J., Marsden, T., de Roest, K., Sevilla-Guzman, E., Ventura, F. (2000), 'Rural Development: From Practices and Policies Towards Theory', *Sociologia Ruralis*, Vol. 40, No 4, pp. 391-408.

Ward, W.A. and Hite, J.C. (1998), 'Theory in Rural Development: An Introduction and Overview', *Growth and Change*, Vol. 29, pp. 245-58.

Warleigh, A. (2003), 'Informal Governance: Improving EU Democracy?', in T. Christiansen and S. Piattoni (eds), *Informal Governance in the European Union*, Edward Elgar, Cheltenham, pp. 22-35.

Whatmore, S. (2002), *Hybrid Geographies: Natures, Cultures, Spaces*, Sage Publications, London.

White, L. (2001), 'Effective Governance Through Complexity Thinking and Management Science', *Systems Research and Behavioural Science*, Vol. 18, pp. 241-57.

Zahariadis, N. (2003), 'Complexity, Coupling and the Future of European Integration', *Review of Policy Research*, Vol. 20, No 2, pp. 285-310.

Chapter 5

European Union Social Policies: The Conundrum of Policy Integration

Theodoros Iosifides and Helen Briassoulis

Introduction

Social problems, such as poverty, social exclusion, crime, malnutrition, and illiteracy, from the neighborhood up to the global level, reveal the unequal distribution of social welfare among individuals and social groups over and across space. They result from the dynamic interplay of autonomous socio-economic forces and public policies acting at various spatio-temporal levels and, in their turn, they cause other socio-economic and environmental problems (Becker and Vanclay, 2003). Globalization and its consequences have exacerbated the incidence of social problems and have revealed their close and multiple ties with economic and environmental problems. The contemporary understanding of social problems and of associated policies recognizes their complexity, their critical role in the achievement of sustainable development and the necessity of combined and integrated public interventions that transgress formal political boundaries, specific spatial levels and narrow policy areas (OECD, 1996; Blakemore, 1998; Byrne, 2001).

In human-environment systems, a frequent, puzzling question is whether social problems cause economic adversity and environmental degradation or the other way around (Blaikie and Brookfield, 1987). This dilemma is more pronounced in Less Favoured Areas (LFAs) where socio-economic disadvantages combine with environmental marginality to reduce available livelihood opportunities, rendering the sustainability of their development more precarious (Spilanis et al., 2003). In desertification-sensitive areas, in particular, the debate on the dynamically evolving, complex, causal relationships among environmental, social and economic forces has influenced importantly both the definition of the phenomenon and the proposed avenues to address it (Blaikie and Brookfield, 1987; Perez-Trejo, 1994; Thomas, 1997; Reynolds and Stafford-Smith, 2002).

Traditional social policies, European Union policies included, prove to be increasingly ineffective in alleviating social problems and fostering sustainable development, especially in LFAs, because they ignore their complexity and do not fully account for the interrelatedness of the social, economic and environmental determinants of social welfare and of the associated policies (Papadopoulos and Liarikos, 2003). To achieve sustainable development and cope with social problems effectively and efficiently, the integration of social with economic and environmental policies is imperative. Policy integration (PI) in the present context is conceived as substantive and procedural compatibility, congruence and coordination among the objects, goals, actors, procedures and instruments of social, economic and environmental policies.

The topic of policy integration is little researched in the case of social policies currently. This chapter aims to make a modest contribution in this direction by examining the present state and future prospects of the integration of EU social with economic and environmental policies on the basis of selected criteria. Combating desertification in the LFAs of the Mediterranean EU member states (MS) is used as an illustrative example. The chapter comprises five sections. The second section negotiates the relationships between social welfare, social policies, sustainable development and the need for PI in the European Union, focusing on LFAs sensitive to land degradation and desertification. The third section presents the current situation in the general EU social policy area and in selected social sectors. The fourth section examines the present state of integration of EU social with economic and environmental policies. Summarizing the analysis, the last section discusses the conundrum of policy integration and its implications for combating desertification in Mediterranean Europe, and offers future research directions.

Social Welfare, Social Policies, Sustainable Development and the Quest for Policy Integration

Social Welfare, Social Policies and Sustainable Development

Social welfare refers to meeting basic social needs[1] such as education, health, housing, social security, income generation and employment. Four major types of needs are usually identified; normative, felt, expressed and comparative (Pinch, 1997). Normative needs concern those defined by policy makers while felt needs are those historically and socio-culturally constructed by various social groups. Social pressures on state or other authorities to satisfy some of the felt needs lead to the notion of expressed needs, while comparative needs relate to the measurement of the characteristics of those who receive social assistance (Pinch, 1997). Contemporary social welfare, both at the global and the European level, is influenced by major socio-economic changes, related mainly to globalization that has two important characteristics; first, the growing, and highly asymmetrical, interdependence of economies and societies and, second, the proliferation of a

---

[1] What exactly constitutes a 'basic' social need depends on historical, social and cultural factors.

liberalized, open market system of economic organization which tends to take its 'pure' form with market forces dominating the economic arena without the 'barriers'[2] of previous periods.

The liberalizing globalization process has caused socio-economic and political restructuring and changes worldwide. These include accelerated flows of short-term investment based on speculative currency trade, policies aimed at further reducing barriers to trade (deregulation), increased shares of transnational corporations in global production and trade, global interconnectedness of production, increased movement of people worldwide, changing power balances between nation-states and supranational organizations, and proliferation of new forms of communication and information technologies. The social welfare implications of these changes include increased social inequalities, both within and between countries, rising impoverishment, vulnerability of people to social risk, and accelerated exclusion of people, communities and regions from the benefits of globalization (Deacon, 2000).

In this broader global context, the character of social policy making has undergone important changes following the economic crisis of the mid-1970s and the retreat of the fordist model of mass production and consumption that was strongly associated with state interventionism and the development of a universalistic social policy regime. The gradual transition to more flexible types of economic organization, associated with the adoption of neoliberal policies, has resulted in increased pressures to dismantle the universalistic social policy regimes in favour of selectivity and residualism[3] (Petmezidou-Tsoulouvi, 1992).

In the European Union, general social welfare and particular social problems have worsened under these conditions. European social policy has proven unable to address satisfactorily and comprehensively persistent and strongly interrelated problems such as income inequalities, long-term unemployment, social exclusion, poverty, and social security and protection, which affect directly and importantly the cohesion of European societies. It is true, however, that the social situation would have been worse in the absence of EU social policy interventions whose impacts depend on the social policy model of each MS and on broader socio-economic conditions (Ferrera et al., 2000; Bernhagen, 2000).

Social welfare is multifariously related not only to economic but also to environmental welfare. Through its influence on population size, structure and well-being, it influences the level of pressures on natural resources and the environment. Employment creation, higher wages and social assistance affect income levels and distribution and, consequently, resource consumption and levels of waste (Huby, 1998). Obviously, social policies alone cannot address satisfactorily multidimensional and complex social problems. Instead, the broader pursuit of sustainable development that treats explicitly the interrelatedness of the economic, environmental and social dimensions of welfare provides the proper

---

[2] Such as state intervention and regulations related to economic organization.
[3] See section *"The Evolution of the EU Social Policy Area"*.

framework fostering the coordinated application of economic, environmental and social policies.

The complexity of social welfare issues and the problems it presents for effective EU social policy towards sustainable development are examined next for the case of European Union LFAs, especially those sensitive to land degradation and desertification.

The Desertification-Sensitive LFAs of the EU – Social Welfare, Social Policies and the Quest for Policy Integration

Based on particular biophysical, social and economic criteria,[4] the European Union defines three categories of LFAs:

(a) Mountainous areas, where altitude and slope account for reduced land productivity, characterized by high cultivation costs, decreased cultivation periods and relative difficulties for agricultural mechanization;

(b) Areas of specific disadvantages, where the preservation of agriculture is necessary for the conservation of natural resources, the protection of the environment and the coastal zone, and the enhancement of their tourist potential;

(c) Other disadvantaged areas, characterized by low land productivity, low income and weak demographic base, frequently threatened by depopulation and land abandonment (Papadopoulos and Liarikos, 2003).

EU policies for LFAs aim at balancing the negative economic impacts of permanent physical disadvantages, reversing rural depopulation trends, and protecting the rural environment. Two sets of specific measures are provided; equalizing compensations per animal head or per acreage, and subsidies for investment plans (10 per cent higher in value than those given in other areas) and for agricultural improvement plans (12.5 per cent higher for young farmers) (Spilanis et al., 2003).

The low level of social welfare and the reduced attractiveness of LFAs owe to their physical disadvantages, high living and production costs, lack of sufficient social infrastructure, high dependence on the primary sector, lack of alternative employment opportunities, especially for women and the young, and outmigration and depopulation (Spilanis et al., 2003). Especially in the northern Mediterranean, LFAs located in semi-arid and dry sub-humid regions suffer from land degradation and desertification that affects over 60 per cent of the southern European territory (UNEP, 1992).

The human causes of desertification originate in unsustainable ways of production and consumption, associated with particular patterns of socio-economic organization and practice, which strain environmental resources and media (Perez-

---

[4] Established under Directive 75/268 and Regulations 950/97 and 1257/99.

Trejo, 1994). In the northern Mediterranean, important human driving forces of land degradation and desertification include population dynamics, social inequality, poverty, migration and urban development (Blaikie and Brookfield, 1987; UNEP, 1992). The common perception is that population pressures intensify and exacerbate land degradation and desertification. Nevertheless, high absolute population numbers, or population increase, do not necessarily lead to land degradation. Desertification occurs often in geographic areas with low population pressures, and periods of population decline may coincide with the exacerbation of the problem (Blaikie and Brookfield, 1987). This is so because the relationship between population pressure and desertification is mediated by several other factors that include land sensitivity and fragility, land use and settlements patterns, social and economic conditions, mode of production, cultivation practices, etc. (Perez-Trejo, 1994). There seems to be no simple, linear and clear causal links between population growth and decertification or a stable and static 'carrying capacity' of land beyond of which the problem starts to worsen. The relative absence of a direct causal relationship must not lead to the neglect of population pressure as a determinant of land degradation as frequently it may catalyse the intensification and severity of the problem (Blaikie and Brookfield, 1987).

Social inequality and poverty, especially rural poverty, lie at the heart of the debate about the relation between social conditions and desertification. Both absolute and relative poverty can be seen as the result of social inequality (UNESC, 1995) and has been viewed as both a cause and as a consequence of land degradation and desertification. Evidently, their links are more complex, with no clear, linear causal relationship between the two (UNEP, 1992).

The direction of causality between social inequality and poverty and desertification is case- and context-specific. Frequently, the poor rely directly on their own limited natural resources and/or complement their income by exploiting common property resources (UNEP, 1992). Human pressures on land and the environment increase when poverty leads to the use of intensive cultivation methods in sensitive areas. The broader implications of poverty, such as lack of access to education and training, information, health and social provisions, etc. must be taken into account also. Living in conditions of social deprivation, the poor find it very difficult or even impossible to understand, accept, and implement sustainable land management practices. Migration to other rural areas or urban centers and changes in livelihood strategies with shifting employment from agriculture to other sectors, like petty trade, services, mining, construction, etc., are common responses to such situations (UNRISD, 1994). The responses of the poor contribute to land degradation in more indirect ways also. For example, rural-rural migration may increase population pressures in other areas; rural-urban migration produces excessive population concentrations in urban centers, hence, higher demand for food and agricultural products that generates pressures for intensification of production in rural areas. Overall, the poor are not always the passive victims of the degradation-poverty spiral.

Migration and population mobility correlate variously with problems of land degradation and desertification. Migration can be seen either as a driving force or as a consequence of these phenomena or as a process helping to combat them. The relationship between the two is complex and non-linear mediated by factors such as local socio-economic conditions and change, urban-rural dynamics, public policies, institutional dynamics, etc. (UNESC, 1995).

The brief account of the social determinants of desertification reveals their complex, multi-level, multi-causal, non-linear, context- and scale-specific relationships to the phenomenon. Fully understanding and coping with such complex problems requires the adoption of a systems approach that views natural and social phenomena as 'open systems' characterized by non-equilibrium and non-linear causality chains (Holling, 2001). Broader, national or global socio-economic developments and processes interact with local and regional conditions and processes through diverse forward and backward linkages. Consequently, individual, unrelated and uncoordinated policies cannot adequately address complex problems, necessitating their integration to achieve simultaneously social, economic and environmental goals.

Social policies can play a crucial role in promoting sustainable development as they impinge on population structure, mobility and dynamics, income distribution, and social inequalities, factors which are intricately related to economic and environmental conditions and performance. However, these interdependencies are not adequately reflected in EU and MS social policies whose lack of explicit relationships and coordination of with other policy areas results, in many cases, in ineffectiveness in addressing complex problems which persist and continue to worsen. This chapter examines the state and prospects of integration between EU social, economic and environmental policy areas as a prerequisite for fostering sustainable development especially in LFAs prone to land degradation and desertification.

Social Policies of the European Union

The Evolution of the EU Social Policy Area

The European Union is characterized by a relative absence of a common social policy framework, as it is the case, for example, with agriculture (Common Agricultural Policy) or the environment. Nevertheless, several directives and guidelines on social policy issues shape a distinct quasi-European social policy framework, setting mainly minimum social standards. The gradual involvement of European institutions in social policy matters and the evolution towards a European social policy framework are briefly outlined in the following.

The 1957 Treaty of Rome contained several points on social provision such as the free movement of workers (articles 45-81), the right of establishment, particularly of the self employed (articles 52-58), the freedom of service provision (articles 59-66), the functioning and rationale of the European Social Fund (articles 123-128) and economic and social cohesion (articles 130A-A30E) (Gold, 1993). The difficulty to move from these general and isolated provisions to direct policy measures or to formulate a coherent common social policy is mainly due to three reasons. The first concerns the tension that exists between supranational bodies (European Commission, European Parliament, European Court of Justice) and national interests; the second is the differing social policy frameworks of the MS, which often transcend the traditional left-right divide[5] and the third is the different views of employers and trade unions regarding the formulation of a common EU social policy (Gold, 1993).

Gold (1993) identifies *four stages* in the evolution of European social and labour policies up to the mid 1990s. The relative absence or neglect of social policy harmonization efforts[6] in the Community characterizes the first stage (1958-1972), although social provisions were made in many cases. During the second stage (1972-1980), the Social Action Program was adopted in 1974 which contained some 40 different initiatives, related mainly to the promotion of full and better employment, improved living and working conditions and worker participation. Furthermore, a series of directives were passed concerning employment protection, employee participation, equal treatment for men and women, and health and safety matters. During the third stage (1980-1987), several other directives were passed (on health and safety and equal treatment) but its distinct feature was the evolution of the concept of 'subsidiarity' at the Community level in general.[7] This notion was, to some extent, a drawback from the concept of a European social policy and was introduced mainly because certain governments, concerned with higher labour costs, and due to political and socio-economic differences, were against unified measures. The fourth stage (1987-1992) was characterized by the adoption of the Social Charter (Community Charter of the Fundamental Rights of Workers) at the Strasbourg Summit in 1989, which was accompanied by an Action Program containing 47 proposals on initiatives in different areas of social policy. In 1991, at the Maastricht European Council, an 'agreement on social policy', which included the Social Action Program, replaced the Social Charter, avoiding the need to incorporate its provisions into the Treaty (Gold, 1993). Finally, in 1997, at the Amsterdam Summit, the Agreement was incorporated into the body of the Treaty and applied to all 15 Member States.

Although European nation building is closely associated with the adoption of a strong social policy dimension, the EU social policy area is characterized by relatively clear tendencies towards greater residuality and selectivity (Nieminen, 1995). The major challenges the European Union identifies for the future are related to global economic competition, especially with the United States and

---

[5] Mainly, the Roman-German, the Anglo-Irish and the Nordic system.

[6] Unified social measures and common rules at the European level.

[7] Meaning that the Community would take action only in cases which cannot be dealt efficiently at the national level.

Japan, globalization of trade and production, competitiveness, economic growth and the impact of new technologies on work and demographic ageing (Nieminen, 1995). Social policy is conceptualized primarily as a tool to support economic growth and competitiveness; i.e. it is considered a productive factor (European Commission, 2000), thus losing its autonomy in covering certain social needs, and becoming subordinate to economic concerns. Although economic and social issues and processes are interlinked and interdependent, this is not reflected in the relationships among the respective policy areas, which are not mutually (and equally) supportive and reinforcing (Vikstrøm, 1996).

These trends signify a relative departure, or major pressures for such a departure,[8] from the traditional *universal/institutional social model*, a model of an all-inclusive welfare covering the social needs of extended segments of the population, which take the form of social rights, and a movement towards a *residual and selective social policy model*. The latter assumes minimum state intervention that is a residual to the 'normal' function of the market. Its main purpose is to prevent extreme social deprivation and exclusion in order to avoid social unrest and conflicts. Social benefits are means-tested[9] and in many cases the recipients are stigmatized. Furthermore, institutions or organizations other than the state are important in providing social services such as the family, voluntary organizations, charities, etc. (Petmezidou-Tsoulouvi, 1992).

The Social Policy of the European Union: Brief Presentation

Currently, the social policy area of the EU contains actions and measures concerning selected sectors such as employment and labour market, health care, social security and protection, migration, social exclusion and poverty. The actions and provisions of the European Union related to general social welfare policy are considered below; notably the Social Action Program 1998-2000 and the Social Policy Agenda (SPA) 2000-2005 as these two general frameworks influence to a great extent almost all other sectoral policies and measures.

The Social Action Program 1998-2000 was based on the first medium-term program of 1995-1997 and on the 1999 Treaty of Amsterdam. Its goals were to create employment, support the development of skills and promote the free movement of workers in order to meet the challenges of the changing world of work in an era of globalization and rapid technological developments as well as to create an inclusive and cohesive society. The follow-up of the Social Action Program, the SPA 2000-2005, acts as an umbrella initiative that affects almost all aspects of social policy in the Union and forms a framework that requires all key actors to join their forces to advance economic, employment and social policy goals. These include the promotion of social cohesion, the realization of full

---

[8] For example, the German Agenda 2010 on welfare and social protection reform.
[9] Subject on inadequacy of means of subsistence which is tested and evaluated mainly by state agencies.

employment through the creation of more jobs and adaptation to the new work environments, the modernization of social protection and the enhancement of social dialogue. Two important points are in order here. First, under the Agenda, the European Union recognizes social policy as a productive factor, and second, the Agenda stresses the need to establish links between economic, employment, social protection, social cohesion and environmental policies in the context of a new and dynamic policy framework towards sustainable development (European Commission, 2000).

The SPA is formulated and implemented by the following major actors (European Commission, 2000):

- The European Commission (DG-Employment and Social Affairs), which makes all relevant proposals using its rights of initiative;
- The Council of Ministers and the European Parliament, which have responsibility over specific measures;
- The national governments and the regional/local authorities who are the main implementers of actions included in the Agenda;
- The European Economic and Social Committee (EESC);
- Social partners (mainly employers and employees organizations such as the European Trade Union Confederation, the European Association of Craft, SMEs, the European Centre of Enterprises with Public Participation and of Enterprises of General Economic Interest, and the Union of Industrial and Employers Confederation of Europe), that play a crucial role in creating consensus and forming meaningful agreements;
- Non-governmental organizations (NGOs), which are concerned with the development of inclusive policies and the promotion of equal opportunities for all.

The principal procedure used for the implementation of the Agenda is the open method of coordination (OMC) (O'Connor, 2003). The OMC comprises four major stages (Sakellaropoulos, 2001): (a) the European Commission in cooperation with the MS sets guidelines and certain short, medium and long term goals and objectives, (b) specific qualitative or quantitative indicators or targets associated with the identification of 'best practices' from different cases and countries are set, (c) national and regional goals, in accordance with the general guidelines and objectives, and the appropriate measures to achieve them are defined and (d) periodic review and progress evaluation of the process.

The main EU social policy instruments are:

(a) Legislation, such as decisions and directives, issued by the European Commission and implemented by the MS, aiming at the development of social standards where appropriate;
(b) Financial initiatives and programs funded through the European Social Fund (ESF);

(c) Community initiatives (ADAPT,[10] NOW,[11] HORIZON,[12] EQUAL,[13] etc.);
(d) Communication in the form of regular reporting based on policy analysis, research results and social dialogue, a process through which trade unions and employer organizations consult each other and influence EU policy making (European Commission, 2000). The Maastricht Treaty made possible for social partners to conclude European level framework agreements on employee's rights, which may take the form of directives.

The ESF is the main social policy financial instrument. Its mission is to support measures for the prevention and combat of unemployment, the development of human resources, the promotion of social inclusion and of equal opportunities for men and women, sustainable development, and economic and social cohesion. The programs funded through the ESF are planned by each MS separately together with the European Commission and are implemented through a wide range of provider, public and private sector, organizations (European Commission, 2000).

Major Social Policy Areas of the European Union

Four major social policy areas, which concentrate the bulk of policy activity of the European Union, are examined more closely in this section; the European employment strategy, actions to combat social exclusion and poverty, social security, and migration.

The *European Employment Strategy* aims at coordinating the employment policies of the MS. Its objectives include achieving a high level of employment in the economy and for all groups in the labour market, moving away from a passive fight against unemployment towards promoting sustained employability and job creation, favouring new approaches to work organization so that EU firms are able to cope with economic change while providing for security and adaptability, and providing equal opportunities for all to have access to and participate in the labour market (Sakellaropoulos, 2001).

The Luxembourg European Council (November 1997) launched the so-called 'Luxembourg process' aiming at employment policy coordination at the European level. This process develops as a rolling program of annual planning, monitoring, examination and re-adjustment. The European Council, based on Commission proposals, sets the Employment Guidelines at the beginning of each year. Each country draws up a National Action Plan to implement these Guidelines. The Council and the Commission jointly examine and evaluate this Plan in December of each year. The Council, based on the conclusions of the heads of state and government, approves the set of Employment Guidelines for the following year

---

[10] On promoting employment and helping employees to adapt to industrial change.
[11] On improving employment opportunities for women.
[12] On integration of people with disabilities in the workplace.
[13] On combating social inequality and exclusion.

whose four pillars are employability, entrepreneurship, adaptability, and the promotion of equal opportunities (Ardy and Begg, 2001).

As regards *combating social exclusion and poverty*, the most important actions of the Community concern (European Commission, 2001d):

- Financial support given through the Structural Funds; following the recent reforms, the Funds will increasingly promote social inclusion for the period 2000-2006;
- Community initiatives, such as EQUAL;
- The Commission's proposals on 'e-Europe - The Information Society for All' and on 'Strategies for jobs in the Information Society';
- Community initiatives which contribute to social inclusion such as the Framework Programs for Research, the Commission's Framework for Action for Sustainable Urban Development, education (SOCRATES), training (LEONARDO DA VINCI) and youth programs, the 'Second Chance Schools' scheme, proposals to combat discrimination, gender equality policies and the Community strategy for disability;
- The European Business Network to Promote Social Cohesion;
- Formulation of common objectives, principles, methods, benchmarking, assessment, monitoring and other support techniques and actions related to social exclusion and poverty for all MS;
- Development of operational tools under the new provisions of the Amsterdam Treaty; in particular, a multi-annual program of operational support for co-operation and a framework instrument to promote integration of people excluded from the labour market.

As regards *social security,* and following the Lisbon Summit, the Commission set up principles and objectives for discussion with the MS in order to maintain the adequacy of pensions, ensure intergenerational fairness, strengthen solidarity in pension systems, maintain a balance between rights and obligations, ensure that pension systems support the equality between men and women, ensure transparency and predictability, make pension systems more flexible in the face of societal change, facilitate labour market adaptability, ensure consistency of pension schemes within the overall pension system and ensure sound and sustainable public finances (European Commission, 2002).

Finally, as regards *migration*, the European Council at its meeting in Tampere (October 1999) set the political guidelines and some concrete objectives for the development of a common EU *Migration Policy* in the following key areas (European Commission, 2001c):

- Close cooperation with countries of transit and origin to address political, human rights and development issues;
- A common European asylum system based on the full and inclusive application of the Geneva Convention;

- A vigorous integration framework to ensure fair treatment of third country nationals, granting rights and obligations comparable to those of EU citizens.

The main process for the intended development of a Common migration policy is again the OMC. The guidelines approved by the Council in 2002 concerned mainly the management of migration flows, admission procedures of economic migrants, integration of third country nationals and cooperation with other countries, especially with countries of origin (European Commission, 2001c).

Compared to other policy areas, such as the economic, monetary, trade and environmental policy that are in the forefront of the Union's priorities, the EU social policy area remains 'underdeveloped' in the sense that there are no observable trends towards the creation of a coherent, 'Europeanized' policy regime encompassing all interdependent social issues. Currently, EU social policy making is characterized by an incremental approach to dealing with social issues, high fragmentation and the absence of a clear commitment towards meeting specific and binding, European-wide goals and targets. This owes to major differences among MS as regards the types and level of social provisions, the dominance of the neoliberal political philosophy, especially after the Maastricht Treaty and through the Economic and Monetary Union process, the prevalence of the subsidiarity principle and the application of the OMC in social policy making. The result is a very weak integration of the EU social with the economic and environmental policy areas, a subject to which the discussion turns next.

The Integration of the EU Social, Economic and Environmental Policies

Methodological Approach

The tight interrelationships and linkages among the social, economic and environmental dimensions of socio-environmental phenomena and of the related policy problems stand in strong contrast to the relatively weak connections among the respective policy areas in the European Union, raising the question of whether the EU policy system can deliver sustainable development ultimately. This section presents a coarse analysis of the current state and future prospects of integration between the social and the economic as well as between the social and the environmental policies of the European Union as they stand now, from both a theoretical and an applied perspective. The methodological approach adopted for this analysis is outlined in the following.

In a theoretical perspective, the thematic and the conceptual dimensions of PI are examined. The *thematic dimension* negotiates relationships among the objects, goals and actors of the policies or policy areas studied; in other words, the definitions of the policy problems each adopts, the aspects of reality each considers and the theories about them, which, in their turn, shape the policy goals and determine the policy actors that are implicated. The sustainable development imperative asks that sectoral policies address the interrelatedness and interdependence of the social, the economic and the environmental aspects of policy problems, avoiding their compartmentalization, characteristic of the traditional, sectoralized policy making. The same holds true for social policies; ideally they should account for other than the social aspects of reality as well as for particular sectoral concerns. The more multidimensional is the definition of the object of a policy, the more probable is that this policy is thematically integrated with other policies, drawing on common (integrated) theories about the problem in question. In this case, integration among policy goals and of the policy actors implicated follows naturally.

The *conceptual dimension* of PI involves a deeper analysis of the relationships among the definitions of concepts that are explicitly or implicitly used in different policy contexts; such as, social impacts, welfare, social security, social exclusion, social deprivation, risk exposure, vulnerability, etc. Socio-culturally determined variations in the value systems of policy actors generate different constructions of the various aspects of reality and different meanings of the same concept which carry over to differences in problem definitions, policymaking approaches and proposed solutions (for the same problem) among policy areas. The quest for conceptual integration is the unending effort to achieve intesubjectivity through the use of a common vocabulary and the development of common understanding and shared meanings as the essential basis for achieving satisfactory PI.

In an applied perspective, PI is examined along the *procedural dimension* that concerns the structural and procedural aspects of the relationships among policies; more specifically, relationships among policy actors, procedures and instruments through which PI materializes. Organizational arrangements and procedures that facilitate, accommodate and secure communication and cooperation among economic, social and environmental policy actors and promote their participation in joint decision making as well as congruent and coordinated use of policy instruments on the same and across spatial/organizational levels are indications of, at least, *intent* for PI.

Similarly, an essential procedural requirement for PI is that all actors possess the right to participate in decision-making within and between policy areas. As argued in Chapter 2 (this volume) the procedural dimension of policy integration is tightly related to the thematic and conceptual dimensions. In a procedural sense, the analysis examines whether the administrative and institutional apparatuses in different policy areas enable or promote the thematic integration of social with economic and environmental policies; in other words, whether the policy system is coordinated and functions coherently, producing social, economic and environmental benefits while minimizing the respective costs.

The analysis of the integration of social with economic and environmental policies adopts particular criteria drawing on the dimensions discussed before (Table 5.1).

Table 5.1   Criteria for the analysis of policy integration (PI)

Criteria related to the thematic and the conceptual dimension	
Policy object	• Comprehensive and compatible problem definitions • Congruent and consistent theories about the policy problem in question • Shared core belief systems across policy sectors
Goals and objectives	• Political commitment and leadership for policy integration • Common, congruent and compatible policy goals, objectives and targets
Actors and actor networks	• Existence of common (and/or complementary) formal and informal actors among policies
Criteria related to the procedural dimension	
Procedures	• Formal/institutionalized relationships among policy actors and actor networks (communication, cooperation, coordination) • Consistent, compatible and coordinated procedures and rules of decision-making in competent administrative bodies • Informal interaction among policy actors and actor networks • Interaction among formal and informal policy actors • Administrative reform (restructuring) in favour of PI • Provisions for implementing PI requirements
Instruments	• Existence of a legal framework for PI among the policies analyzed • Common, compatible, consistent and coordinated legal and institutional instruments • Use of financial mechanisms and fiscal incentives • Use of integrative instruments; such as, legal, economic, financial • Use of planning and management instruments for PI • Common (or compatible and consistent) data and information bases, monitoring programs and infrastructure • Common assessment and evaluation methodologies, and tools (PI indicators) • Use of communication instruments for PI • Education and training services for civil servants, bureaucrats, etc. on PI issues

Source: Compiled from various sources (ex. Lenschow, 2002; Persson, 2002; Peters, 1998)

The Integration of EU Social and Economic Policies

The economic policies of the European Union The economic policy area of the European Union covers mainly the fields of economic and monetary union, internal market, competition, taxation and foreign trade. The Common Agricultural and the Regional Policy are more specialized economic policies which are not considered here.

On January 1st 1999, the European Central Bank (ECB) assumed the responsibility for the economic and monetary union of all MS that adopted the single currency (euro), the primary objective being to maintain price stability; i.e. an annual increase in the Harmonized Index of Consumer Prices (HICP) of below two percent over the medium term. The main macro-economic policy instrument used is the interest rate mechanism, which is determined centrally for the Euro area by the ECB. In contrast to general economic policy, monetary policy is formulated for all MS within the Euro area (EMI, 1997). Policymaking is centralized, overseen by the governing council of the ECB. The decisions are made in accordance with the principles of an open market, competitive economy, favouring an efficient allocation of resources.

The overarching goal of EU economic policy, according to the decisions taken at the Lisbon European Council in 2002, is for Europe to become the most competitive and dynamic knowledge-based economy, capable of sustainable, non-inflationary growth with more and better jobs and greater social cohesion. These general goals are incorporated in the Broad Economic Policy Guidelines (BEPGs) and guide the coordination of the economic policies of the MS. The BEPGs are proposed annually by the EcoFin Council and constitute the framework for the close coordination of four other special procedures and their respective objectives:

a) The Stability and Growth Pact (SGP), which sets certain monetary and budgetary limits (inflation rate below two per cent over the medium term and budget deficit below three per cent of the gross national product);[14]
b) The Employment Guidelines, on labour market actions. The aforementioned Luxembourg process extends to the entire social policy area as a more appropriate way to address the structural problems of the European labour market;
c) The Cardiff process, which aims at the coordination of structural reforms for improving the functioning of product and labour markets;
d) The Cologne process, on wage developments (macroeconomic policy dialogue with the active involvement of social partners).

Internal market and competition policies are implemented centrally by the European Commission aiming at the gradual development of a common market and of an economic environment characterized by fair competition. The taxation policy of the European Union aims at minimizing the externalities of the national tax systems, especially in the field of competition and circulation of capital, services, goods and people. The foreign trade policy, implemented by the

---

[14] The budgetary policies of individual MS are constrained also by the ceiling on deficit and debt ratio determined by the provisions of SGP, which contain specific review and sanction procedures; namely, the excessive deficit procedure. In the case of 'excessive deficit' of a MS (a budget deficit above three per cent of Gross Domestic Product) the Council can penalize it, in the form of a non-interest bearing deposit with the Union which can be converted into a fine if the deficit is not 'corrected' within two years.

Commission, aims at trade liberalization and the development and application of common foreign trade rules.

The principal actors involved in EU economic policy making include:

- The European Central Bank;
- The European Commission;
- The European Council;
- The European Parliament;
- The individual MS that participate in the formulation but especially the implementation of the BEPGs;
- The European Economic and Social Committee (EESC);
- The social partners at the European level;
- Non-Governmental Organizations (NGOs).

EU economic policy making is based, on the one hand, on the subsidiarity principle and, on the other, on close coordination of the economic policies of the MS on the basis of generally agreed rules at the European level (the BEPGs and other procedures). The existence and functioning of the SGP limits to a great extent the flexibility of economic policy making of the individual MS.[15]

In addition to the interest rate mechanism, economic policy instruments include single market legislation, competition laws, common foreign trade measures and common rules of compulsory coordination of agricultural production and market organization of individual MS. There are no common policy instruments as regards labour market, wage developments and the regulation of capital and product markets.

In a broad sense, the EU economic policies espouse the neoliberal rationale. Neoliberalism is a contemporary variation of the classical liberalism of the 19th century, promoting the abolition of 'barriers' to the functioning of the free market, such as government intervention, tariffs, social and labour regulation, etc. It advocates the commodification of almost every aspect of socioeconomic life, deregulation and minimization of the economic role of the state, privatization of state-owned enterprises, budgetary discipline and anti-inflationary policies, public expenditure and taxation cuts and it is notoriously hostile to the notion of 'public goods' (Martinez and Garcia, 1997). International organizations, such as the International Monetary Fund (IMF) and the World Bank, are among the major proponents of neoliberal ideas and of the respective economic policies.

As regards the EU monetary and economic policies, Pollack (1999, p.268) stresses that:

...from Rome to Maastricht, the fundamental thrust of the treaties has been neo-liberal, in a sense that each of the Community's constitutive treaties facilitated the creation of a

---

[15] See section *"Procedural integration of EU social and economic policies"*.

unified European market, while setting considerable institutional barriers to the regulation of that same market.

The adoption of the Stability and Growth Pact led to the gradual avoidance of non-monetarist interventionist economic policy actions and instruments and to the reinforcement of neoliberal supply-side economic policies (Bernhagen, 2000). Thus, the 'neoliberal policy consensus' replaced the demand-side, neo-Keynesian macroeconomic policies[16] of the 1960s and 1970s (Bernhagen, 2000, p.6). Major economic processes directly connected with recent changes include accelerated flows of short-term speculative foreign investment, longer-term foreign direct investment, gradual reduction of global trade barriers, increased role of transnational firms in global production and trade, and accelerated elite and labour mobility (Deacon, 2000). According to UNDP (1999), these developments have led to increased social inequality both within and between countries and have intensified socio-spatial exclusion.

Thematic and conceptual integration of EU social and economic policies
Concepts and values which traditionally are associated with a strong and inclusive social state, such as social solidarity and social equality are changing meaning and adapt to the necessities of applying neoliberal economic policies in the European Union (Nieminen, 1995). From the concept of 'equality of outcome', which strongly relates to the universalistic model and to efforts to reduce social inequality through income redistribution policies, emphasis shifts towards concepts of 'equality of condition' and of 'equality of opportunity', which are linked to notions of free market, individual freedom and entrepreneurship (Turner, 1986; Nieminen 1995). Within this conceptualization, social inequality poses no real threat to society as far as there are opportunities of inclusion, mainly in the form of some kind of employment; hence, the overriding emphasis on 'social exclusion' rather than on 'poverty'. "Thus while poverty simply exists, 'exclusion' is created. And if it is created, it can be reversed. While the poor may always be with us, the excluded can be integrated" (Wickham, 2000, p.6). The combination of neoliberal economic policies and selective social policies leads to a form of 'management' of social exclusion and deprivation and not to a comprehensive resolution of problems of poverty and social and income inequality (Cahill, 1994; Streeck, 1999, 2001).

It appears that the principal goals of EU social policies (promotion of social cohesion and inclusion) and of economic policies (creation of a competitive and dynamic, knowledge-based economy) are only rhetorically compatible mainly because of two reasons. The first is the complementary and subordinate position of

---

[16] These are macroeconomic policies supporting the demand for goods and services, as the major driver for economic growth, through extensive state intervention and regulation of the economy. They are compatible with the universalistic-institutional model of social policy.

social policies in relation to economic policies and the second relates to the neoliberal content and character of economic and monetary policy making of the Union. The subordinate position of social policies weakens considerably the simultaneous achievement of social and economic policy objectives. Certain linkages between economic and social policy objectives are discernible only within the 'loose' and much debated context of sustainability. The European Commission (2001a, p.10) acknowledges:

> ...development has an economic, a social and an environmental dimension. Development will only be sustainable if a balance is struck between the different factors that contribute to the overall quality of life.

However, within the dominant neoliberal economic policy paradigm of the EU, sustainability is viewed as a set of necessary, although limited, adjustments allowing the continuation of the current patterns of economic organization and development (Rammel and Bergh, 2003) The adoption of such an approach:

> ...offers an incrementalist agenda that does not challenge any existing entrenched powers and privileges. In this sense, the mantra of sustainable development distracts from the real social and political changes which are required to improve human well-being, especially of the poor, in any significant way (Robinson, 2004, p.376).

Furthermore, the focus on 'efficiency' and the associated market-based incentives may lead to unsustainable long-term socio-economic patterns, and especially to negative distributional effects (Rammel and Bergh, 2003). Thus, the overt congruence of the goals and objectives of EU social and economic policies within the sustainability framework proves to be superficial and rather dubious.

Social policy objectives and targets, set by the Council or the Commission, constitute mainly propositions for national action, not associated with binding EU-level targets, sanctions and related procedures (Ardy and Begg, 2001). Meeting these objectives rests with the national governments, according to the subsidiarity principle, and on 'soft' coordination methods. Although certain quantifiable targets exist (e.g. increasing overall employment by 70 per cent up to 2007 and increasing the female participation rate to 60 per cent), they differ substantially from corresponding economic policy targets or ceilings which are centrally imposed by the European Council and are associated with certain review and sanction procedures in the case of not meeting them. Although the BEPGs constitute a policy coordination framework, based on the principle of subsidiarity, they are strongly influenced by the SGP, which is a relatively rigid and compulsory macroeconomic and budgetary framework.

Within the current economic policy framework and the high priority of economic goals, such as macroeconomic stability and competitiveness, the subordinate role of social policies in the Union becomes gradually evident. The economic policy objectives and the dominance of the neoliberal paradigm, influence the EU and national social policy goals, changing their emphasis from combating social problems and promoting social integration and development to 'managing' social polarization and exclusion (Cahill, 1994; Streeck, 1999). The difference between comprehensively 'managing' and combating social problems draws from the divergent value systems underlying neoliberal and redistributive approaches to socio-economic development respectively.

The alleged political commitment of the Union to social and economic policy integration can be identified in several instances in official documents such as in the conceptualization of social policy 'as a productive factor' and the adoption of strategic economic policy goals at the Lisbon European Summit. Within this framework, and with the adoption of the SPA, the European Commission (2000, p.6) "sets to ensure the positive and dynamic interaction of economic, employment and social policy and to forge the political agreement to mobilize all key actors to work jointly towards the new strategic goal". The 1999 Cologne European Council "consolidated the European employment strategy and created the basis for a Community employment policy which takes account all the economic factors that affect the employment situation".[17] This consolidation resulted in the European Employment Pact (Cologne process). The Pact reemphasizes the importance of price stability and controlled pay increases for employees and, thus, leads to a certain kind of economic, employment and social policy integration within the dominant neoliberal policy priorities and consensus.

The integration between EU economic and social policies is, nevertheless, asymmetric. Economic, or 'market-making',[18] policy has become gradually Europeanized, comprising in many instances strict rulings and sanctions, while the social, or 'market-correcting', policy has to be agreed through intergovernmental processes, on the basis of, in most cases, unanimous consensus (Scharpf, 2002; Mosher and Trubek, 2003). These remarks raise important questions as regards the content and the ideological-conceptual basis of economic and social policy integration in the Union. If the ultimate goal is the minimization of the 'autonomy' of the economic element and the promotion of an all-inclusive and cohesive society, then neoliberalism and the associated highly asymmetrical social and economic policy relations cannot support the achievement of this goal (Mariolis and Stamatis, 1999; Bernhagen, 2000).

Turning to policy actors, the EU economic and social policies appear to have several 'common' actors, such as the individual MS, the Commission, the Council, the social partners, and various NGOs. Nevertheless, in reality these are different agencies which are separately responsible for each policy area.[19] There are no joint formal bodies for promoting pan-European social and economic policy goals in

---

[17] http://europa.eu.int/scadplus/leg/en/cha/c00002.htm (p.3).

[18] A set of economic policies aiming at the formation of an open market economic environment (Scharpf, 2002).

[19] More in-depth research is needed to identify relationships among policy actors and assess whether these suggest some form of integration between social and economic policies.

mutually congruent and integrated ways. The EESC could play a potentially important role in this direction. However, its present role is consultative and its major tasks are to advise the European Parliament, the Council and the Commission, to foster the involvement of civil society in the European integration project, and to strengthen the role of civil society organizations in non-EU countries promoting 'structured dialogue' in common areas of interest (EESC, 2003). Currently, the EESC includes 222 members divided into three groups; the employers group, the employees group and the various interests group. The employers group includes public and private sector representatives from industry, small businesses, chambers of commerce, wholesale and retail trade, banking and insurance, transport and agriculture. The employees group includes mainly national or pan-European trade unions while the various interests group comprises representatives of NGOs, cooperatives, voluntary associations and of the scientific and academic community. The EESC advises the European Union on a broad spectrum of issues, such as agriculture, rural development and the environment, economic and monetary union, social cohesion, employment, social affairs, citizenship, external relations, the single market, transport, energy, infrastructure and the information society (EESC, 2003). A more positive contribution of the EESC to social and economic policy integration presupposes considerable changes in the conceptual preconditions of EU policy making, a reversal of the subordinate character of social in relation to economic policies and a greater degree of Europeanization of EU social policies.

Procedural integration of EU social and economic policies   EU economic and social policy making employs particular policy coordination methods to ensure that MS policies are in agreement among them and with the EU goal of economic and social cohesion. The close coordination method applies to economic policies. "The activities of the member states and the Community shall include...the adoption of an economic policy which is based on the coordination of member states economic policies in order to achieve the Community objective of high level of employment and sustainable non-inflationary growth" (Wickham, 2000, p.16). The OMC applies to social policies. The two policy coordination methods are characterized by a fundamental difference that underlies the divergence between economic and social policies. Economic policies exhibit a better balance between 'Europeanization'[20] and subsidiarity, mainly because several crucial monetary and economic policy decisions are centralized and subject to strict review and sanction procedures.[21] On the contrary, binding, target-meeting procedures or centrally induced legislative change, signifying major moves towards 'Europeanization', are

---

[20] "The emergence and development at the European level of distinct structures of governance, that is, of political, legal, and social institutions associated with political problem solving that formalize interactions among the actors, and of policy networks specializing in the creation of authoritative European rules" (Cowles et al., 2001, p.3).
[21] These policies include the monetary, budgetary, internal market, competition, trade and the agricultural policy.

absent from social policy making. Furthermore, the trend towards subsidiarity and decentralization is reinforced in recent years, as there is an observable departure from the social standard and directive style of social policy provisions towards 'loose' coordination and 'policy learning'[22] procedures (Scharpf, 2002) that have several limitations and drawbacks (Ardy and Begg, 2001).

In contrast to economic policy, there is no clear commitment and a common political vision as regards EU social policy (Wickham, 2000). But

> ...if there is no sanction (or, as it was the case with the EMU convergence criteria, a reward) for failing to adopt suitable measures, let alone meeting targets, the attempt to co-ordinate could prove to be empty" (Ardy and Begg, 2001, p.12).

Additionally, the legitimacy of drawing social policy guidelines by unaccountable officials and national government representatives and the use of the OMC to avoid 'hard law' initiatives in EU social policy have been seriously questioned (Sciarra, 2000; Ardy and Begg, 2001; Mosher and Trubek, 2003). These issues should be taken into account and addressed if the 'policy learning' promoted through the OMC is to produce concrete and binding policy outcomes at the European level.

The divergence in the mode of implementation of EU economic and social policies and the resulting differences in their 'Europeanization' trends, flowing from the principal postulates of neoliberalism and the 'residual' and subordinate character of social policy, does not favour formal interaction among policy actors or the adoption of consistent and compatible procedures and rules of joint and balanced social and economic policy making at the level of the EU. Furthermore, administrative reform in favour of policy integration and the establishment of procedures for implementing policy integration requirements presupposes different economic priorities and goals as well as different conceptual linkages between economic and social policies. It presupposes a view of economic policy goals, such as growth and competitiveness, as means for social cohesion and development and not as ends themselves. Furthermore, it requires the reduction of the autonomy of national social policies through pan-European and market compatibility regulation (Ferrera et al., 2000). The absence of these prerequisites detracts not only from the procedural integration of economic with social policies but also from efforts to move towards the creation of a unified, high standard social policy regime for Europe as a whole (Storey, 2004).

The above-mentioned divergence influences also strongly the use of economic and social policy instruments and their relationships. Economic policy instruments are more 'Europeanized' while the role of individual MS in the development of social policy instruments remains crucial. For example, the main monetary policy instrument is the interest rate which is imposed centrally by the independent ECB.

---

[22] A process of 'learning' from best practices of individual MS in adopting and implementing social policy proposals and guidelines.

The main macroeconomic and public finance policy instruments are commonly agreed rules derived from the SGP, the Excessive Deficit Procedure and the BEPGs which, although they conform to the subsidiarity and policy coordination principles, embody certain binding limits and ceilings associated with sanctions and fines. Regulatory and legislative instruments are used by the internal market, competition and external trade policies. Only, the labour market, capital and product market policy areas are less Europeanized and are partly influenced by the requirements of the BEPGs.

On the other hand, the main instruments of EU social policies, financial incentives and programs funded through the SFs, are less 'Europeanized'. Regulatory and legislative instruments are used at the MS level. Currently, the political and ideological preconditions for the weakening of the subsidiarity principle and the gradual 'Europeanization' of social policy together with considerable changes in economic policy directions and goals are missing. Thus, the development of integrative, common legal and institutional instruments for the EU social and economic policy areas cannot be fully achieved at present. For the same reasons, particular financial mechanisms, fiscal incentives, planning or management instruments or specialized tools promoting comprehensive economic and social policy integration at a pan-European level are absent.

A potentially important PI instrument could be a unified, integrated impact assessment, covering all economic, social and environmental consequences of EU policies. Currently, sector-specific impact assessments are in use, the most important of them relating to business, trade, the environment, health, gender mainstreaming and employment. To overcome this fragmentation, the Commission has proposed the adoption of Sustainability Impact Assessment (SIA) (DEFRA-IEEP, 2002). The SIA is characterized by comprehensive coverage of the economic, social and environmental impacts of EU policies, the consideration of short and long term, as well as global and local impacts, and the definition of the 'problem' or situation under assessment in a way which helps to understand its underlying causes and its economic, social and environmental dimensions (DEFRA-IEEP, 2002). The adoption of SIA would definitely be a progress in itself, but the degree of its contribution to the promotion of PI towards stronger versions of sustainability, depends on broader conceptual and socio-political preconditions and developments as discussed before.

Integration of EU Social and Environmental Policies

The environmental policies of the European Union The environmental policy of the European Union includes, on the one hand, horizontal provisions of a general, procedural character (horizontal environmental policy) and, on the other, policies related to specific environmental media, sectors, or types of pollutants. The most important, from the point of view of desertification, are the water resources and biodiversity policies, the former being among the first sectors covered by the EU environmental policy. Currently, there is no common soil protection policy, which is an important consideration for combating desertification. Several soil protection-related issues are variously covered in other environmental and non-environmental policies.

Although the first environmental directives were adopted in 1975, specific references to the environment did not appear in the EU Treaties until 1987. Until then, action could be taken in the context of the more general powers of Articles 100 and 235. The Single European Act of 1987 introduced Title VII on 'Environment' confirming that environmental and research policies were formally matters of Community competence and obligation. Articles 130r, 130s and 130t explicitly enabled the introduction of environmental legislation that required unanimity of the Council of Ministers. The Maastricht Treaty of 1992 removed the right to veto, so that now qualified majority voting applies to most environmental legislation, except in some reserved cases.[23]

The environmental policy of the EU was formulated in the framework of the successive Environmental Action Programs (EAP). The 4th, but especially the 5th EAP, introduced explicitly the goal of sustainable development. Since the beginning of 2000, and especially in the context of the 6th EAP, the Union promotes a more strategic approach to environmental policy making, calling for the active involvement and accountability of all sections of society in the search for innovative, workable and sustainable solutions to environmental problems.

The overarching goal of the EU environmental policy is the achievement of sustainable development (Article 2 of the 1999 Amsterdam Treaty). Aiming to ensure optimum environmental improvement in different regions of the Union, its main goals include (Krämer, 1992; Johnson and Corcelle, 1997):

a) The preservation, protection and improvement of the quality of the environment;
b) The protection of human health;
c) The prevention and reduction of pollution;
d) The conservation of natural habitats and of wild fauna and flora of EU importance;
e) The prudent and rational utilization and sustainable management of natural resources;
f) The de-coupling of economic growth and environmental degradation.

More specific objectives are (Krämer, 1992; Johnson and Corcelle, 1997):

a) The application of integrated resources management;
b) The integration of environmental concerns into sectoral policies;
c) Ensuring the implementation of existing environmental legislation;
d) Formation of a network of protected areas (NATURA 2000);

---

[23] Such as legislation of a primarily fiscal nature or affecting choice between energy sources and legislation relating to land use planning or water resources management.

e)   Public participation, consultation and information.

EU environmental policy making is guided, among others, by the precautionary, the preventive action, the rectification of damage at source, the 'polluter pays' and the 'consumer pays' principles. Article 6 of the Amsterdam Treaty introduced the environmental integration principle requiring that environmental protection requirements be integrated into the definition and implementation of Community policies and activities, with a view to promoting sustainable development.

Formal EU environmental policy making involves the 'institutional triangle' of European Commission (DG-Environment, mainly), Council and Parliament as well as the European Court of Justice and the Court of Auditors. Other important formal actors are the Committee of the Regions,[24] the European Ombudsman,[25] the European Investment Bank[26] and the European Environmental Agency.[27] To facilitate environmental policy integration (EPI), each Directorate-General has nominated an 'integration correspondent'. National formal actors, mostly policy implementers, are the Management Authorities (e.g. for protected areas) and River Basin Authorities (RBA) (responsible for water resources policy implementation). Environmental NGOs are important semi-formal actors, among several other European and national/regional NGOs, participating both in policy formulation and implementation.[28] Finally, various pressure groups, such as Industry, Farmers, Hunters, Forest Owners and other Associations, seek to influence environmental decision making.[29]

Although regulation remains the principal style of Community environmental policy, other approaches are increasingly being introduced and used in conjunction with regulation. Direct involvement of the Union in environmental protection in MS takes the form of providing financial support for environmental protection programmes and projects. Decentralized, participatory decision making approaches are promoted, drawing on the subsidiarity principle, as it is the case with the Water Framework Directive (WFD) where the ultimate responsibility of securing appropriate water resources management rests with the MS (RBAs) and the local interests involved. Market approaches to pollution control are also proposed in

---

[24] It expresses the opinions of regional and local authorities on regional policy, environment and education.

[25] It deals with complaints from citizens concerning maladministration of an EU institution or body.

[26] It finances public and private long-term investments.

[27] It provides environmental information and data.

[28] Important among them are the EEB (European Environment Bureau), the WWF (World Wildlife Fund), the RSPB (Royal Society for the Protection of Birds), the Birdlife International and Water Pact, and the Greenpeace.

[29] It is common for such pressure groups to establish offices in Brussels to facilitate lobbying.

several policy areas. Voluntary approaches, targeting the third sector mainly, are gaining ground while persuasion, through the use of Commission Communications, Green or White papers on particular subjects, environmental education, and the like, herald a gradual shift in EU environmental policy making towards less 'command-and-control' approaches.

EU environmental policy is implemented through institutional, legislative, financial, economic and communication instruments, provision of infrastructure, and funding of research programmes. The EAPs constitute the broad institutional frameworks setting the general environmental policy goals and objectives and specifying various means for their achievement. The legislative instruments include (a) *regulations* that are directly applicable and binding on all MS, without the need for national implementing legislation, (b) *directives* that are transposed to the national legislative order, and are binding on MS as to the objectives to be achieved within a certain time frame while leaving the choice of form and means to be used to the national authorities, (c) *decisions* that are binding in all their aspects for the entities to which they are addressed and (d) *recommendations* and *opinions* that are not binding. Directives are the preferred legal policy instruments of the EU, the recent trend being towards framework directives that delegate decision-making power and accountability to national and sub-national levels. Important among them are the Environmental Impact Assessment (EIA) and the Strategic Environmental Assessment (SEA) directives as well as the WFD.

The economic instruments include various types of environmental levies and resource pricing following the principle of full cost recovery of resource use. The financial instruments usually cover the costs of constructing environment-related facilities (e.g. waste treatment), research facilities, and the implementation of directives (such as the Habitats and Birds directives). The Structural Funds, the Cohesion Fund and the Community Initiative LEADER may also finance nature protection activities. Finally, communication instruments, such as the voluntary schemes Eco-label and the EMAS[30] and environmental education programmes are used to facilitate the transition to more sustainable production and consumption modes.

The 'sustainability' and 'sustainable development' discourse frames the contemporary environmental policy of the European Union. The 5th EAP (1993-2000) marks the shift from a uni-dimensional emphasis on the environmental dimensions of development towards holistic, multidimensional approaches that emphasize the critical role of human interventions and the importance of the socio-economic and institutional dimensions of environmental change. It accepts that unsustainable patterns of human consumption and behaviour are the driving forces of environmental problems. These have to change in the long term through, among others, the integration of environmental concerns in decision making. It contains a clear commitment to a policy towards sustainable development, linking environmental protection to socio-economic well-being and growth for the present

---

[30] Environmental Management and Audit System.

and the future generations, the maintenance of the overall quality of life and the continuing access to natural resources and the avoidance of lasting environmental damage (Baker et al., 1997). As a result, more integrated approaches towards environmental problem solving are advanced. Several horizontal environmental policy provisions reflect this new comprehensive theoretical framework such as the SEA and the Integrated Pollution Prevention Control (IPPC) directives. Similarly, the WFD adopts a holistic approach to water resources problems considering both the biophysical and the socio-economic factors (water use and consumption patterns) affecting water quality and quantity. This is reflected in the ecosystem approach and the integrated river basin management it espouses.

Thematic and conceptual integration of EU social and environmental policies The EU Sustainable Development Strategy (EUSDS) provides the broad framework for examining the linkages between social and environmental policies as "...in the long term, economic growth, social cohesion and environmental protection must go hand in hand" (European Commission, 2001a, p.2). The Commission identifies poverty, social exclusion, population ageing and mobility, among the six major priority topics of the EUSDS[31] (European Commission, 2001b). Furthermore, it proposes specific actions to improve policy coherence towards sustainable development, which mainly concern the Common Agricultural, Fisheries and Transport policies and the Cohesion policy.

Sustainable development is a highly ambiguous and contested concept, taking different meanings in different socio-political contexts. Although the theoretical demand for the simultaneous satisfaction of economic, social and environmental goals may be construed as implying their equivalence, in practice, goal trade-offs are inevitable in order to come up with viable, acceptable and implementable solutions. In particular, "the social dimension has been commonly recognized as the weakest 'pillar' of sustainable development, notably when it comes to its analytical and theoretical underpinnings" (Lehtonen, 2004, p.199).

Societal value systems influence considerably which goals are prioritized and societal preferences determine their ultimate ranking. This is why the literature, informed by the economic analysis of environmental problems mainly, suggests four versions of sustainable development and sustainability: very weak, weak, strong and very strong sustainability (Pearce et al., 1994; Baker et al., 1997; O'Riordan and Voisey, 1998). Each version makes different assumptions about the potential (and desirability) for substitution among goals and their relative weighting. Very weak and very strong sustainability represent two extremes; the former gives low priority to environmental over economic and social goals, paying lip service to policy integration while the latter assigns supremacy to the achievement of environmental goals at any (economic or social) cost, backed by strong international and national environmental institutions and a formal shift to

---

[31] The other topics being climate change and clean energy, public health, management of natural resources, land use and territorial development.

sustainability accounting at all levels (O'Riordan and Voisey, 1998). Strong sustainability implies a situation where there are no substitution possibilities of natural for human capital. Consequently, it prioritizes the stability or even the augmentation of natural capital, aided by binding policy integration, international agreements, green accounting and creation of 'civic' income for social use (O' Riordan and Voisey, 1998). On the contrary, weak sustainability, although giving some weight to environmental goals and supporting formal policy integration, considers strongly the costs of attaining them. Under a weak sustainable development regime, expansive economic growth is considered possible, aided by substantial restructuring of economic incentives, while achieving a minimum of environmental protection (Redclift, 1993). The environment is valued in monetary terms primarily, with the use of cost-benefit and accounting techniques, ignoring to a great extent its non-pecuniary cultural, social or spiritual dimensions.

The approach of the Union to sustainable development comes closer to the weak sustainability variant as revealed by, e.g., the promotion of 'sustainable growth',[32] the gradual adoption of market-based policy instruments, and the low degree of policy integration pursued, except for certain procedural provisions (e.g. Article 6 of the Amsterdam Treaty). Furthermore, this approach does not take into account the unequal distribution of environmental costs and benefits between different social groups and regions at global and European levels. This raises the question of 'environmental justice', which is closely related to social justice and to the rationale of social policy making, and concerns mainly the allocation of costs and benefits of environmental degradation and protection and the distribution of economic, social and political power as regards access and utilization of natural resources (Siwakoti, 2004). Although essentially not separate from social justice, being rather one of its components, this indivisibility is not reflected in the conceptual framing and practice of social and environmental policy making. Beyond formal rhetoric, policy practice shows an inclination towards improving the 'management' of social and environmental problems separately within a model of 'weak' sustainability instead of promoting their comprehensive and joint resolution through the promotion of policy integration and 'stronger' versions of sustainable development (Robinson, 2004).

If the aim of the integration of social and environmental policies is not simply the 'management' of problems, through adjustments and incremental measures, but their definitive mitigation, then the combination of selective and 'residual' social policy making with environmental policy making of the 'weak' sustainability variant appears inadequate. This is reinforced by declarations in favour of social cohesion and sustainable development without compromising the dominant socio-economic ideology and policy making model. Thus, the European Commission "…welcomes recent tax cuts in member states and calls for a further reduction in taxation by 2005 to 1%" (ETUC et al., 2002, p.2). However with reduced taxation and, consequently, public and social spending, the need to allocate sufficient

---

[32] A contradictory term according to several authors.

financial resources to meet social and environmental goals comprehensively cannot be satisfactorily addressed.

The EUSDS includes social issues, such as poverty and social deprivation, together with economic and environmental considerations and goals. However, their linkages are very weak, especially those between social and environmental goals, reflecting the continuation of treating the respective policy areas in relative isolation (ETUC et al., 2002). Breaking this isolation requires that the environmental and the social policy areas adopt a value system supporting 'strong' versions of sustainability whose absence accounts for the loose linkages between social and environmental issues and is reflected in the political commitment of Article 6 of the Amsterdam Treaty and the associated Cardiff Integration Process (see Chapter 1, this volume). Thus, integration aims at incorporating environmental concerns in EU policies in the areas of agriculture, regional development, energy, enterprise, fisheries, internal market, research, trade and external relations, transport, and economic and financial affairs. Social policy areas are not included in this framework yet and major issues such as income inequality and distribution, employment, social exclusion, poverty and migration are not linked to environmental processes and concerns. Thus, the environmental policy integration project is disconnected from the social processes and structures within which it is naturally embedded and its social implications are deceptively considered external to the economy-environment connection; the residual and subordinate treatment of social issues is manifested again in this case.

The weak thematic and conceptual integration of the EU social and environmental policy areas carries over to the absence of common actors or dedicated joint administrative bodies vested with powers to handle the complex relations between social and environmental processes and elaborate integrated socio-environmental policies.

Procedural integration of EU social and environmental policies   EU environmental policy making is largely regulatory. Nevertheless, persistent pressures to find compromises based on the lowest common denominator in most cases leads to the formulation of 'minimal' regulations and directives (Johnson and Corcelle, 1997). These encourage a kind of environmental management congruent with the current socio-economic and political philosophy, its central feature being marginal or incremental adjustment requirements (Baker et al., 1997; Krämer, 1992). Despite the regulatory character of EU environmental policy, certain moves towards new models of environmental governance favour 'policy-making without legislating' (Héritier, 2002). These include definitions of targets, exchange of information on best practices, monitoring, publications on performance and voluntary accords. They are characterized by non-binding targets, the use of 'soft law', subsidiarity and participation of diverse social actors (Héritier, 2002). Furthermore, the growing tendency towards the use of environmental 'framework directives' increases the implementation powers and flexibility of individual MS, compromising the definition of specific targets and of associated explicit measures or actions for meeting them.

These new forms of environmental governance resemble the OMC used in EU social policy making. Indeed the vast majority of policy actions associated with these forms of governance can be found in the social and environmental policy areas (Héritier, 2002). Despite this resemblance, however, EU environmental policy making differs substantially from social policy making, where the subsidiarity principle is still dominant and the resistance to its Europeanization greater. Environmental policies are more 'Europeanized' (Cowles et al., 2001) compared to social policies where the OMC keeps them highly nationalized.

The continuation of the 'softening' tendency of EU environmental policies might further obstruct efforts to integrate them with social policies, among others. Especially in the case of environmental policies concerning phenomena with important social implications, such as land degradation and desertification,

> ...there is an inherent conflict between targets that can easily be agreed upon, and carry few decision-making costs, on the one hand, and the effectiveness of those targets, on the other. If targets are formulated in a general way, which make them politically easy to achieve, then they often have 'no bite' when implemented, and they remain without much effect. Against this background it has also been pointed out that these soft instruments do not purvey legal certainty and that frequently, there is a lack of accountability for the impact of applying these instruments (Héritier, 2002, p.19).

As a consequence of the above, no evidence exists currently for the development of consistent, compatible or coordinated decision making procedures in common administrative bodies dealing with problems with strongly interlinked social and environmental dimensions. Furthermore, there is no evidence of formal and institutionalized interaction and joint decision making between, for example, the European Commission DGs on social affairs and on the environment. Administrative restructuring favouring the integration of EU social and environmental policies does not seem to be contemplated either. The official documents of the Union acknowledge rather clearly the limited progress towards the integration of social with environmental concerns:

> Sustainable development implies a society-wide approach to policy design. Sustainability must be placed at the core of the mandate of all policy makers. *This means more than tagging on environmental and social objectives to existing policies.* Achieving these objectives should be as relevant to judging the success or failure of a policy as its sectoral targets. Otherwise integration and sustainability risk becoming buzzwords to which policy-makers pay lip service only. *Integration must mean something more than minor adjustments to 'business as usual' if sustainable development is to move from rhetoric to reality.* This needs political commitment and leadership.[33] (European Commission, 2001b, p.49).

---

[33] Emphasis added.

This lack of real commitment derives mainly from the prevailing adherence to the 'weak sustainability' model within which social and environmental concerns, such as social and environmental justice, are loosely linked, if at all. This conceptual framework accords with the overall neoliberal economic consensus of the Union that favours a 'management' approach towards environmental and social problems deriving from the constant pursuit of the 'ultimate goal' of greater competitiveness.

The lack of integration between EU environmental and social policies is reflected in the relationships among their instruments too. The EU environmental policy still uses various kinds of 'command and control' instruments mostly while social policies delegate to a great extent the regulatory activity to individual MS. Furthermore, and for the time being, there are no moves towards the creation and use of financial incentives or planning and management instruments for environmental and social policy integration. Persuasive or communicative instruments such as information exchange, public awareness-raising, and education and training programs could provide a basis for better understanding the connections between social and environmental issues (Gysen et al., 2002). Currently, the European Union funds environmental education and training projects through various initiatives and programs, such as Leonardo Da Vinci, but the main responsibility for the development of such schemes lies with the individual MS (European Commission, 2001a). The development of holistic and integrative instruments, such as the proposed Sustainability Impact Assessment,[34] could enhance PI. Nevertheless, SIA's effectiveness depends on the coherence of its theoretical and conceptual basis and on the degree of thematic and procedural penetration of social policy concerns in European policy making.

The inadequate integration of EU social and environmental policies implies that socio-environmental problems continue to be tackled mainly as 'technical' challenges and their management remains compartmentalized, treating their social and environmental dimensions separately.

Social Policies: The Conundrum of Policy Integration

Although selective, general, and based on limited evidence, the present analysis uncovered the conundrum of the integration of the EU social with the economic and environmental policy areas. The sustainability rhetoric, as formally expressed in the EUSDS (EC, 2001a) is not sufficiently capitalized yet in integrating the three policy areas. The current combination of neoliberal EU economic policies, selective social policies and environmental policies of the 'weak' sustainability variant does not portray an integrated view of reality, creating instead a context of policy 'disintegration', oriented towards 'crisis management' rather than towards the essential resolution of contemporary complex socio-environmental problems

---

[34] See section "*Procedural integration of EU social and economic policies*".

such as desertification. This section summarizes the main points of the analysis, discusses the implications of the lack of PI for combating desertification in the LFAs of Mediterranean Europe, suggests actions to come to terms with the policy integration conundrum, and offers directions for future research.

Currently, the integration between the EU social and economic policy areas is relatively weak and its future prospects do not appear bright mainly because of the 'subordination' of social goals to the dominant economic goals and the associated pressures towards a selective and residual social policy regime,[35] and because of differences in the degree of 'Europeanization' between the two policy areas. The lack of thematic and conceptual integration affects their procedural integration as well. Economic policies are more binding and closely coordinated while social policies are mostly non-binding and loosely coordinated.

The current integration between the EU social and environmental policy areas appears even weaker and their future prospects less bright than it is the case with economic policies. The social dimensions of environmental problems are loosely added on environmental policies within the 'weak' sustainability paradigm where a 'management' approach to social and environmental problems prevails. Characteristically, the Cardiff integration process leaves out social policies from the list of sectoral policies considered.[36] In terms of 'Europeanization' and centralization of policy making, the EU social and environmental policies differ considerably with the latter being more 'Europeanized' and regulatory, although a 'softening' of EU environmental governance is on the rise. The weak substantive linkages and the procedural differences of the two policy areas do not favour their effective operational integration.

The implications of the lack of adequate integration among all three EU policy areas for combating desertification in the sensitive LFAs of Mediterranean Europe are rather obvious. Land degradation and desertification are not only complex problems characterized by strongly interlinked environmental, social and economic dimensions over and across space but their resolution requires local and regional interventions coordinated from above and from below. Integration of the EU policies is crucial because these set the overall framework for policy action at lower levels (Lehtonen, 2004). In addition, the survival and prosperity of

---

[35] As Lucas (2001, p.2) masterfully observes "...the EU's over-riding priority of ever increasing international trade and competitiveness is seriously undermining its aspirations to achieve greater sustainability. Its unwavering support for economic globalization means that it is unable to address the problems that inevitably follow in its wake ... (if) the EU continues to put its corporate-led, deregulated, neo-liberal agenda above social justice and sustainable development, the result will be further marginalization and exclusion of growing number of its citizens".

[36] This might change in the future, however, as "In October 2002, the Environment Council called upon the European Council to invite the Council formations responsible for education, health, consumer affairs, tourism, research, employment and social policies to develop strategies to promote sustainable development by integrating environmental considerations into their existing policies and action" (European Commission, 2004, p.6).

environmentally disadvantaged, desertification-sensitive LFAs depends critically on their socio-economic vitality because their residents will eventually carry out those actions needed to stop or reverse land degradation while striving for acceptable livelihoods. Social policies are, therefore, particularly important in the context of supporting the sustainable development of these LFAs that have a limited repertoire of development alternatives.

The current inadequate integration of EU social with economic and environmental policies does not guarantee either the short or longer term effectiveness of whatever EU social policies are implemented in these areas or their essential contribution to combating desertification. The poor substantive and procedural linkages among all three policy areas rather exacerbate social problems, such as unemployment, poverty, and population ageing, and weaken social cohesion. Given the bio-climatic and socio-economic constraints facing the desertification-sensitive LFAs, the current policy mix favours mostly developed areas with greater competitive advantages, thus, intensifying spatial competition, encouraging outmigration and land abandonment, and discouraging the development and application of sustainable alternative livelihood strategies and options. Because the welfare problems of these LFAs are simply 'managed' rather than resolved, they remain 'locked' in low development states, deprived of the possibility to break the vicious circle of underdevelopment and exploit the available economic, social and environmental policy measures and initiatives, combining them effectively towards the protection of their sensitive environmental resources and the satisfaction of their development needs. Higher level lack of PI constrains effective lower level planning action.

To improve the current situation, efforts should concentrate on both the substantive and the procedural aspects of the integration of all three EU policy areas and, more importantly, on the consistency between the two. The substantive, and more fundamental, change required is the gradual abandonment of the neoliberal consensus and the adoption of a common theoretical outlook. Societal and policy learning processes, using available communication and education instruments, should be initiated to facilitate the fusion and transformation of belief systems across all policy sectors. Shared values and conceptions of socioeconomic welfare will most probably lead to congruent and compatible goals and objectives. The critical prerequisite for the materialization of the above is political commitment and leadership for policy integration that is desperately missing at present.[37]

The procedural arrangements needed to move on the PI project should be based ideally on, and should be facilitated greatly by, the fulfilment of the substantive changes proposed. However, because substantive changes have a long gestation period, procedural changes can be introduced independently, wherever possible and feasible, to provide the necessary policy integration 'hardware' (Jordan, 2002).

---

[37] As bridging official rhetoric and policy practice requires a thorough reorientation of EU's priorities (Deacon, 2000).

Administrative restructuring in favour of PI on and across spatial/organizational levels is necessary to provide the requisite capacity to deal with cross-cutting policy issues and engage in *joint* problem-solving. The policy actors associated with the three policy areas should be linked through formal communication, cooperation and coordination procedures. Influential, informal policy actors should be encouraged to participate and communicate more openly with formal actors.

The establishment of an official, comprehensive PI framework is necessary to provide the enabling context for the required actions. The legal, institutional, economic, financial and other instruments used in the different policy areas should be harmonized and streamlined, guided by the common theoretical framework and outlook referred to before. Integrative instruments, such as EIA, SEA, SIA, taxation and budgeting, should be improved and promoted. The use of communication, education and training instruments is particularly important for effecting both short-term and long-term changes in attitudes and beliefs that are the *sine qua non* of the success of any PI endeavor.

This chapter offered a very general analysis of selected aspects of the integration of EU social with economic and environmental policies in the perspective of addressing complex socio-environmental problems. The future research agenda is long and demanding given the number and variety of the policies that have to be thoroughly analyzed to provide practical advice and policy support. Multiple quantitative and qualitative methodologies should be employed to capture as many aspects of the complex socio-environmental reality and the corresponding policy domains.

One research stream concerns the deeper analysis of the theoretical and conceptual differences and commonalities among the objects of social, economic and environmental policies and their links to their goals and objectives, actors, and instruments. The study of the associated actors and actor networks at the EU level as well as their relationships to their national and subnational counterparts is necessary to provide essential knowledge needed to craft meaningful and effective linkages among policies. Research on the relationships among the procedural aspects of different policies represents another vast research domain given the multiplicity and variety of existing and potential procedures and instruments. Finally, the role of non-policy factors on the current state of PI should be explored to identify meaningful ways to manipulate them to reinforce future PI efforts.

Combating desertification is an ideal test bed for the study of the implications of the current state of PI among EU social with economic and environmental policies on the management of complex policy problems at the local and the regional level and the contribution of better integration to alleviating this problem. Without such focused and case study-based research it will not be possible to provide meaningful practical policy support.

References

Ardy, B. and Begg, I. (2001), *The European Employment Strategy: Policy Integration by the Back Door*, Paper prepared for the ECSA 7th Biennial Conference, Madison, Wisconsin.

Baker, S., Kousis, M., Richardson, D. and Young, S. (eds) (1997), *The Politics of Sustainable Development. Theory, Policy and Practice within the European Union*, Routledge, London.

Becker, H.A. and Vanclay, F. (eds) (2003), *The International Handbook of Social Impact Assessment*, Edward Elgar, Cheltenham.

Bernhagen, P. (2000), *Convergence, Stability and the Perpetuation of the Neoliberal policy consensus: Using EMU as External Discipline*, Department of Political Science, Trinity College, Dublin.

Blaikie, P. and Brookfield, H. (1987), *Land Degradation and Society*, Routledge, London.

Blakemore, K. (1998), *Social Policy, An Introduction*, Open University Press, Buckingham.

Byrne, D. (2001), 'Complexity Science and Transformations in Social Policy', *Journal of Social Issues*, 1(2), Electronic Issue.

Cahill, M. (1994), *The New Social Policy*, Blackwell, London.

Cowles, M., Caporaso, J. and Risse, T. (eds) (2001), *Transforming Europe: Europeanization and Domestic Change*, Cornell University Press, Cornell University Press, Ithaca and London.

Deacon, B. (2000), *Globalization and Social Policy: The Threat to Equitable Welfare*, United Nations Institute for Social Development (UNRISD), Geneva.

DEFRA-IEEP (Department for Environment, Food and Rural Affairs-Institute for European Environmental Policy) (2002), *Sustainability Impact Assessment*, Seminar Proceedings, British Embassy, Brussels.

EESC (European Economic and Social Committee) (2003), *The EESC: A Bridge Between Europe and Organised Civil Society*, Office for Official Publications of the European Communities, Luxembourg.

EMI (European Monetary Institute) (1997), *The Single Monetary Policy in Stage Three. Specification of the Operational Framework*, Frankfurt.

ETUC (European Trade Union Confederation), EEB (European Environmental Bureau), Social Platform (2002), *Making the Economy Work for Sustainable Development*, Brussels.

European Commission (2000), *Social Policy Agenda*, Communication from the Commission, COM (2000)379 final, Brussels.

European Commission (2001a), *A sustainable Europe for a better World: A European Union Strategy for Sustainable Development*, Communication from the Commission, COM(2001)264 final, Brussels.

European Commission (2001b), *Consultation Paper for the Preparation of a European Union Strategy for Sustainable Development*, SEC (2001) 517.

European Commission (2001c), *Communication from the Commission to the Council on an Open Method of Coordination for the Community Immigration Policy*, COM (2001) 387 final, Brussels.

European Commission (2001d), Joint Report on Social Inclusion, Brussels, (http://europa.eu.int/comm/employment_social/soc-prot/soc-inc/join_rep_en.htm).

European Commission (2002), *Social Protection in Europe 2001*, DG General for Employment and Social Affairs, Unit EMPL/E.2.

European Commission (2004), *Integrating environmental considerations into other policy areas – a stocktaking of the Cardiff process*, Commission Working Document, COM(2004)394 final, Brussels.

Ferrera, M., Hemerlijck, A. and Rhodes, M. (2000), *The Future of Social Europe: Recasting Work and Welfare in the New Economy*, Report for the Portuguese Presidency of the European Union, (http://www.iue.it/SPS/People/Faculty/CurrentProfessors/PDFFiles/RhodesPDFfiles/report.pdf).

Gold, M. (ed.) (1993), *The Social Dimension. Employment Policy in the European Community*, Macmillan, London.

Gysen, J., Bachus, K. and Bruyninckx, H. (2002), *Evaluating the Effectiveness of Environmental Policy: An Analysis of Conceptual and Methodological Issues*, Paper presented at the European Evaluation Society Seville Conference, Spain, Seville.

Héritier, A. (2002), *New Models of Governance in Europe: Policy-Making without Legislating?*, Max Planck Project Group, Common Goods: Law, Politics and Economics, Bonn.

Holling, C.S. (2001), 'Understanding the complexity of economic, ecological and social systems', *Ecosystems*, 4, pp. 390-405.

Huby, M. (1998), *Social Policy and the Environment*, Open University Press, Buckingham.

Johnson, S.P. and Corcelle, G. (1997), *The Environmental Policy of the European Communities*, Kluwer Law International, London.

Jordan, A. (2002), 'Efficient hardware and light green software: Environmental policy integration in the UK', in A. Lenschow (ed.), *Environmental Policy Integration: Greening sectoral Policies in Europe*, Earthscan, London, pp. 35-56.

Krämer, L. (1992), *Focus on European Environmental Law*, Sweet and Maxwell, London.

Lehtonen, M. (2004), 'The environmental – social interface of sustainable development: capabilities, social capital, institutions', *Ecological Economics*, 49, pp. 199-214.

Lenschow, A. (ed.) (2002), *Environmental Policy Integration, Greening Sectoral Policies in Europe*, Earthscan, London.

Lucas, C. (2001), *The Future of the European Union – Radical Reform or Business as Usual?*, The Greens, European Free Alliance, London.

Mariolis, T. and Stamatis, G. (1999), *EMU and Neoliberal Policy*, Ellinika Grammata, Athens (in Greek).

Martinez, E. and Garcia, A. (1997), *What is Neoliberalism? A Brief Definition for Activists*, National Network for Immigrant and Refugee Rights, (http://www.corpwatch.org/issues/PID.jsp?articleid=376).

Mosher, J.S. and Trubek, D.M. (2003), 'Alternative approaches to governance in the EU: EU social policy and the European employment strategy', *Journal of Common Market Studies*, 41, 1, pp. 63-88.

Nieminen, A. (1995), 'EU Social Policy: from Equality of Outcome to Equality of Opportunity', Paper prepared for the COST A7 workshop "Welfare Systems and European Integration", Tampere, Finland, August 24-27, (http://www.helsinki.fi/~arniemin/eu_sospo.htm).

OECD (1996), *Globalization: What challenges and what opportunities for government?* Organization for Economic Cooperation and Development, Paris.

O'Connor, J.S. (2003), 'Welfare state development in the context of European integration and economic convergence: situating Ireland within the European Union context', *Policy and Politics*, 31, 3, pp. 387-404.

O'Riordan, T.R. and Voisey, H. (eds) (1998), *The Transition to Sustainability*, Earthscan, London.

Papadopoulos, A. and Liarikos, K. (2003), 'Towards what kind of rural development in less favoured areas in Greece', in Proceedings of the Conference: *LFAs and Development Strategies*, Department of Environmental Studies and Department of Geography, University of The Aegean, Mytilini, Greece (in Greek).

Pearce, D., Atkinson, G.D. and Dubourg, W.R. (1994), 'The economics of sustainable development', *Annual Review of Energy Environment*, 19, pp. 457-74.

Perez-Trejo, F. (1994), *Desertification and Land Degradation in the European Mediterranean*, European Commission, Directorate-General XII Science, Research and Development, Luxembourg.

Persson, . (2002), *Environmental Policy Integration, An Introduction*, Stockholm Environment Institute, Stockholm.

Petmetzidou-Tsoulouvi, M. (1992), *Social Inequalities and Social Policy*, Exantas, Athens (in Greek).

Pinch, S. (1997), *Worlds of Welfare. Understanding the Changing Geographies of Social Welfare Provision*, Routledge, London.

Pollack, M. (1999), 'A Blairite treaty: neoliberalism and regulated capitalism in the treaty of Amsterdam", in K.H. Neunreither, and A. Wiener (eds), *European Integration after Amsterdam: Institutional Dynamics and Prospects for Democracy*, Oxford University Press, Oxford.

Rammel, C. and Bergh van den, J. (2003), 'Evolutionary policies for sustainable development: adaptive flexibility and risk minimizing', *Ecological Economics*, 47, pp. 121-33.

Redclift, M. (1993), 'Sustainable development: needs, values, rights', *Environmental Values*, 2, 1, pp. 3-20.

Reynolds, J.F. and Stafford-Smith, M. (2002), *Global Desertification: Do Humans Cause Deserts?*, Dahlem University Press, Berlin.

Robinson, J. (2004), 'Squaring the circle? Some thoughts on the idea of sustainable development', *Ecological Economics*, 48, pp. 369-84.

Sakellaropoulos, T. (2001), *Transnational Social Policies in the Era of Globalization*, Kritiki, Athens (in Greek).

Scharpf, F.W. (2002), 'The European social model: coping with the challenges of diversity', *Journal of Common Market Studies*, 40, 4, pp. 645-70.

Sciarra, S. (2000), 'Integration through co-ordination: the employment title in the Amsterdam treaty', *Columbia Journal of European Law*, 6, 2, pp. 209-29.

Siwakoti, G.K. (2004), *Eco-Justice: Defending the Fourth Generation's Right*, INHURED International, Nepal, (http://www.inhured.org/ecojustice.htm).

Spilanis, I., Kizos, A. and Iosifides, T. (2003), 'Economic, social and environmental dimensions of development in less favoured areas', in Proceedings of the Conference: *LFAs and Development Strategies*, Department of Environmental Studies and Department of Geography, University of The Aegean, Mytilini, Greece (in Greek).

Storey, A. (2004), *The European Project: Dismantling Social Democracy, Globalizing Neoliberalism*, Paper presented at the Conference 'Is Ireland a Democracy?', Sociology Department, National University of Ireland, Maynooth.

Streeck, W. (1999), *Competitive Solidarity: Rethinking the "European Social Model"*, Max-Planck Institute for the Study of Societies, Cologne.

Thomas, D.S.G. (1997), 'Science and the desertification debate', *Journal of Arid Environments,* 377, pp. 599-608.

Turner, B.S. (1986), *Equality*, Ellis Harwood and Tavistock Publications, Chichester.

UNCCD (United Nations Convention to Combat Desertification) (1994), http://www.unccd.int/convention/text/convention.php#begin.

UNDP (United Nations Development Program) (1999), *Human Development Report 1997*, United Nations, New York.

UNESC (United Nations Economic and Social Council) (1995), *Poverty Eradication and Sustainable Development. Commission on Sustainable Development*, Third Session, 11-28 April, 1995.

UNEP (United Nations Environment Program) (1992), *World Atlas of Desertification,* Edward Arnold, London.

UNRISD (United Nations Research Institute for Social Development) (1994), 'Environmental Degradation and Social Integration', Briefing Paper No 3, *World Summit for Social Development*, November, 1994.

Vikstrøm, J. (1996), *The Future of the Welfare State-Some Socio-Ethical Considerations,* The Finnish Institute in London, 29 November, 1996. (http://www.evl.fi/arkkipiispa/welfare.htm).

Wickham, J. (2000), *The End of the European Social Model: Before it Began?*, EurUnion, Dublin, (http://www.ictu.ie/html/publications/ictu/Essay2.pdf).

Chapter 6

Integration of EU Water and Development Policies: A Plausible Expectation?

Giorgos Kallis, Helen Briassoulis,
Constantinos Liarikos and Katerina Petkidi

Introduction

Water, an essential life support resource for humans and ecosystems, is inextricably linked to development. Since antiquity, differences in availability and proximity to sources of water have accounted for differences in the socio-spatial structure of human settlements and modes of production (see, for example, Wittfogel, 1957; Mumford, 1961). The available quantity of water determines the development potential of all human activities (housing, farming, energy production, etc.). Water quality is crucial for public health while the state of the aquatic environment is becoming an important determinant of regional attractiveness and location factor for investments. Although water resources are renewable, they cannot be exploited indefinitely without eventually degrading. The mismanagement of water has long-term impacts on the sustainability of a region. In the semi-arid and sub-humid areas of Mediterranean Europe, the incidence of desertification owes, to a considerable extent, to serious shortages of water resources due to climatic and hydrologic variability and changes, development policies dissociated from water availability, mismanagement of water resources, and increasing water consumption.

Water management in Southern Europe, as in other arid regions of the world, has been based on an expansionist, short-sighted, single-use logic following the dominant supply-driven economic development model that considers water as merely a passive resource 'to be developed', a 'fuel' for economic growth (Worster, 1985). A patchwork of often inefficient waterworks, built to meet, or even generate, demand has led to overuse of water and deterioration of its quality. In recent years, regional and interregional conflicts among competing use(r)s have intensified. Furthermore, the ecological importance of water has been neglected. Critical aquatic ecosystems, including some of Mediterranean's most important wetlands have been damaged, sometimes irreversibly, by waterworks, water abstraction and wastewater discharges. Ecosystem degradation has important 'knock-on' effects on the natural environment and is an important contributor to desertification (EEA, 1998).

Water resources are complex systems, parts of broader nature-society systems, serving multiple, human and non-human uses. The presence of several agencies with separate responsibilities over various aspects and uses of water has contributed to a fragmented approach to water management and a relative blindness to the numerous interdependencies between water quantity and quality and, more broadly, between water resources, human activities, uses of land, ecosystems and society.

The need for 'integration' in water resource management has been widely recognized in policy documents starting with the 1992 milestones of the International Conference of Dublin on Water and the Environment and the UN Conference on Environment and Development (UNCED) in Rio resulting in the freshwater chapter of Agenda 21. Integration is the reason behind the major reform of European Union's (EU) water policy, marked by the 2000 Water Framework Directive (WFD). The success of the WFD, and by projection, the combat of the water-related causes of desertification, will much depend, however, on its integration with other policies directly or indirectly influencing water use.

EU regional and agricultural policies are major drivers of development and of land use change in the desertification-prone, mainly rural, areas of southern Europe. EU funds account for the bulk of public investments in these areas. Several past EU regional and agricultural development programmes, designed without taking into account water resource constraints, have caused detrimental impacts on water quantity and quality (WWF, 2003). Recent revisions of EU development policies attempt to account better for water-related, among other environmental, issues.

For sceptics, evidence from the ground, where Community-financed developments in many cases continue to undermine water management goals, suggests that this quest for integration is elusive. For optimists, recent policy changes, such as the WFD and the revisions of the Common Agricultural Policy (CAP) and the Structural Funds (SFs), mark important progress in the process of integration. This chapter comes to examine the current status and future prospects of integration between EU water and development policies in the light of past experience and recent policy changes, seeking to identify whether it can be plausibly expected to materialize in the future and bring the desired results; namely, contribute to sustainable development. For the southern, desertification-sensitive EU regions, the question becomes if this integration can contribute to combating desertification and alleviate the associated development problems.

Section 2 reviews concepts of integration in water resources management, examines the characteristics of integration between water and development policies and presents guiding criteria for assessing policy integration. Section 3 presents the main features of the EU water policy, focusing on the WFD, as well as the key features of EU regional and agricultural policies, especially as they relate to water management. In Section 4, the degree of integration between the three policies is assessed. The concluding section returns to the principal question of this chapter.

Water, Development and Policy Integration

Integrated Water Resources Management (IWRM)

In traditional, fragmented, water resources management, separate agencies are responsible for different functions of water resources (e.g. drinking water supply, wastewater management, energy production, irrigation, etc). Managers see themselves mostly in a neutral role, managing the natural system to supply water to meet externally determined demand (GWP, 2000). Integrated Water Resources Management (IWRM) instead refers to the joint consideration and management of interrelated factors such as supply and competing demands, water and land resources and uses, human and ecological needs, upstream with downstream and in-stream with off-stream uses, groundwater and surface water sources, water supply and wastewater management, water quantity and water quality. IWRM is defined as:

> ...a process that promotes the co-ordinated development and management of water, land and related resources, in order to maximize the resultant economic and social welfare in an equitable manner, without compromising the sustainability of vital ecosystems (GWP, 2000, p.22).

The river basin[1] is advocated as the most appropriate spatial unit for IWRM as the different uses and functions of water within its territorial limits are interdependent. It is, therefore, an appropriate spatial unit for planning the allocation of water resources, waterworks or pollution control measures to alternative use(r)s. The hydrological, instead of the political-administrative, boundaries of river basins help to overcome the boundary problems that plague water resources management (Moss, 2003).

River basin management has been a popular concept among engineers and water planners since the early 20th century. Countries such as Spain, the Netherlands, and England and Wales have a long history of river basin organization and management. What is new, however, is the emphasis on environmental protection, ecological needs and quantity-quality interactions.

---

[1] "The area of land from which all surface run-off flows through a sequence of streams, rivers and, possibly, lakes into the sea at a single river mouth, estuary or delta" (CEC, 2000, p.6).

Water Management and Development: from IWRM to Policy Integration

An extensive scientific literature advocates IWRM at the river basin level and deals with its implementation requirements (Jaspers, 2003; Jonch-Klausen et al., 2001). Most studies stem from an engineering perspective, emphasizing the need to extend management decisions to account for important factors, previously thought of as 'external'.

IWRM, however, is not an issue of water resources management alone. Water management is part and parcel of the development process that determines demand for and impacts on water resources. For example, urbanization in the arid areas along the Mediterranean coast has created demands to transfer water from distant places. There is a reciprocal, positive-feedback, co-evolutionary relationship between water management and development, or supply and demand; an "egg-chicken" situation where it is difficult to determine what comes first (Vlachos, 1982). Public investment and subsidization of large waterworks for most of the 20th century went hand-in-hand with the dominant, Fordist model of economic growth, leading to urbanization, industrialization and agricultural intensification. Roberts and Emel (1992) suggest that water problems should not be seen as a "tragedy of the commons" over open access water sources, but as conflicts over "competing development opportunities". This perspective emphasizes that a holistic approach should address water management and economic development processes together. Sustainable water resources management is an inseparable part of the sustainable development process.

Water managers and analysts, with a predominantly natural science/engineering background, are not at ease with extending analysis and prescriptions to broader issues of political economy, a weakness reflected in the techno-managerial emphasis of the pertinent IWRM literature. It is unlikely, however, that techno-managerial interventions in the operational aspects of water resource management alone will suffice to bring forth the necessary changes, unless linked to (and supported by) broader shifts in the development policies and processes that generate water demand and water problems (GWP, 2000).

Public policies, together with autonomous societal forces, are important determinants of socio-economic development. Some policy documents recognize this importance for water resources management and emphasize the need for a holistic approach where water policy is integrated with economic and sectoral policies (GWP, 2000). There is scant theoretical and empirical work, however, both to conceptualize and operationalize the integration of policies advocated. The term integration has been used in various, partly *ad hoc*, ways in water policy contexts. For example, Nijkamp and Dalhuisen (2002) refer to the need for integration between spatial planning and water policies and Zalidis and Gerakis (1999) to the integration of soil criteria into water policy. Both studies focus on selected aspects of policy integration only and do not offer theoretical insights into its conditions and requirements. OECD (1990) examines administrative integration for IWRM and offers guidelines to national authorities to integrate water with other sectoral policies. This is, however, an empirical study, based on best-case examples of administrative integration from OECD countries, lacking an overall theoretical perspective. Bressers and Kuks (2002) offer a more relevant analysis from a policy integration perspective, as part of a European research project on integrated institutional regimes for water resources management. Their emphasis, however, is on governance structures and property rights rather than on policies *per se*, which is the subject of this chapter.

A Framework for Analysing the Integration of Water Resources and Development Policies

Following the conceptual and methodological approach for the analysis of policy integration (PI) presented in Chapter 2 of this volume, PI is considered to occur when policies take into account their effects on each other. It refers to "a process of sewing together and coordinating various policies, both across (vertically) and along (horizontal) levels of governance, modifying them appropriately if necessary, to create an interlocking, non-hierarchical, loosely coupled, multi-level, policy system that functions harmoniously in unity" (Chapter 1, this volume p.37). For the analysis of the integration between water and development policies, the following interlinked criteria are used:[2]

1. *Substantive integration.* Different policies possess different logics and embody different theories. This is especially true for environment-related policies, such as the EU WFD, and development-related policies, such as the EU regional policy and the CAP, that have dissimilar origins and very different historical trajectories. Unearthing these underlying theories (ideologies) and examining the degree of (or, trends towards) conceptual convergence/integration is a key analytical task in assessing progress towards PI. Such an analysis sheds light on the essential compatibility of policy goals and approaches and the larger issues hidden in implementation conflicts. This chapter emphasizes the different conceptions of development underlying each policy, particularly, their pre-analytic theory of nature-economy relationships.

2. *Integration of policy goals and objectives.* Eliciting the degree of PI requires an analysis of the goals, objectives, and targets of the policies considered to assess whether they are common or compatible or, at least, agree with one another. Furthermore, PI *per se* may be set separately as an explicit policy goal. Statements indicating political commitment or stipulation of specific integration goals, targets, and timetables can be seen as evidence of PI.

3. *Integration between actors and actor networks.* Policy formulation and implementation involves particular networks of formal and informal actors.

---

[2] Their separate treatment serves analytical purposes only.

Actors shared by these networks can introduce indirectly a degree of coherence between different policies. Another issue is whether actors belong more to 'policy communities' rather than to 'issue networks', implying more consensus and continuous interaction (Bressers and Kuks, 2002).

4. *Procedural integration* refers to the existence of provisions that integrate the implementation procedures and the instruments of the policies considered at and across the different spatial/organizational levels on which they apply. Examples of moves towards procedural integration include regulations for cross-compliance, statutory responsibilities to consult water actors when drafting regional development plans, joint committees, new agencies or administrative divisions with a coordinative mission. Particular integrative instruments and related procedures might also serve to integrate policies (e.g. Environmental Impact Assessment, green taxation, etc.).

This chapter is based on desk research including study of policy documents and scientific literature, plus selected interviews with experts and policy makers. A pitfall of this approach is that policy rhetoric is often very distant from implementation practice. Incorrect conclusions may be reached by looking at policy documents alone or by consulting policy implementers only. Evidence from the actual working of policies and decisions is used where relevant to demonstrate whether and how provisions work on the ground. This chapter focuses on the EU level, but at some points the discussion unavoidably refers to experience from national/local policy implementation practice in southern Europe. Implementation instruments and processes, however, may differ considerably among Member States (MS), particularly since recent EU policies rely heavily on subsidiarity and decentralization. Commenting on regional policy, Roberts (2003, p.2) notes that "whilst in theory a single system for regional development is present throughout the EC, in practice 15 or more regional systems exist".

EU Water and Development Policies

EU Water Policy

Overview Water is the most extensively regulated sector by EU's environmental policy (Kallis and Nijkamp, 2000). The first EU water directives (1975-1980) catered to protecting uses important for public health (drinking, bathing, fish and shellfish harvesting) and to controlling the discharge of dangerous chemicals to surface and ground waters. This was a standards-based legislation, setting EU-wide quality or emission limit values for specific substances/parameters. As the EU extended its competence from public health to environmental protection, emphasis shifted to the control of important polluting activities at their source. The 1991 Wastewater Directive mandated secondary biological treatment for all urban settlements with population over 2,000 inhabitants plus additional treatment for discharges in eutrophication-sensitive areas. The 1991 Nitrates Directive followed a more decentralized approach, asking for the implementation of action programmes including good farming practice codes in nitrate-sensitive rural zones (Kallis and Nijkamp, 2000).

Other environmental directives had also important implications for water. The 1991 Plant Protection Directive linked the authorization of pesticide products to compliance with drinking water standards. The 1996 Integrated Pollution Prevention and Control Directive introduced a common pollution permit system for industrial installations based on Best Available Technologies, granting authorizations on the basis of the total environmental performance of the installations, including effluents and their impacts on water. Finally, the 1979 Birds Directive and the 1992 Habitats Directive require the MS to identify sites of European importance, many of which are aquatic ecosystems (rivers, lakes, etc), wetlands, or other water-dependent terrestrial ecosystems, and to draft special management plans to protect them.

The 2000 WFD aims to combine the various pieces of legislation into a coherent whole as well as to extend regulation to account for the ecological quality of waters. Within this framework, surface waters and groundwater are to be protected using a common management approach and following common objectives, principles and basic measures across the EU (Moss, 2003). It adopts the 'ecosystem approach' to environmental management that conceives humans and their artefacts as parts of broader ecosystems including also natural elements (Smith and Maltby, 2003); human activities are partly constrained by the limits of the ecosystem within which they operate. It emphasizes a holistic approach to nature-society interactions recognizing that ecosystems function as integrated entities and not in pieces, providing a broad range of goods and services whose production is the result of a healthy ecosystem. Management interventions aim to preserve or increase the capacity of ecosystems to produce desired benefits (Jewitt, 2002). The WFD sets procedural rules and guidelines for organization, planning and management at the river basin level. River basin plans and programmes are to provide the platform to achieve ecological and pre-existing standards (Kallis and Butler, 2001).

EU water policy is primarily a *water quality* policy. Water directives are part of EU environmental policy, decided by the Ministers of Environment of the MS and executed by DG-Environment of the European Commission. The Maastricht and the Amsterdam Treaties have separated "the management of water resources" from environmental policy, establishing in this way a division in competences between *water quality* and *water quantity* management (Kraemer, 1998). Unanimity in the European Council is needed for decisions concerning water resources management, whereas majority voting suffices for environmental policy. Southern MS have defended this division in the Treaty on the grounds that, for them, water management is not only an environmental but also a national development issue and they should maintain a veto power. In designing the WFD, the EC had to balance on a thin line between "integration" (which necessarily requires joint consideration of quantity and quality issues) and non-interference with the management of water quantity. The Directive proclaims that the "control of quantity is an ancillary element in securing good water quality" (CEC, 2000, preamble 19). However, many WFD provisions affect significantly water quantity management (Kallis and Butler, 2001).

Goals of EU water policy   In line with the Community's environmental policy goals, as expressed in its Environmental Action Programmes, the overarching goal of EU water policy is to contribute to sustainable development in the EU. This includes, but is not limited to, the goal of environmental protection. Older Directives, now subsets of EU water policy, pursued two principal goals, now subsumed under the sustainable development goal: public health protection and establishment of a 'level-playing field' of environmental standards to reduce distortions in competition (Kallis and Nijkamp, 2000).

The principal aim of the WFD is to "maintain and improve the aquatic environment in the Community" (CEC, 2000, preamble 19). More specific objectives include: prevention of further deterioration and enhancement of aquatic ecosystems, dependant terrestrial ecosystems and wetlands; sustainable use of water; enhanced protection and improvement of the aquatic environment; prevention and reduction of groundwater pollution; provision of sufficient, good quality water as needed for use, and mitigation of droughts and floods (CEC, 2000, Article 1). The overriding target is to achieve at least a 'good' status for water bodies in terms of ecological and chemical parameters (plus water quantity for groundwater) and to prevent the deterioration of aquatic ecosystems (ibid, Art. 4). Bathing areas, drinking sources, habitats and euthorphication/nitrate-sensitive zones are to be registered as 'protected areas', incorporating, hence, the standards of the respective directives in the WFD structure (ibid, Art. 6).

Water policy instruments   The main water policy instrument is regulation through directives. This is not accompanied by specific supporting financial instruments. Certain water-related projects[3] may be financed through the Life programme of DG-Environment. EU water policy contains a mix of 'command and control' and procedural instruments, with a definite trend towards assigning the latter a greater role. Command and control instruments can be further subdivided into quality/emission standards for particular substances/parameters[4] and uniform technological requirements.[5] In comparison, procedural regulation prescribes processes (and not standards or measures) that competent authorities should implement. The main procedural requirement of the WFD is the delineation of River Basin Districts (RBDs), the designation of River Basin Authorities (RBAs)

---

[3] E.g. innovative environmental demonstration projects.
[4] Such as those for drinking water, bathing water, dangerous chemicals, etc.
[5] Such as the mandate for secondary treatment plants or BATs in industrial installations.

and the preparation of River Basin Plans (RBPs) and programmes of measures to achieve the Directive's goals. Other important procedural requirements include: establishment of monitoring programmes; licensing schemes for abstractions, impoundments or discharges; processes for public information, consultation and participation; and reporting to the EC (CEC, 2000). Requirements for planning, monitoring and reporting are precisely defined in the WFD and other water directives.

Other than complying with the procedural requirements, the national competent authorities have considerable discretion in choosing the organizational form of the RBAs, the design of participatory processes, or any other means to achieve the Directive's goals as long as they conform to its general principles. The minimum requirement is to implement existing legally-set instruments.[6] In addition, the WFD introduces a number of mandatory new instruments to be implemented by MS and RBAs without prescribing their details, however. These include costing/pricing, zoning of designated areas, abstraction and discharge permitting and authorization of water quality-impacting activities. When the aforementioned 'basic measures' do not suffice to achieve the environmental objectives of the directive, the RBAs will have to implement additional measures, such as stricter permit standards, zones for good farming practices beyond the nitrate sensitive areas, water demand management programmes, etc.

Policy implementation processes The DG-Environment has the responsibility to ensure that MS comply with the legal requirements of the water Directives. Monitoring and reporting, however, are responsibilities of each MS. The EC can check whether certain implementation deadlines or procedural requirements are respected[7] but it is less able to evaluate and judge both the content and the accuracy of the voluminous data provided by MS for specific standards.

The WFD is a milestone in the Union's environmental policy (Kallis and Butler, 2001). Recognizing the limits of the top-down, command-and-control approach, it adopts a more flexible and cooperative implementation strategy. Many of its core requirements were not defined in the legal text. A so-called Common Implementation Strategy (CIS) is meant to lead to EU-wide standards and guidance for the implementation of the WFD (EC, 2001). The most important output of the CIS is the agreement on the EU-wide set of parameters/standards differentiating water status classes (high, good, moderate, poor, bad) by type of water body. The EC has set up working groups with the participation of national delegates, experts and civil society/NGO representatives to prepare non-binding guidance on various implementation-related tasks such as planning processes, pricing, economic evaluation, participation, monitoring, etc. (EC, 2003a).

---

[6] E.g. secondary treatment, good agricultural practices in nitrate sensitive zones, industrial installation permits, etc.

[7] E.g. for designation of protected areas, establishment of plans and monitoring programmes, reporting, etc.

The framework approach of the WFD blurs the border between compliance and non-compliance. The Directive is (deliberately) ambiguous over whether MS should achieve the good ecological status objectives or should simply 'aim to' achieve them, meaning that they are obliged to implement the necessary measures and procedures, but are not culpable if these do not achieve eventually the objectives. Furthermore, the Directive allows generous derogations (e.g. when costs are 'excessive', when waters are 'significantly modified', or when the reason for damage is a 'sustainable human activity'). Compared to the straightforward command-and-control directives, the ambiguity of such terms makes it much more difficult to judge on non-compliance (Kallis and Butler, 2001).

Implementation processes are highly variable and depend on the institutional organization of the water sector in each MS. The minimum procedural requirements of the WFD will soon have to apply all over the EU. Still these leave considerable freedom in the allocation of powers between different tiers of government, the choice of legal and economic instruments, the planning process and content, etc., implying potentially great differences in implementation even in neighbouring regions of the same country (Moss, 2003).

EU Regional Policy[8]

Regional policy aims to achieve social and economic cohesion in the Union by "reducing disparities between the levels of development of the various regions and the backwardness of the least-favoured regions or islands, including rural areas" (CEC, 1999a, preamble 1). To achieve harmonious development, support is provided to help the least well-off regions rise to the challenge of the common market and the monetary union. For the period 2000-2006, three specific regional development objectives target the development of lagging regions, of regions facing structural adjustment problems and the development of human resources and the modernization of education (CEC, 1999a).

EU regional policy is delivered through financial instruments, the SFs, which include: the European Regional Development Fund (ERDF), the European Social Fund (ESF), the Guidance Section of the European Agriculture Guarantee and Guidance Fund (EAGGF) and the Financial Instrument for Fisheries Guidance (FIFG). The Cohesion Fund (CF) provides assistance for environmental and transportation infrastructure in four lagging countries of the EU (Greece, Portugal, Spain, Ireland), and is also included under the SFs, albeit implemented independently. Finally, the SFs include a number of Community initiatives, such as the INTERREG, LEADER+, URBAN, and EQUAL, targeting cross-border cooperation, innovative rural development, urban regeneration, and employment, that account, however, for a tiny proportion of the overall regional support budget.

Community funding is allocated and managed on the basis of rules and processes specified in regulations concerning eligibility criteria for areas and

---

[8] See also Chapter 3, this volume.

beneficiaries; procedural rules for applying, granting, monitoring and evaluating funding; enforcement of investment standards and quality controls; auditing, etc. Other procedural requirements include publicization of programme reporting, organization of public-targeted events and consultation processes (CEC, 1999a).

MS applying for structural support propose, negotiate and agree with the EC a Community Support Framework (CSF), which is made up of different Regional and Sectoral Operational Programmes (OP). Regional OPs concern the development of a specific region, while Sectoral OPs target specific sectors (e.g. transport) or issues (ex. environment). Instead of submitting and negotiating a CSF, a MS or a region can submit a Single Programming Document (SDP), which is the equivalent of an independent Operational Programme.

MS submit their CSFs or SPDs in a predefined format and then agree on the programme details setting out the budgetary envelope, aims, objectives, priorities, eligible measures, monitoring and evaluation procedures and general implementation requirements. Responsibility for programme implementation and monitoring rests with the MS, which should establish a management authority (MA) and a Monitoring Committee (MC), for the whole CSF and the OPs included. The MC is composed of MS officials, EC representatives (no voting rights) and economic, social and other partners, and is responsible for overseeing programme implementation and approving all major management decisions. Programme effectiveness is evaluated before programme commencement (*ex-ante*), in the middle of the implementation period (*interim*) and after programme completion (*ex-post*). Although evaluations are meant to provide information on management effectiveness and on programme implementation results, their majority is limited to management and financial issues, with the evaluation of impacts and results been systematically neglected.

The EC influences the direction of regional development in the MS by defining the funding priorities and the eligibility criteria of the SF, by negotiating and approving the proposed programmes, by participating in the MCs and through its competence on approving major programme changes. However, its role in the actual selection of projects by the MS (other than 'major' SF, or CF projects) and the monitoring of their implementation is limited. It may intervene through formal processes in cases where gross mismanagement of funds is suspected, but it does not have the capacity to follow closely the implementation of separate projects against predefined goals/indicators.

EU regional policy has several implications for water resources management. Many SF-supported infrastructure projects (e.g. highways) have impacted negatively on water resources and aquatic ecosystems (e.g. damage to wetlands). Development patterns, indirectly driven by the Funds (e.g. urban or tourism development), increase water demand and cause water pollution, hence, intensifying pressures on scarce resources. Furthermore, the Funds have supported infrastructure works such as water transfers, dams, networks, etc., indirectly subsidizing the cost of water and serving to maintain unsustainable supply-side management practices. Certain projects[9] have had important ecological impacts and have contributed to the overuse of local water resources. However, following recent reforms (see below), the Funds have started targeting also alternative IWRM or environmental conservation projects.

EU water policy may potentially impinge on the cohesion and development of lagging regions, the main objective of regional policy. The WFD foresees a gradual elimination of subsidies and cross-subsidies of water users. Unless counterbalanced by other measures, this may impact negatively on cohesion by worsening the condition of poorer users or regions (WRc, 2001). Furthermore, EU water Directives entail significant implementation costs and investments (for drinking and wastewater treatment, establishment of monitoring infrastructures, funding of new agencies and planning procedures, etc.). Relative to the more developed regions, lagging regions are typically in worst 'initial positions' for implementing water directives. On the other hand, the relatively better condition and higher assimilative capacity of their environment may reduce investment requirements. The verdict, therefore, of the impact of EU water directives on cohesion is indecisive (WRc, 2001).

EU Agricultural Policy

The CAP is the most important policy of the EU in expenditure terms (55 per cent of EU budget in 1998 to decrease to 46 per cent by 2006). Upon its formulation, in 1958, the CAP purported to assure fairly-priced food supplies for European peoples and at the same time to provide for sufficient incomes for farmers. In the face of vast overproduction and environmental problems, its attention gradually shifted towards a more spatial and less-sectoral outlook and the development of rural space. This shifting attention to rural development, the second pillar of the CAP, was put forward in the 1996 Agenda 2000 reform and affirmed by the current Rural Development Regulation (RDR), which provides for the coordination between rural and regional development instruments and the implementation of a spatially defined rural policy.[10] If the proposals for a new RDR are finally approved and put forward, the rural development element of the agricultural policy will be strengthened, but direct coordination with regional policy will be disrupted; rural development policy will be financed by a specially-set instrument, alongside the first pillar of the CAP (direct price and income support).

The main instruments of the CAP are price and financial supports. Procedural regulations govern the allocation and monitoring of funds. Until 1992, the CAP market policy relied almost exclusively on price support mechanisms for products, combining high border protection, export subsidization and intervention buying at

---

[9] E.g. irrigation programmes in rural areas contributing to the exhaustion of groundwater reserves. Although irrigation projects cannot be funded formally under the SF, such projects are commonly funded under 'water management' measures and funding lines.

[10] See, chapters 3 and 4, this volume.

guaranteed prices in the internal market. Price support and guarantees have stimulated capital investment and intensification of production. Market and price support are provided by the Guarantee section of the EAGGF, the financial instrument of the CAP, which is managed by DG-Agriculture. Support is decided on a per product basis through Common Market Organizations (CMOs). About 60.5 per cent of total support of the Guarantee section to producers concerns market prices and the remaining 39.5 per cent is direct payments (Chatelier and Daniel, 2001). The Guidance section is managed through national CSFs and follows the implementation procedures described previously.

Rural development policy is defined in the Rural Development Regulation (RDR)[11] (CEC, 1999b). The related expenditures account for about 10 per cent of the total CAP budget. There are two basic strands of rural development measures: (a) measures for agricultural sector restructuring and for supporting areas with natural deficiencies (agri-environmental, young farmers, compensatory allowances, etc.) and (b) measures for the development of the rural economy, targeting sectors other than agriculture, including forest management, early retirement, agro-tourism, Less Favoured Areas (LFAs) and areas with environmental restrictions. For Objective 1 regions, the latter are financed through the Guidance section of the EAGGF and are designed and implemented under the aegis of the regional development regulation. They take the form of OPs integrated in the relevant national CSFs. The former are detached from regional policy implementation and their application is independently managed by agricultural authorities; they follow the programming principle, though, and are prescribed in seven year Single Programming Documents. The rural development measures and the associated implementation procedures and requirements are differentiated on the basis of the classification of assisted regions.[12] The RDR defines principles, administrative and financial provisions, formats for the preparation of the aforementioned programmes, public participation procedures, etc.

In the rural regions of the southern MS, agriculture is by far the largest consumer of water and one of its most important polluters. Pesticides, fertilizers and heavy metals from sewage sludge and manure application affect groundwater and surface water quality. Land drainage from farming causes wetland destruction. Intensive water abstraction for irrigation lowers groundwater tables and reduces surface water flows. Soil sediments modify the morphology of water courses and wetland ecosystems (EC, 2003b). These pressures are due to agricultural intensification driven primarily by the price/market support measures of the CAP, that are mostly approved on the basis of land area and production levels, thus, encouraging intensification and spatial expansion of production. The CAP has sustained and intensified the production of water-intensive crops in water-short Mediterranean regions. CAP subsidies indirectly favour irrigated crops through

---

[11] See, also, Chapter 4, this volume.
[12] See, Chapters 3 and 4, this volume.

higher payments,[13] thus providing strong incentives for continued use of irrigation over rainfed agriculture. Compensatory allowances are allocated using land area criteria. Funding for the modernization of enterprises is oriented, almost exclusively, towards the utilization of capital-intensive techniques. Furthermore, the CAP (and other regional funds) have subsidized investments in waterworks (reservoirs, irrigation networks) that have maintained a low cost of water for users, and may have played a role in the overuse of water resources.

Environmental, and especially water-related considerations, have been, at least partly, behind the gradual reform of the CAP from price support to single, decoupled income payments per farm, and from market support to structural adjustment and rural development programmes. Investments in agricultural holdings, training, LFAs and agri-environment measures (AEMs) have the potential to contribute positively to water resources management (EC, 2003b). The budget of these measures, however, is still very small compared to price supports.

EU water policy may bear important implications for agricultural and rural development. The 'polluter pays' and the 'user pays' principles mark an (indirect) change in the property rights of farmers. Land ownership is decoupled from the, until recently *de facto*, right of farmers to use their land and its resources (including groundwater) without limitations. If the WFD is implemented properly, farmers will have to obtain permits for using water or for applying agrochemicals that pollute water courses. The reduction of subsidies and the increase in the proportion of cost recovered by charges put forward by the WFD will most probably increase the cost of water, and, hence, production costs for most farmers. The WFD and the Nitrates Directive also demand significant farm level investments to implement good agricultural practices. Leaving aside the debate of whether such investments will pay off in the longer-term,[14] in the short-term at least they increase production costs and strain especially smaller farmers who operate under marginal profits (if not losses).

Analyzing the Integration of Water with Development Policies

Substantive integration

Three aspects of substantive integration are examined below; policy problem and theories, policy object and policy approach/intervention style.

Policy problem and theories   Referring to *the* theory (or theories) underlying a certain policy can be misleading. Several (even contradictory) theoretical ideas may coexist in a single policy. Policies evolve over time, accumulating several different ideas. Implementation also matters significantly; political and ideological

---

[13] As they continue to be paid in proportion to productivity per hectare.

[14] E.g. if farms become competitive in "niche" organic food markets.

conditions, reflecting particular theoretical positions, influence the "translation" and realization of policies. Moreover, theories are not monolithic. Names attached to certain theories often do injustice to the plurality of ideas and the diverse viewpoints held by different scholars. Being aware of the limits of generalizations, the following discussion attempts to identify broad trends in some very basic theoretical underpinnings of the three policies, based not on an empirical analysis of the theoretical ideas that influenced policy makers when drafting or implementing the policies, but on a 'reading' and understanding of policies and their relations to broad theoretical paradigms, especially those concerning the development-environment interface.

Early EU water, regional and agricultural policies had very different origins. Table 6.1 summarizes the broad problems addressed and remedies put forward by each policy. The problems addressed by the three policies were disconnected, indicating the strong policy sectorialization at the early periods of the EU. The three policies also were implicitly endorsing very different views of problem-response relationships. Whereas water policy viewed economic development as the source of the problem (pollution) that had to be controlled, for regional and agricultural policy it was the response to their problem.

Table 6.1   Perceived problems and remedies in early EU water and development policies

EU Policy	Problem	Remedy
Water	Health impacts from pollution Market distortion and 'race to the bottom' from different national quality standards	EU-wide water quality standards
Regional	Regional disparities, lack of cohesion	Supporting the economic development of laggard regions
Agricultural	Food sufficiency	Supporting the economic development of the agricultural sector

With the danger of oversimplification, one might relate early EU water (and more generally environmental) policy to what the literature calls 'ecological', 'limits' (de Graaf et al, 1996, Norgaard, 1995) or 'closed economy' (Daly, 1999) paradigm. This paradigm sees human activity bound within environmental limits which when superseded have impacts that may limit development (Kallis, forthcoming).

In comparison, early regional and agricultural policies were characterized by a more 'economistic' paradigm, emphasizing output growth without sufficient recognition of natural resource or environmental limits.[15] From this perspective, natural resources are considered important factors in the production and development processes. They are not seen, however, as being intrinsically limiting (as in the 'closed economy' paradigm); with sufficient technology and investments, resources can be 'developed' to contribute to economic growth (see, for example, the substantial provisions of early regional or CAP funding for capital investments in waterworks).

Recent policy developments, however, indicate some converging tendencies in the theoretical underpinnings of the three policies. EU water policy has moved away from its exclusive focus on qualitative environmental standards to embrace the broader goal of contributing to sustainable development. In policy terms, a pragmatic multi-dimensional interpretation of sustainable development has prevailed, emphasizing the combination of multiple objectives; economic, social and environmental. For example, Engwegen and McLaren (1998) define sustainable development as a new model of development which aims to pursue three objectives in such as a way as to make them mutually compatible; first, sustainable, non-inflationary economic growth, second, social cohesion through access for all to employment and a high quality of life and, third, enhancement and maintenance of the environmental capital on which life depends.

This interpretation of sustainable development is central in the notion of IWRM and underlies the WFD. Right from its preamble, the WFD declares that it establishes the principles of a sustainable water policy and commits to a sustainable management of freshwater. The Directive is explicitly designed to combine environmental objectives (status objectives), social objectives (participation) and economic objectives (economic evaluation and cost-benefit justification of investments, pricing for efficiency).

Perspectives on economic development as reflected in regional and agricultural policies have also changed significantly, moving away from an exclusive emphasis on output and GDP growth towards a 'new development' theoretical paradigm of a more qualitative, multi-functional and multi-purpose development process. Quality of the environment (including the aquatic) is recognized as a social welfare factor, instead of being subsumed under gross production output. For example, the SF regulation states that "the quality of the natural ... environment ... contributes to making regions economically and socially more attractive" (CEC, 1999a, preamble 6). The shift of EU regional funding from basic infrastructure to structural investments (education, innovation, etc.) parallels the shift in regional development theorizing towards qualitative, 'non-infrastructure' development.

A similar trend is noted in the theoretical framing of agricultural policy with the adoption of the concepts of 'rural development' and 'multifunctionality of agriculture'. The former reflects a shift from an emphasis on agricultural product growth to the development of rural space as a whole, and in particular the

---

[15] However, regional policy quickly evolved to account for environmental concerns (although its main corpus continued to be focused on GDP growth in laggard regions).

fulfilment of social goals such as reducing poverty and inequality.[16] The latter,[17] which has influenced several EU policies emphasizes the economic, social and environmental benefits from well-functioning rural economies. The Berlin European Council on the reform of CAP has declared that the aim is to "secure a multi-functional, sustainable and competitive agriculture throughout Europe, including regions facing particular difficulties ... [and to] ... maintain the landscape and the countryside, make a key contribution to the vitality of rural communities and respond to consumer concerns and demands regarding food quality and safety, environmental protection and maintaining animal welfare standards".[18] These views on development embraced by regional and agricultural policy indicate a convergence with the conception of sustainable development as a multi-functional process combining economic, social and environmental factors.

Unlike, however, the older, straightforward output-based notions of development, the meaning and practical interpretation of sustainable development are much more ambiguous and contested. The expectation that different objectives can be compatible is no guarantee that they *are* really compatible, not in all instances, at least. Indeed, real life is replete with situations where there are no "win-win" possibilities and hard trade-offs and sacrifices have to be made by some for others to reap benefits (Kallis and Coccossis, 2004).

Decisions over water transfers in southern Europe, for example, have tested in many cases the limits of commitment of the EU to sustainability and integration. A water transfer benefits the recipient region but imposes costs on the source region. Whereas agriculture in the recipient region may benefit from the increased supply of water (contributing to the avoidance of desiccation and land abandonment, and, therefore, potentially reversing phenomena of desertification), the source region incurs ecological damages and reduced economic opportunities. Striking trade-offs between the economic, social and environmental goals of different regions, or between people and ecosystems, is not a matter of rational analysis and objective decision making; it involves deeper issues of justice and of reconciling different value systems. The failure of the EC to deal effectively and unambiguously with such confrontational projects is not only a matter of lack of administrative coordination between the various DGs but owes to the very substantive limits of concepts such as sustainable development and integration.

Agricultural supports is another illustrative example of important trade-offs and conflicts between economic, social and environmental goals and outlooks. Economic and environmental goals emphasize the need to make polluters and users (in this case, agricultural producers) pay the full cost of resource use. Several issues, however, arise and limit the implementation of this theoretically sound rule. Paying the full cost of water may have important income and distributive

---

[16] See, Chapter 4, this volume.
[17] Referring to the "achievement of multiple food and non-food objectives in the most cost-effective manner, taking into account direct and indirect costs" (OECD, 2001).
[18] Berlin European Council, Presidency Conclusions, 24-25 March 1999.

implications on farmers, which, in turn, raises some important questions of social justice. In the past, farmers were providing, at least partly, a social service in guaranteeing national food security, enjoying, thus, many 'privileges'. Now that food security is no longer essential, removing supports may be to an extent both unfair and risky (what if global markets collapse in the future?). Furthermore, it is not always clear to what extent farmers should pay for cleaning up a river, or instead urban water users who wish to enjoy higher levels of drinking water quality and safety should cover this cost. As a result, in practice, the issue of whether the polluter should pay or should be compensated not to pollute remains unresolved and decided on a case-by-case basis.

The above two examples illustrate that beneath the surface of some a-politicized definitions of concepts such as 'sustainable regional development' or 'rural development' may still lie different, irreducible values, perspectives or world-views on *how to* settle trade-offs between economic efficiency, environmental protection and social justice. Inevitably, these resurface when making decisions or implementing policy; it is then that the integrative spirit of sustainable development faces its test, and often breaks down. Procedural mechanisms enabling to deliberate and coordinate the settling of different objectives and trade-offs become thus important.

Substantive integration is an important goal and the convergence of EU water and development policies, even if only in rhetoric, marks an improvement in the perspective of their integration. The above discussion suggests, however, that a full integration in the form of a single theory of development guiding water, regional and agricultural policies is elusive. More importantly, it is not necessarily desirable. Pluralism and dialogue among different theories may be a more effective means towards policy integration as will be argued in the concluding section of this chapter.

Policy object Early water, agricultural and regional policies had very different objects: resource, sector and territory, respectively. Water policy focused on the quality of the resource for specific uses, agricultural policy on the development of the sector and on farmer income support, and regional policy on regional economic development. These very different policy objects evidently made integration efforts more difficult.

Recent policy changes, however, have brought about a *territorialization* of water and agricultural policy, bringing them closer to regional policy and creating more opportunities for synergies in implementation (Moss, 2003). Water policy operates now at the river basin scale. Agenda 2000 emphasizes regional cohesion as a key goal of agricultural policy and commits to the structural support of (rural) regions facing difficulties. Rural development policy marks a gradual (although hesitant) shift of the CAP from a sectoral to a territorial approach.

This convergence is not complete yet; important elements of divergence still exist. RDR accounts for a minor portion of CAP outlays; farm-based supports absorb the majority. Moreover, not all RDPs are region-based; MS may design national-sectoral plans open to proposals from individual farmers.

A second case of divergence is between regional and water policy. Regional policy employs administrative regions based on the NUTS system. River basins instead follow hydrological boundaries. The two policies, therefore, address different spatial systems of reference, a misfit not necessarily negative. Moss (2003) observes: "the replacement of existing institutional units by institutions oriented around biophysical systems will inevitably create new boundary problems and fresh mismatches". Perfect spatial system compatibility is elusive. Moss (2003) argues in favour of a constructive "institutional interplay", whereby the problems of imperfect matches are overcome by formal and informal processes and mechanisms which facilitate coordination and integration.

Policy approach/intervention style The changes in the objects of all three policies identified above are paralleled by a similar trend in changing policy styles from top-down, command-and-control to bottom-up, procedural[19] and network-based[20] approaches (Moss, 2003). Early EU water policy was based on the establishment of EU-wide regulatory standards with which MS had to comply. The WFD instead regulates primarily procedures (river basin planning and management, participation, licensing, etc.), following a network-based, multi-partnership implementation approach (e.g. Common Implementation Strategy, River Basin Councils, etc.).

These policy changes at the EU level are driven by the binding subsidiarity principle. Subsidiarity, a mandate for decentralization which maintains an important coordinative role for the EU, is a compromise between 'federalists' and 'anti-federalists' in the debate over the future of the EU. Recent environmental and water policy analyses recognize that there are no uniform policies applicable in all contexts and that it is more effective to adapt goals to local conditions and needs (Ward et al., 1997; Collier, 1997).

All three policies contain clear statements of commitment to the principle of subsidiarity. The WFD states that "decisions should be taken as close as possible to the locations where water is affected or used" (CEC, 2000). River basin management expresses the decentralization approach of the Directive. So does the provision to differentiate ecological quality standards depending on the type of water body. Similarly, the RDR states that "rural development policy should follow the principle of subsidiarity ... [and be] ... as decentralized as possible ... [with] ... emphasis on participation and a bottom-up approach" (CEC, 1999b). Regional policy has long made this decentralization shift through its programming structure, whereby powers of decision and implementation are delegated to MS and regional authorities. The SF regulation calls for "partnerships to be strengthened"

---

[19] Regulating procedures, not results.
[20] Flexible partnerships between higher and lower level actors.

and for a "decentralized implementation of the operations of the SFs by MS" (CEC, 1999a).

In terms of PI, the decentralization trend has both positive and negative aspects. The EU has assumed a supervisory role, limited to procedural regulation and coordination, in all three policies. The positive aspects of partnership-based implementation have been documented both for the SFs (Roberts, 2003) and for water resources management (Moss, 2003), in the case of northern, more developed MS, however. On the other hand, a more centralized control of policy implementation could create a more solid ground for the integration of EU policies. Decentralization and the devolution of EU powers to lower organizational levels weaken an already weak EU capacity to control actual outcomes, hence threatening integration during implementation. This is particularly disquieting for the southern MS, where the old "modernization" paradigm still dominates, new development and environment ideas have yet to trickle down and there are concerns that the lack of effective monitoring and control by the EU distorts the proper implementation of both water standards (Ward et al., 1997; Kallis and Butler, 2001) and regional development programmes (WWF, 2003).

Integration of Policy Goals and Objectives

Two issues are examined with respect to this criterion: (a) whether the goals of the different policies are congruent and compatible and/or exhibit commonalities and (b) whether integration of EU water policy with other policies is a policy goal.

The previous discussion of substantive integration showed that the early goals of all three policies were highly divergent, addressing environmental limits (water policy), social and economic cohesion (regional policy) and food sufficiency/sectoral output growth (CAP). Recently, a hesitant convergence is noted among their goals under the rhetoric of multi-functional, sustainable development and the injunction of Article 6 of the Amsterdam Treaty,[21] asking for the integration of environmental considerations into sectoral policies, although no separate reference to water as such is made.[22] However, this convergence is rather superficial, stated in the abstract and far from complete. Important underlying, partly irreconcilable, differences remain as to how each policy perceives and operationalizes the grand goal of sustainable development as well as the actual relationships of its goals with those of other policies.

All three policies take up and refer to the goal of integration. The WFD stresses the importance of "further integration of protection and sustainable management of

---

[21] Representing the highest level official, global commitment to the goal of policy integration.

[22] However, Article 6 embodies a vertical notion of policy integration, namely, environmental policy integration (EPI), while the horizontal notion of integration among policies is not addressed.

water into other Community policy areas such as energy, transport, agriculture, fisheries, regional policy and tourism" (CEC, 2000, preamble 16). Moreover, it recognizes that water policy should take into account "the economic and social development of the Community as a whole and the balanced development of its regions as well as the potential benefits and costs of each action" (ibid, preamble 12). Regional and agricultural policies also refer to the need for integration of environmental goals (which include water). The SF regulation states that in "its efforts to strengthen economic and social cohesion the Community also seeks to promote the harmonious, balanced and sustainable development of economic activities ... and a high level of protection and improvement of the environment" (CEC, 1999a, preamble 5). However, sustainable development *per se* is not the goal of regional policy. The RDR defines as its objective "the preservation and promotion of a high nature value and a sustainable agriculture respecting environmental requirements" (CEC, 1999b, Art. 2). Rural development goals are also tuned to those of regional policy; rural development measures should "take into account the specific targets of Community support" set in the SF regulation (ibid, Art. 1).

Summing up, it seems that the goals of one policy are 'added' on, rather than 'fused' with, the goals of another. Calls for integration are not taken up rigorously by specific policy provisions or policy implementation plans with defined objectives, targets and programmes/timetables of implementation.[23] EPI is stressed mostly and not horizontal integration of environmental, water policy in this case, with development policies. In particular, the integration of the WFD with development policies is not an explicit policy goal, a fact owing partly to its appearance at a latter date (2000) than the other two policies.

Integration between Actors and Policy Networks

The EU policy process is best described as "a multi-national, neo-federal system, extremely open to lobbying by a wide variety of organizations with an unpredictable agenda setting process creating an unstable and multi-dimensional policy-making environment" (Richardson, 1997, p.140). A rather messy amalgam of interrelationships between formal institutions and non-governmental actors exists. EU policy networks evolve continuously, their composition changing as EU expands its competences in new areas and as the effects of EU policies are realized on the ground, motivating actors to intervene at the European level. Even relationships among the key EU institutions are not settled and are redefined with each round of Treaty reform.

A general structure of EU policy networks comprising formal EU institutions, economic actors, non-governmental organizations and experts is valid more or less for all EU policies. However, the nature and composition of the networks of different policies vary widely. EU water policy networks have been described as

---

[23] Based on the principle of subsidiarity, these are left to the discretion and choice of MS.

open rather than closed and more 'issue-driven' rather than part of a stable 'policy community' (Kallis and Nijkamp, 2000; Richardson, 1997). This, however, reflects more the pre-WFD period, where different actors and coalitions were motivated around each of the water directives, with their different scope and area of intervention. A more stable, albeit still open, water policy network emerged around the WFD, reflecting its broader coverage and implications, the long period (almost seven years) that its preparation and negotiation lasted, and the development of the Common Implementation Strategy (Kaika, 2003). EU water policies include a wide range of non-institutional actors from industry, trade unions, citizens organizations, etc.[24] The role of scientific experts is important in view of the extensive use of *ad hoc* committees by the EC to support its legislative functions, and especially the establishment of several working groups, as part of the CIS, that have been particularly open to newcomers and environmental NGOs.

These trends confirm Richardson's (1997, p.142) prediction that:

...the more there is an attempt to coordinate environmental policy (including water policy) with other sectors such as regional development and agriculture, the more the process will shift from a narrowly-based set of actors to a more extended network of actors.

Several development actors (agricultural interests, trade unions, regional authorities, industries, etc.) have taken an active role in the process of formulating and implementing the WFD.

However, while the WFD policy process seems to provide an arena for more interaction among actors and, hence, favouring PI, this seems a one-sided opening, not followed by analogous developments in regional and agricultural policy networks which, especially the latter, remain closed. The agricultural policy network forms around DG-Agriculture, the Council of Ministers of Agriculture and the major Farmers Unions. Both at the MS and the EU level, these actors have been exceptionally resistant to change. Environmental interests and NGOs have a marginal role (Lenschow, 2002). The rigidity of the network has limited opportunities for fundamental CAP reform in practice despite the radical rhetoric of EC policy statements (Daugbjerg, 1999). The same can be said for the regional policy network. Governments have a much tighter grip on the process, than in water policy, as the funds allocated provide an important source of revenue for national economies. Environmental NGOs intervene actively, especially when the funding of environmentally controversial projects is at stake, but their role is mediated mainly through the arenas of environmental policy and DG-Environment. At the institutional level, water and regional/agricultural policies differ in that

---

[24] Some of the most active actors are EUREAU (the European Association of Water Suppliers), ECPA (European Crop Protection Association), CEFIC (European Chemical Industry Council) and UNICE (Union of Industrial and Employer's Confederation of Europe), BEUC (European Bureau of Unions of Consumers), WWF and EEB (European Environmental Bureau).

while in the former decisions need only majority voting in the Council and co-decision with the EP (i.e. the EP has the right to veto), in the latter unanimous agreement is required, whereas the EP is constrained to a consultative role.[25] Individual MS and their vested economic interests, therefore, yield much more power in regional and agricultural policies and are able to tailor provisions to their demands. In comparison, in water policy more drastic reforms in favour of environmental goals have higher chances of success, since reactant MS cannot veto the decision; furthermore, the pro-environment EP, to which NGOs have better access (Richardson, 1997), has considerable power. Indeed, hard bargaining by the EP has been decisive in 'saving' some of the more concrete commitments of the WFD to environmental objectives (Kallis and Butler, 2001; Kaika, 2003).

Lack of integration among the policy networks at the EU level partly reflects the lack of coordination in national administrations. Ministers of Environment represent MS in water policy, whereas Ministers of Finance in regional policy and Ministers of Agriculture in the CAP. Ministers of Environment typically have a weak role in the national cabinets, especially in the southern MS. Drastic reform (such as the WFD) agreed at the EU level meets the opposition of other sectoral policies at the national level of implementation. As Richardson (1997) notes for water policy "it is relatively easy to sign up to new regulations, in the knowledge that there are so many opportunities for policy erosion at the implementation stage that it is not worth the risk of being seen as bad European by opposing the process of European integration". This is particularly true for the WFD which includes many ambiguous definitions leaving many legal loopholes that can be exploited by MS unwilling to shoulder implementation costs (Kallis and Butler, 2001).

Actors involved in policy implementation at the national and regional level are different from those involved in EU policy making. There is often limited interaction between policy formulation and implementation actors, especially in water policy, reducing the degree of cross-scale integration and contributing to implementation failures. However, this is subject to change as implementation actors come to realize the importance of EU policies and organize more at a European level (Richardson, 1997) and the EU expresses its intention to promote dialogue among actors from all spatial levels (EC, 2004a).

The network-based approach of recent EU policies creates more opportunities for environmental NGOs and under-represented interests to make their voice heard during implementation. Regional policy regulations, for example, provide for the participation of NGOs in the MCs of the OPs, while the proposed reform of the SFs establishes the creation of wide partnerships (potentially including NGOs) for the designation of plans and programmes (EC, 2004b). The new proposed regulations also set out to institutionalize an annual consultation of the Commission with NGOs as regards programme implementation (CEC, 2004). Similarly, the WFD demands the active participation of all interested stakeholders in the formulation and implementation of the RBPs. This might lead to an opening-

---

[25] It can suggest amendments, but these are not necessarily adopted.

up of both regional development and water policy networks, which in some southern MS are notoriously closed, dominated by government, engineers, construction and producer interests. Reality, however, is still far from rhetoric. In the implementation of SF, the inclusion of stakeholders and social partners in consultation and monitoring procedures is limited and construed in a way that leaves much to be desired. This has been clearly identified in the ex-post evaluation of the SF for the 1994-99 programming period (ECOTEC, 2003), and for the case of environmental NGOs there are indications that this continues to be the case during the current programming period too (WWF Adena, 2003; Liarikos, 2004).

Procedural Integration

Three issues are singled out for examination here: cross-compliance, planning provisions, and financial and economic instruments.

Cross-compliance   'Cross-compliance', referring to compliance of EU-funded actions with environmental (including water) policies, provides an important mechanism for integrating water policy provisions in regional and agricultural policy. Article 12 of the SF regulation states that "operations financed by the Fund ... shall be in conformity with ...Community policies and actions ... on environmental protection". Article 26 declares that 'major projects' (exceeding 50 MEuros) will be judged by the EC, among other factors, on the basis of whether they comply with other Community policies (CEC, 1999a). Likewise, the EAGGF regulation refers to cross-compliance with environmental regulations, demanding that MS set out verifiable standards entailing compliance with general mandatory environmental requirements and good farming practices in their RDPs (EC, 2002). Although the cross-compliance provision does not specifically address water policy, it is a definite improvement over past practices, where projects funded by the CAP or the SFs constituted some of the most important violations of Community law in the southern MS.

CAP reforms agreed in 2003 and due to be implemented beginning January 1st 2005, introduce a single decoupled payment per farm. Cross-compliance, which was a voluntary requirement and only for set-aside payments since 1992, hitherto becomes compulsory for all payments; all farmers receiving direct payments from CMOs should comply with all statutory EU water standards (EC, 2003b).

In principle, DG-Regio and DG-Agriculture should not fund projects that contravene existing water directives. The role of DG-Environment is to inform these DGs on projects that do not comply with environmental regulations. The main instrument for checking the impacts of projects on the environment (including aquatic and wetlands) is the legal requirement for an Environmental Impact Assessment (EIA).[26] Specific administrative procedures have been

---

[26] As defined in the 1985 Directive (amended in 1997).

established allowing the EC services to assess the environmental compatibility and conformity of the evaluation process in the EIA conducted in the MS (Nychas, 1998). An EIA is demanded *ex-ante* by the EC as a pre-requisite for funding 'major' and CF projects. In certain major and controversial water projects (e.g. big dams, transfers, etc.), the EC has also asked some MS for a Strategic Environmental Assessment (discussed below).

The SF regulation demands explicitly the incorporation in OPs and regional plans of

> ...an ex-ante evaluation of the environmental situation of the region concerned, in particular of those environmental sectors which will presumably be considerably affected by the assistance; the arrangements to integrate the environmental dimension into the assistance and how far they fit in with existing short and long term national, regional and local objectives ... ; the arrangements for ensuring compliance with the Community rules on the environment. The ex-ante evaluation shall give a description, quantified as far as possible, of the existing environmental situation and an estimate of the expected impact of the strategy and assistance on the environmental situation (CEC, 1999a, Art. 41b).

Ex-post monitoring of environmental indicators following OP implementation is also foreseen (ibid, art. 36). Similarly, RDPs should include an "appraisal of the expected ... economic ... impacts" and a definition of quantified indicators for evaluation (CEC, 1999b).

The 2001 SEA Directive has strengthened the provisions for ex-ante environmental evaluation of programmes (CEC, 2001), asking that an environmental report accompanies government plans and programmes, containing relevant information on the likely significant effects of implementing the plan or programme together with an examination of reasonable alternatives. Water management is one of the areas for which SEA is mandatory. The Directive requires a comprehensive assessment of possible environmental impacts, especially with respect to standards set by EU legislation. Importantly, SEA is mandatory for "plans and programmes co-financed by the European Community" and should be prepared and approved before the formal submission and approval of the plan.

The breadth of 'cross-compliance' has been expanded in practice by the EC which, in some cases, has made general support to a regional OP conditional upon compliance with environmental directives. DG-Regio has withheld funds from some regions that were not properly implementing the Habitats, Nitrates and Wastewater directives. This, however, has concerned mainly gross breaches of Community law and has been implemented in a partly *ad hoc* and inconsistent manner by the EC (WWF, 2003).[27]

---

[27] A 1992 Court of Auditors report was highly critical of the lack of coordination between DG-Environment and DG-Regio, producing many examples of contradictory policies being pursued by the two DGs, some concerning water (Court of Auditors, 1992 quoted in Richardson, 1997).

Evidently, the afore-mentioned provisions in recent SF and RDR regulations contribute to the integration of water management objectives into development policies. Still, there are many gross deficiencies as the actual coordination mechanisms between DG-Environment, DG-Regio or DG-Agriculture are far from effective. Drawing on the example of the controversial Spanish National Hydrological Plan (NHP),[28] WWF (2003) observed that there are "differences in analysis between DGs and delays as regards drawing clear conclusions and decisions to initiate infringement proceedings". In recent years, several controversial grand scale waterworks and transfers, similar to the NHP, have been petitioning funding from the SFs or the CF (e.g. the Acheloos diversion in Greece or the Alqueva dam in Portugal). The EC has responded by consultation procedures between DG-Regio and DG-Environment and the establishment of committees to assess whether the works contravened Community law. SEAs were demanded from applying MS and assistance from external experts in assessment was sought. However, these processes were far from transparent or accountable to outsiders and the public. Key procedural issues, such as the selection of experts, the setting up and decision rules of committees, consultation and resolution of differences between experts or DGs, appraisal of the scientific content of the SEAs, etc., are unregulated and developed by the EC on an *ad hoc*, case-by-case basis. This situation far from satisfies the criterion of procedural PI.

The implementation of ex-ante environmental evaluation tools is often far from satisfactory. Studies on SF implementation in Greece document that ex-ante environmental profiles in the CSF and the OPs are limited to simplified descriptions of the state of the environment in the region and a definition of very general environmental goals that are obviously insufficient for guiding policy implementation (Liarikos, 2004). The legal requirement for a mandatory SEA marks a definite improvement. However, the guidelines provided for environmental profiling and for conducting a SEA of the plans are optional and not mandatory, the exact procedures to be followed are not specified precisely neither any standards for assessing the validity of the process (c.f. the EIA directive which provides a very specific process blueprint). Research on past major water projects requiring a SEA reveals many problems with the practice of assessment such as consideration of limited and pre-defined alternatives, political manipulations, poor use of science and limited participation (Antunes et al., 2002). The more 'programmatic' (rather than project) nature of OPs or RDPs may make the application of SEA even more difficult and subject to manipulation.

Much of the responsibility to check cross-compliance is delegated to MS following the subsidiarity principle. Often southern MS see environmental rules as constraints imposed by Brussels, rather than as considerations to be integrated in development programmes. 'Tricks', such as breaking major projects into several

---

[28] Initial plans (now dropped) included several dams and a major inter-basin transfer, probably contravening both the WFD and the Habitats and Birds directives, and were submitted to DG-Regio for co-financing (del Moral et al., 2002).

smaller ones to avoid the requirement for an ex-ante EIA/SEA, are often used (WWF, 2003). 'Whistle-blowing' from environmental groups can draw the attention of the EC in cases of law violations. However, due to the limited resources of both EC and environmental NGOs, attention focuses only on 'big' and emblematic cases (such as the Spanish NHP). Several smaller regional projects, however, may contradict the goals of water policy.

Whereas cross-compliance with quality standards is relatively straightforward, checking cross-compliance with a framework directive such as the WFD is a more complex task. For example, the impact of a single farm on the status of a water body cannot be isolated from the impacts of other polluting activities. Indeed, the WFD is not included in the statutory obligations upon which farmers' compliance should be appraised.[29] Cross-compliance will, therefore, need to refer to standards or restrictions set by national laws transposing the WFD or by the specific standards and rules set by the RBAs and RBPs. In many southern MS, however, there are concerns that laws and plans will never go so far as to pose real limitations on farmers. It is not clear whether and how cross-compliance provisions can be applied by the EC since it has neither a reference framework nor an objective assessment mechanism. In theory, the requirement of 'good farming practice' can provide a benchmark upon which to judge general farm compliance. Still this is much less specific than exact qualitative or emission standards; clear guidelines are needed on what constitutes 'good practice' and for a variety of cases/contexts.

Essential PI requires much more than mere regulatory cross-compliance, however. WWF (2003) notes for the Nitrates directive[30] that "work on the cross-compliance mechanism has resulted in funds being withheld in certain cases but did not lead to a comprehensive strategy to integrate the directive's requirements with regional development plans". Development policies produce broader socio-environmental change. The SEA and the environmental profiling requirement, as they stand now, seem limited in addressing the longer-term implications of development policies for the aquatic environment.

Planning provisions   The decentralization and territorialization of EU policy implementation and the emphasis on programming (SFs and RPDs) and planning (WFD) suggest that programme/plan preparation is a key mechanism for procedural PI. The EU cannot control the outcome of these activities, but sets procedural requirements that can foster integration, such as the mandates for public consultation in the design of regional and rural programmes and for including environmental actors in committees and project partnerships. However, until now participation of environmental actors in the design and implementation of regional programmes has been limited to formal actors, such as Ministry of Environment representatives. Their role in the process has been marginal, reflecting their limited

---

[29] The Nitrates directive, however, is (EC, 2003b).
[30] One of the least implemented water directives in the EU.

power in national administrations. Consultation with environmental and citizens groups has been *ad hoc* at best (for the case of Greece, see Liarikos, 2004).

River basin planning provides a key platform for PI. As the preparation of RBPs has just started, it is early to assess the extent to which development and water management goals will be integrated in the process. The following discussion explores some of the opportunities and potential problems for PI.

The WFD is designed so as to integrate water-related goals into development decisions. Integrated river basin planning and management are the main vehicles for this. First, all new major developments in a river basin that may affect the condition (status) of water bodies and aquatic ecosystems should be authorized and included in the RBP (CEC, 2000). Clear justification should be given when being exempted from the requirements of the WFD. Authorizations are explicitly needed for water abstractions or impoundment works (dams, transfers, etc.); controls may be applied if these affect negatively the water body or result in groundwater over-abstraction causing 'significant' damage to dependant terrestrial ecosystems. Similar authorizations and controls can be required of polluting activities or changes that may affect the hydromorphology of a river (e.g. drainage works, land use changes, etc.). Most of the physical interventions funded through the CSF or the CAP will require such authorization and inclusion in the RBPs. This comes on top of requirements for cross-compliance with existing water and environment directives.

Second, river basin planning is an open process. The broad participation of stakeholders provides a procedural arena in which various water management and development interests[31] can debate policies and decisions. Many MS have established permanent multi-stakeholder river basin Councils (with varying degrees of decision-making power) and also foresee wider consultation processes for the authorization of plans or specific decisions/authorizations (EEB, 2004).

The reverse integration, i.e. of development policy goals into water management decisions, is less explicit. Permitting derogations[32] from environmental objectives on cost grounds is an indirect form of integration. The EC recognized that the WFD could not go as far as to institutionalize water (and the aquatic environment) as *the* limiting factor to development. Strictly adhering to a 'no deterioration' principle would practically put an end to all physical development projects. A more pragmatic approach was to group desired environmental objectives in broad classes ('status'), prohibiting deterioration of status (not individual standards) in principle, and allowing a certain degree of freedom to RBAs in judging when derogations can apply on economic grounds (Kallis and Butler, 2001). The WFD provides that MS may aim to achieve less stringent environmental objectives when their achievement is 'infeasible' or

---

[31] E.g. water utilities, irrigation associations, water and environment agencies, development agencies, farmers' and industry representatives, environmental NGOs, etc.

[32] These include lower standards, authorization of damaging projects, deviation from the full costing principle, etc.

'disproportionately expensive'. Similarly, waterworks or other physical interventions with negative impacts on the aquatic environment are allowed when the reasons are of 'an overriding public interest' and/or the benefits of the new modifications to human health and safety or to sustainable development outweigh the benefits to the environment and society from the environmental objectives. 'Sustainable human activities', a term whose interpretation is left wide-open,[33] can also be allowed even if environmental standards are violated, as long as the benefits of the projects outweigh the costs of violation.

The stance of the EC is that, according to the principle of subsidiarity, deciding on derogations and the development-environment trade-offs they entail cannot be resolved from the top; they have to be debated at the river basin level with the procedural tools provided by the WFD, benefiting, hence, from a procedural integration of different actors and their perspectives. There are, however, concerns on how the actual procedures (e.g. Councils, public consultation, etc.) will work in practice. For example, the Spanish NHP, despite its clear breach of some water and environment directives, was approved by both national and river basin water Councils, reflecting the dominant influence of the composition of the Councils (government official and economic interests) (Kallis, forthcoming). "Conversation" between policies and competing interests, that is, depends on the real democratization of the decision process.

Apart from constraints, there are also potential positive synergies between river basin planning, SF programming and RDPs. Several environmental measures funded by the SF or the CAP, for example, can contribute to the achievement of river basin goals (e.g. AEMs) and (should) be included as part of the river basin "programme of measures". Good farming practice codes are also a shared instrument between agricultural and water policy. In agricultural policy, they are used to judge the compliance of farmers receiving support. In water policy, they are mandatory in nitrate-sensitive zones and supplementary in river basins where agriculture is a main source of pollution. A certain harmonization between the requirements of codes as part of the two policies might be needed to avoid contradictions or duplication of efforts (EC, 2003b).

Despite the obvious interrelationships and synergies among the three planning processes there is no concrete EC initiative to push forward such procedural integration. A discussion paper by DG-Environment for the integration of the CAP and the WFD (EC, 2003b) suggests that competent authorities for rural and regional development should be involved in the drafting of RBPs and vice versa. Representatives should also have a more permanent position in related bodies (e.g. representatives of OPs partnerships having statutory position in RBAs and vice versa). The EC, however, concedes that it cannot prescribe the details of such integrative procedures because of the subsidiarity principle (EC, 2003b). Current experience with the incorporation of environmental goals and actors in SF

---

[33] Thought to refer, for example, to projects for navigation, clean energy production including hydropower, etc, although this is nowhere specified in the text.

implementation in southern MS suggests that such cross-representation of actors will be very difficult unless the EC provides clear guidance and a certain degree of commitment (WWF, 2003; Liarikos, 2004). There is a certain imbalance so far, as economic actors seem to dominate in the river basin planning process, whereas the opposite is not true for water and environment actors in regional and rural development planning.

Financial and economic instruments EU regional and agricultural policies provide important complementary financial resources for the implementation of water policy, especially in southern MS, as the WFD does not have its own financial instrument. The SFs can be used to co-finance some of the measures required to achieve "good ecological and chemical status" (WWF, 2003). Already, the SFs and the CF finance some investments that have positive impacts on water quality (e.g. construction of sewerage treatment plants). EC (2003b) identifies RDR measures that can contribute to the achievement of the WFD objectives: "investment in agricultural holdings" can subsidise procurement of water-friendly equipment;[34] "setting up of young farmers" can provide aid to comply with WFD standards; "training" can be used to educate farmers on good farming practices; "early retirement" aid can be given to reassign agricultural land with negative impacts on water resources to non-agricultural uses; compensation can be given to farmers in LFAs facing restrictions due to WFD standards. The most useful measures for implementing the WFD, however, are the AEMs (EC, 2003b). These can contribute to good farming practices through reduced use of agro-chemicals and water, protection of aquatic ecosystems, growing of catch crops and buffer strips along surface waters, etc.

However, the use of CAP and SF instruments to support WFD goals is incidental and much depends on the will of MS, regional authorities or individual farmers. PI can be enhanced by referring explicitly to the WFD in SF, the RDR or guidance documents and 'earmarking' programmes or funds' quota for the implementation of WFD measures (WWF, 2003; EC, 2003b). This would require more cooperation between the respective DGs and perhaps a delegation of powers to DG-Environment (directly or through inter-directorate committees) in the design of funding programmes and the rules and selection of eligible projects.

Past experience suggests that the actual use of funding instruments by MS may create important problems. Water investments, as part of environmental protection up to now, have focused primarily on infrastructure works. These sometimes have negative environmental effects. For example, the CF has financed the repair and enlargement of the aqueduct of Athens, Greece, and the extension of the distribution network to the periphery as an 'environmental investment'. This subsidization may have acted, however, against a longer-term, cautious, demand-side management of water for the city (Kallis and Coccossis, 2003).

---

[34] E.g. drip irrigation, machinery to spread pesticides better, etc.

Even worse, some MS have exploited environmental funds to finance conventional infrastructure works. For example, several small projects have been financed for the regeneration of Lake Karla (Greece) as part of implementing the Habitats directive (and in line with the demands of the WFD for restoration). Critics, however, argue that these projects served to create a new reservoir for irrigating the surrounding area (that otherwise could not get EU support) (Liarikos, 2004). RDP measures such as reforestation have been used in some cases for planting water-intensive trees that do not fit the local landscape and go counter to water management objectives (Georgiadis, 2004). Good intentions for PI at the EU level may, therefore, dissolve during implementation. This points again to the trade-offs between the benefits and drawbacks of subsidiarity and decentralization in the growth-oriented, southern European regions.

Although the WFD has an important procedural orientation, funding is available for physical interventions primarily. Few provisions exist for financing procedures, such as public participation, river basin planning, etc.[35]

Water pricing for cost recovery is an integrative economic instrument provided by the WFD whose inclusion in the RBPs is mandatory. This includes "environmental and resource costs" in order to "provide adequate incentives for users to use water resources efficiently, and thereby contribute to the environmental objectives of the directive" (CEC, 2000, art.9). Disaggregate pricing per user is foreseen, taking into account "social, environmental and economic effects of the recovery as well as the geographic and climatic conditions of the region" (CEC, 2000, art.9). This 'internalization' of water and environmental monetized costs in the cost of development aims to integrate water policy objectives into development policies. The potential tensions between the cohesion goals of CAP/SFs subsidies and the efficiency goals of cost recovery have been discussed before.

Integration of EU Water and Development Policies: A Plausible Expectation?

The picture emerging from the preceding analysis of the current status of integration between EU water and development policies is mixed. Positive signs that the process has started do exist but the difficulties encountered raise questions about the plausibility of expecting an enhanced level of PI in the mid-term (say next ten years) especially amidst changing socio-political conditions. The key findings of the analysis and the prospects for PI are discussed below in the light of recent policy changes. Suggestions for selected necessary improvements are offered.

---

[35] Funding is, however, necessary to support the effective participation of underrepresented groups (WWF, 2003).

Substantive and Goal Integration

There is a notable convergence in the theories and approaches to development underpinning the three policies and, by extension, a greater degree of compatibility among their respective goals. An increasing emphasis on sustainable development in water policy goes well with the focus of regional policy on qualitative, multi-faceted regional development and its renewed interest on environmental quality. There is also an ongoing, albeit hesitant, shift of agricultural policy from sectoral support to a broader, multi-objective rural development emphasising, among others, environmental protection. Decentralization and territorialization are noted in all three policies, opening more opportunities for horizontal coordination and integration at the implementation level.

Much of this progress remains superficial and fails to materialize on the ground, however, especially in 'growth-thirsty' Mediterranean regions. Despite convergence under the apparent inclusiveness of new notions of (sustainable) development, in many cases important trade-offs between the goals of economic growth, social cohesion and protection of the aquatic environment are still made. This becomes evident at the EU level (e.g. financing of water transfers or agricultural supports), and much more, at the implementation level. Making sustainable development an explicit objective of regional and rural development policies would formalize conceptual convergence, but would not necessarily secure more integrated implementation. Much will still depend on the *interpretation* and *operationalization* of sustainable development whose very nature defies monolithic interpretations, requiring instead procedures for mediating and harmonising different objectives and interests. This directs attention to the issue of procedural integration.

Territorialization of the three policies also creates fresh mismatches and institutional misfits. Rather than recommending a deceptively perfect fit of the territorial and administrative levels of reference of the three policies, emphasis should rather shift towards providing flexible yet effective enabling mechanisms of cooperation and coordination between river basin, rural development and regional development planning and the associated administrative services.

In terms of policy approach, decentralization, emphasis on procedural (rather than substantive) requirements and the devolution of implementation responsibilities from the EC to the MS or the regional levels (subsidiarity) could, in theory, allow for more flexible and case-tailored implementation but may not necessarily lead to greater PI during implementation. Reality cautions against such undue expectations; Mediterranean MS often view environmental regulation and related restrictions as a burden imposed by Brussels and may seek to exploit the freedom given in the name of subsidiarity to water down implementation costs. A deficiency that should be addressed is the lack of concrete, detailed and binding guidance on the part of the EC on procedural requirements[36] and the lack of related inspection mechanisms.

Finally, although the need for PI *per se* is recognized in the Union's Treaty and referred to, more or less explicitly, in all three policies, this is not so much a call for a comprehensive, horizontal and vertical, PI, but rather for the narrower EPI. Not one policy operationalizes even the EPI goal into concrete objectives and actions. Existing internal EC documents on PI (e.g. EC, 2003b) could form the backbone for a more explicit and binding document on PI with concrete objectives, actions, implementation time-tables and funding provisions.

Procedural and Actor Integration

New procedural provisions and mechanisms provide enhanced opportunities for PI. Cross-compliance and inter-departmental mechanisms serve to limit the negative impact of EU-financed regional or agricultural developments on the aquatic environment. Such provisions could be further strengthened if cross-compliance applied not only to specific programmes and projects but also to the actual recipient (region, farmer, etc), i.e. binding all EU support to general compliance with environmental (including water) regulation. River basin planning and regional/rural development programming offer opportunities for multi-stakeholder interaction, hence, improving the prospects for actor integration. Moreover, the provision for consultation processes has provided opportunities for a gradual opening-up of previously closed and rigid policy networks.

On the other hand, considerable problems still haunt the proper implementation of the cross-compliance principle. Some EC decisions for financing controversial water projects have generated criticism, however. The lack of an explicit, observable and transparent process of inter-departmental consultation and decision on such controversial projects is a major problem. Clear processes for deciding on cross-compliance should be established at the level of EU institutions, preferably in an official document, including definition of consultation and decision-making rules between DGs, clarification of the role of experts and committees, specific provisions for transparency of the decision-making process, and concrete duties of the EC relating to the justification of decisions.

The same deficit is observed at the implementation level. Few provisions exist to ensure the actual opening-up of planning/programming processes and especially the integration of water/environment objectives into regional and rural development planning. The EC could elaborate and define specific procedural requirements for SFs, CF, RDPs and RBPs as mandatory guidance to MS, accompanying them with strict compliance checks. Furthermore, integration between the different planning processes could be strengthened through the preparation of a guiding working document by the EC suggesting ways and tools to

---

[36] E.g. for drafting river basin or rural development plans, independent mechanisms and indicators to assess the quality of the plans and the planning process.

integrate the river basin, regional and rural development planning processes and outputs.

A notable improvement is the increase in regional and agricultural policy programmes and funds directed to environment-enhancing measures. However, such funding should be directed to more specific and explicit EU regulation-related activities (e.g. funding to support specific implementation tasks of the WFD in river basins) with a more active role of DG-Environment (e.g. in drafting WFD-targeting programmes in SFs and RDPs, approving and monitoring projects, etc.). It is essential to tighten inspection and enforcement mechanisms to ensure that water-related regional or rural development programmes and projects actually deliver on the ground.

Expectations

All three policies are currently in a state of change. The WFD was transposed into national legislation in 2003 and the process of defining ecological standards and preparing RBPs and programmes of measures has just started (to be completed by 2009). The new CAP (2007-2013) should be agreed by 2005. In 2004, a mid-term review of the RDR will take place concluding with the approval of the new regulations in 2006. The next round of SFs for 2006-2013 will soon be decided.

If PI is to be seriously promoted, bolder are urgently needed in the coming policy reforms. These concern both substantive and procedural aspects since substantive differences can only be resolved through procedural mechanisms and vice versa; procedures can operate only on a minimum shared substantive framework of goals and objectives. Changes, therefore, should both promote an integrated vision of sustainable development, especially in rural, water-poor, southern EU regions, and enhance the procedural aspects of streamlining the horizontal and vertical linkages of EU and national level networks of formal and informal actors and of designing appropriate policy instrument mixes.

Achieving such convergence will depend importantly on socio-political conditions. In particular, developments in the European integration process, as reflected in the debate over subsidiarity, and the approach taken towards economic development in relation to sustainability, two highly contested, theoretically and politically, issues, will be decisive. Concerning subsidiarity, the prospects for PI in Mediterranean MS will benefit from a not-too-decentralized, yet flexible policy implementation approach. The capacity of the EC to monitor policy implementation (funds, plans, etc.) and ensure compliance is critical. In terms of development, PI will benefit from greater permeation of the concept of sustainability from environment-only into conventional economic policy spheres. It is essential that sustainable development, however loosely defined, gradually becomes a standard economic blueprint for the EU guiding all policies (and not only environmental ones).

The integration of EU water with development policies will most probably contribute to combating desertification in the sensitive regions of Mediterranean

Europe, at least indirectly through better management of their scarce and variable water resources and the promotion of development patterns fit to the available supply of water resources in the region. However, since the primary emphasis of the WFD is on water quality rather than on quantity, much will depend on the guidance the EU will offer and the requirements it will promulgate for the coordination of the procedures governing the preparation of river basin, rural development and regional plans and programmes. Inevitably, however, the success of any integration effort will be played out at the implementation stage where conflicts among competing interests, jurisdictions and established rules of resource use are resolved and the subsidiarity principle renders predictions pointless and futile!

References

Antunes, P., Videira, N., Kallis, G., Santos, R. and Lobo, G. (2002), 'Integrated evaluation for sustainable river basin governance in Europe', in Proceedings of *Third Iberian Congress on Water Management and Planning*, University of Seville, Seville, pp. 161-6.

Bressers, H. and Kuks, S. (2002), *European water regimes and the notion of a sustainable status: Case study comparison*, EUWARENESS project report, CSTM, University of Twente, Twente.

CEC (2001), Directive 2001/41/EC on the assessment of the effects of certain plans and programmes on the environment, Council of the European Communities.

CEC (2000), *Directive of the European Parliament and of the Council: Establishing a framework for Community action in the field of water policy*, Council of the European Communities 2000/60/EC, (23 October 2000), L 327/1, (22.12.2000), Official Journal.

CEC (1999a), *Council Regulation (EC) No 1260/1999 laying down general provisions on the Structural Funds*, Council of the European Communities Official Journal of the European Communities, L 161/1, (26.6.1999).

CEC (1999b), *Council Regulation (EC) No 1257/1999 on support for rural development from the European Agriculture Guidance and Guarantee Fund (EAGGF) and amending and repealing certain Regulations*, Council of the European Communities Official Journal of the European Communities, L 160/80, (26.6.1999).

Court of Auditors (1992), *Special Report no 3/92 Concerning the Environment, together with the Commission's replies*, Official Journal, 92/C245/01, Vol. 35.

Chatellier, V. and Daniel, K. (2001), 'Direct Payments to European Agriculture: Territorial Impact of Reforms', Paper presented in the 73rd Seminar of the EAAE entitled: Policy Experiences with Rural Development in a Diversified Europe, 28-30 June 2001, Ancona.

Collier, U. (1997), 'Sustainability, Subsidiarity and Deregulation: New Directions in EU Environmental Policy', *Environmental Politics*, Vol. 6, No. 2, pp. 1-23.

Daly, H.E. (1999), *Ecological Economics and the Ecology of Economics. Essays in Criticism*, Edward Elgar, Northampton, MA, USA.

Daugbjerg, C. (1999), 'Reforming the CAP: Policy Networks and Broader Institutional Structures', *Journal of Common Market Studies*, Vol. 37, No 3, pp. 407-28.

De Graaf, H.J., Musters C.J.M., and Ter Keurs W.J., (1996), 'Sustainable Development: Looking For New Strategies', *Ecological Economics,* 16, pp. 205-16.

Del Moral, L., Pedregal, B., Calvo, M. and Paneque, P. (2002), *The River Ebro interbasin water transfer,* ADVISOR Project Report, University of Seville, Seville.

ECOTEC (2003), *Ex-Post Evaluation of the Objective 1 1994-1999: A Final Report to the Directorate General for Regional Policy*, European Commission, Ecotec Research and Consulting, Birmingham.

European Commission (2004a) *Third report on Economic and Social Cohesion.* COM(2004) 107, Luxembourg, Office for Official Publications of the European Communities.

European Commission (2004b), Proposal for a Council Regulation on support for rural development by the European Agricultural Fund for Rural Development (EAFRD), COM(2004) 490 final, European Commission, Brussels.

European Commission (2003a), *Carrying forward the Common Implementation Strategy for the Water Framework Directive – Progress and Work Programme 2003/2004*, Strategic Document, Directorate for Environment, Brussels.

European Commission (2003b), *The Water Framework Directive (WFD) and tools within the Common Agricultural Policy (CAP) to support its implementation*, Working Document, DG ENV.B.1/BB D (2002), Directorate for Environment, Brussels.

European Commission (2002), *Commission Regulation No 445/2002 laying down detailed rules for the application of Council Regulation (EC) No 1257/1999 on support for rural development from the European Agricultural Guidance and Guarantee Fund (EAGGF)*, Official Journal of the European Communities, L 74/1, (15.3.2002).

European Commission (2001), *Common Strategy on the Implementation of the Water Framework Directive*, Strategic Document, Directorate for Environment, Brussels.

EEA (1998) *Europe's Environment. A Second Assessment*, European Environment Agency, Copenhagen.

Engwegen, J. and McLaren, G. (1998), 'Sustainable Regional Development and Community Structural Funds, and the eastern Scotland Experience', in "Proceedings of the International Workshop Series on Sustainable Regional Development. Regions-Cornerstones for Sustainable Development", Graz, Austria, October 28-30, 1998, pp. 87-95.

Georgiadis, N. (2004), 'The use of Pseudoacacia in the implementation of the "forestation of agricultural land" measure of the RDR in the Prefecture of Evros, Greece: Investigation, utilization, potentials and spread of the species', unpublished research report, WWF Greece, Athens, (in Greek).

Global Water Partnership (2000), *Integrated Water Resources Management,* TAC Background papers no 4, Technical Advisory Committee, GWP, Stockholm.

Jaspers, F.G.W., (2003), 'Institutional arrangements for integrated river basin management', *Water Policy*, 5, pp. 77-90.

Jewitt, G. (2002), 'Can integrated water resources management sustain the provision of ecosystem goods and services?', *Physics and Chemistry of the Earth*, 27, pp. 887-95.

Jonch-Clausen, T. and Fugl, J. (2001), 'Firming up the Conceptual Basis of Integrated Water Resources Management', *Water Resources Development*, Vol. 17, No. 4, pp. 501-10.

Kaika, M. (2003), 'The Water Framework Directive: a new directive for a changing social, political and economic European Framework', *European Planning Studies*, 11, 3, pp. 299-316.

Kallis, G. (forthcoming), 'Beyond limits and efficiency, what? Assessing developments in EU water policy', *International Journal of Water*.

Kallis, G. and Butler, D. (2001), 'The EU Water Framework Directive: Measures and Implications', *Water Policy*, 3(3), pp. 125-42.

Kallis, G. and Coccossis, H. (2004), 'Theoretical reflections on limits, efficiency and sustainability: implications for tourism carrying capacity assessment', in H. Coccossis, and A. Mexa (eds), *The Challenge of Tourism Carrying Capacity Assessment: Theory and Practice*, Ashgate, Aldershot, pp. 15-35.

Kallis, G. and Coccossis, H. (2003), 'Managing water for Athens: from the hydraulic to the rational growth paradigm', *European Planning Studies*, 11(3), pp. 245-61.

Kallis, G. and Nijkamp, P. (2000), 'Evolution of EU Water Policy: a critical assessment and a hopeful perspective', *Journal of Environmental Law and Policy*, 3/2000, pp. 301-35.

Kraemer, R.A. (1998), 'Subsidiarity and water policy', in F.N. Correia, (ed.), *Selected Issues in Water Resources Management in Europe*, A.A.Balkema, Rotterdam.

Lenschow, A. (2002), 'Greening the European Union: An Introduction', in A. Lenschow (ed.), *Environmental Policy Integration: Greening Sectoral Policies in Europe*, Eathscan, London, pp. 103-26.

Liarikos, C. (2004), 'Regional Policy in Greece: Brief Overview, Description of the 3rd CSF and an Analysis of its Relevance to Environmental Protection', World Wide Fund for Nature, WWF-Greece, Working Paper 3/2004, Athens (in Greek).

MEDACTION (2004a), *Module 4: Design of a Desertification Policy Support Framework*, Deliverables 35, European Commission, DG-XII, Contract No. ENVK2-CT-2000-00085, (www.icis.nl/medaction).

Moss, T. (2003), 'Regional Governance and the EU Water Framework Directive: Institutional Fit, Scale and Interplay', in W. Lafferty, and M. Narodoslawsky (eds), *Regional Sustainable Development in Europe. The Challenge of Multi-Level Co-operative Governance*, ProSus, pp.181-206.

Mumford, L. (1961), *The City in History*, Fine Communications, New York.

Nijkamp, P. and Dalhuisen, J. (2002), 'Enhancing efficiency of water provision, Theory and Practice of Integrated water management principles', Proceedings of the 5th International Conference EWRA, Water Resource Management in the era of transition, 4-8 September, Athens, pp. 1-17.

Norgaard, R. (1995), 'Metaphors we might survive by', *Ecological Economics*, Vol. 15, pp. 129-31.

Nychas, A. (1998), 'Environmental Assessment of Structural Funds interventions', in "Proceedings of the International Workshop Series on Sustainable Regional Development. Regions-Cornerstones for Sustainable Development", Graz, Austria, October 28-30, 1998, pp. 108-16.

OECD (1990), *Water resource management; integrated policies*, Organization for Economic Co-operation and Development, Paris.

OECD (2001), *Multifunctionality: Towards an Analytical Framework*, Summary and Conclusions, OECD, Paris.

Richardson, J. (1997), 'EU water policy: uncertain agendas, shifting networks and complex coalitions', *Environmental Politics*, Vol. 3, 4, pp. 139-67.

Roberts, P. (2003), 'Partnerships, programmes and the promotion of regional development: an evaluation of the operation of the Structural Funds regional programmes', *Progress in Planning*, 59, pp. 1-69.

Roberts, R.S. and Emel, J. (1992), 'Uneven Development and the Tragedy of the Commons: Competing Images for Nature-Society Analysis', *Economic Geography*, 68 (3), pp. 249-71.

Smith, R.D. and Maltby, E. (2003), Book review: Using the Ecosystem Approach to Implement the Convention on Biological Diversity: Key Issues and Case Studies, IUCN, Gland, Switzerland and Cambridge, UK.

Vlachos, E. (1982), 'Socio-cultural Aspects of Urban Hydrology', in P. Laconte and Y.Y. Haimes (eds), *Water Resources and Land-use Planning: a Systems Approach* Matrinus Nijhoff, The Hague.

Ward, N., Lowe, P. and Buller, H. (1997), 'Implementing European Water Quality Directives: Lessons for Sustainable Development', in S. Baker, M. Kousis, D. Richardson, and S. Young, (eds), *The Politics of Sustainable Development*, Global Environmental Change Series, Routledge, London.

Wittfogel, K.A. (1957), *Oriental despotism; a comparative study of total power*, Yale University Press, New Haven.

Worster, D. (1985), *Rivers of empire; water, aridity and the growth of the American West*, Pantheon Books, New York.

WRc (2001), *Study on the impact of environment-water policies on economic and social cohesion*, WRc, Bucks.

WWF Adena (2003), 'Comparative study for the integration of the environment in the 2000-2006 Structural Funds Programmes in Various Member States of the EU', Report to the Spanish Ministry of Environment by WWF Adena, Project Coordinator: Guy Beafuy (in Spanish, appendices in English).

WWF (2003), *Structural Funds in an enlarged EU. Learning from the past-Looking to the future*, WWF, Brussels.

Zalidis, C.G and Gerakis, A. (1999), 'Evaluating Sustainability of Watershed Resources Management Through Wetland Functional Analysis', *Environmental Management*, Vol. 24, No. 2, pp. 193-207.

Chapter 7

Biodiversity Policies in the European Union: Achieving Synergies with Other Policies?

Vassilis Detsis

Introduction

Policies to address biodiversity conservation date back to the second half of the 19th century when the first protected areas were established in the USA. Since then biodiversity policy grew in importance, widened its scope and about a century later biodiversity issues began to appear in international agreements. A landmark in this process is considered the 1992 UN Conference on Environment and Development held in Rio de Janeiro, the Rio Summit, which dealt simultaneously with a number of environmental issues. A result of the Summit was the Convention on Biological Diversity (CBD) that declared the determination of contracting parties "to conserve and sustainably use biological diversity for the benefit of present and future generations" (CBD Preamble).[1]

This Summit produced two more Conventions: the Convention to Combat Desertification and Drought (CCD) and the Framework Convention on Climate Change (FCCC). It was evident from the beginning that these Conventions might instigate potentially overlapping actions. A joint liaison group for the three Conventions was established and a Memorandum of Understanding was signed between the Secretariats of CCD and CBD (CBD, 2000a).[2] In the CBD "Program of Work in Dry and Subhumid Lands", provisions were made for cooperation with the CCD (CBD, 2000b). The imperative of achieving synergies between actions undertaken in the frame of these Conventions found its way down to subglobal

---

[1] The purpose and approach of the Rio Summit in general and the CBD in particular are not free from criticism. Critics argue that the Conventions bear the stamp of large economic interests ((Haila, 1999; Adger et al., 2001 and references therein) and that the CBD specifically provides the institutional frame for biodiversity trade; biodiversity protection is not necessarily its main goal (Brand and Görg, 2003).

[2] Similar moves to promote cooperation and coordination have been taken with respect to other international agreements (CBD, 1998).

levels in texts that arose as a response to these conventions, such as the EU Biodiversity Strategy (EUBS) (EC, 1998a) and relevant national documents.

Within the European Union biodiversity-specific legislation has existed since as early as 1979 (the Birds Directive). However, biodiversity policy is a recent development compared to other environmental policy sectors, such as pollution control, which were seen as more relevant to the issues surrounding the creation of the Common Market (Ledoux et al., 2000).[3] Initially the EU adopted a mainly regulatory approach to biodiversity policy making (Ledoux et al., 2000; Fairbrass and Jordan, 2001),[4] exemplified by the Birds and the Habitats Directives that aim to bring habitats and species of Community interest to a "satisfactory conservation status". More recently another approach emerged; that of integrating environmental concerns, including biodiversity, into sectoral policies. As a result, four Action Plans appeared in 2001 in the frame of the EUBS aiming to promote sectoral integration of biodiversity concerns (EC, 1998a; EC, 2001).

Unlike the integration of environmental concerns in EU sectoral policies in general, the integration of biodiversity has received little attention in the literature. Some case studies are available addressing, among others, the integration of biodiversity concerns into economic activities at the local level associated with public funding for agri-environmental measures (AEMs).[5] This chapter aims to contribute to the policy integration (PI) discourse in the context of the EU biodiversity policy. Its starting point is that integration between policies is a more appropriate form of PI than that of integrating biodiversity concerns into sectoral policies. This would facilitate coordinated interventions in the appropriate geographic areas and at the proper spatial scales, would avoid the suboptimal use of the limited means provided in the frame of each separate policy and would improve the effectiveness of mitigation efforts. Mitigation of biodiversity decline and desertification are, by definition, environmental issues concerning rural areas primarily. Therefore, policies affecting these areas are relevant in this discussion. Of the several EU policies directly or indirectly related to the rural environment, the EU rural development policy and its integration with the EU biodiversity policy are examined focusing, where needed, on the desertification-sensitive rural areas of

---

[3] The development of a 'pure' biodiversity approach is attributed to the implementation of the CBD in the EU (van Dijk, 2001).

[4] The lack of financial resources in the environmental field in the EU has been a powerful force to resort to regulatory approaches in policy making (Ledoux et al., 2000).

[5] A German case study suggested that success in environmental and economic terms depended on the combination of a number of favourable conditions, including policy measures that could be properly utilized on site (Knickel, 2001). In a UK site, Mills (2002) found that the necessary management measures in the frame of a local biodiversity action plan would have a positive and multiplicative effect on the local economy and on employment. In another UK study, Harrison et al. (1998), studying the perception held by locals of agri-environmental measures targeting biodiversity, point to an important complication arising from the difficulty of quantifying biodiversity and from the difficulty to predict the exact biodiversity-related outcome of distinct management practices.

the Southern EU Member States (MS). The major question addressed is whether the two policies move in the direction of becoming integrated, what reasons might account for the (positive or negative) outcome and what are the implications as regards the combat against desertification.

The next section briefly discusses the determinants of biodiversity decline and its relationship to desertification in the northern Mediterranean. The third section outlines the theoretical and methodological framework adopted for the analysis of policy integration. The fourth section presents the EU biodiversity and the EU rural development policy. The fifth and the sixth section analyze their integration at the general policy level and at the level of a particular rural policy instrument respectively. The last section summarizes the discussion, presents tentative conclusions as regards the integration of PI the two policies and negotiates the implications of the findings for the desertification issue. It also offers some proposals and future research directions.

Biodiversity Decline and its Determinants

Compared to most other parts of Europe, Mediterranean biodiversity is high owing, among others, to the geographic location, geological and climatic history, geomorphology, and the characteristic mosaic of small-scale variable land uses of the region. It has been argued that this small-scale variability, the result of human-nature interactions over time, has produced land use patterns fit to the environmental conditions prevailing in the Mediterranean during its recent history. In the last decades, however, socio-economic changes have put this land use pattern under pressure (van der Leeuw, 1999). New, economically attractive but environmentally unsustainable, forms of land use and natural resources management compete with traditional, more environmentally sustainable, uses of land that are no more economically viable, and have led to biodiversity decline and desertification in the northern Mediterranean. Both phenomena have a set of more or less common determinants, including abandonment of agriculture in marginal hinterlands, intensification of agriculture on productive lands, and coastal urban and tourism development, that are driven by deeper socio-cultural and economic forces (van der Leeuw, 2004; Firmino, 1999; Moreira et al., 2001). The case of agricultural land use is discussed in some detail as it is more pertinent to the theme of this chapter.

Agricultural land management is characterized by the contrast between intensively managed productive land and underutilized or abandoned marginal land (Potter and Goodwin, 1998). This is the result, among others, of changes in agricultural product prices under the influence of market forces, agricultural policies, international trade negotiations, and urbanization trends generated by the search for better livelihood opportunities (in terms of employment, health care, education, etc.). Changes in agricultural land use and management practices are leading to biodiversity decline that has caused widespread concern about the relationship between farming and biodiversity. Some have viewed the CAP reform as an opportunity to reconcile agriculture and the environment (including biodiversity). However, the CAP reform was primarily driven by international trade negotiations, which have questioned the level and form of agricultural subsidies, and not by environmental concerns (Roederer-Rynning, 2003). The ongoing process of trade liberalization will most probably reduce the prices of agricultural products in the future as well as the support for agricultural production. Optimists suggest that this will result in production extensification and increased availability of financial resources for AEMs. However, a drop in prices may not necessarily lead to extensification but it may as well spur further intensification in areas suitable for intensive agriculture (Potter and Goodwin, 1998; Bongaerts, 2000). The productivity of farming is still considered sub-optimal and farming could be further intensified indeed in terms of single crop production (Rabbinge and van Diepen, 2000). Even if extensification (as regards crop management) occurs, its beneficial effects on the environment will depend on the degree of extensification and the kind of farm management adopted.

Reduced prices, on the other hand, will probably accelerate the process of marginal farmland abandonment. Most agricultural land that still hosts valuable biodiversity is exactly land considered marginal from an economic point of view (Bongaerts, 2000). Furthermore, financial resources made available through production support reductions may not be entirely, or even largely, redirected to AEMs, an approach that may not be realistic in the long term altogether (Rabbinge and van Diepen, 2000). Thus, it seems that the current trends of agricultural land use and management practices associated with biodiversity decline and desertification will persist in the near future, at least.

Similar arguments apply to other processes linked to these phenomena, such as urbanization. The socio-economic changes that took place in Europe in the recent past have led to concentration of population and economic activities in major urban centers. In the Mediterranean, these concentrations are particularly intense in the coastal zone. Urban areas, and not rural areas, are now the major place of production. Consequently, the populations of the most European countries, including the southern ones, are not predominantly farmers any more. These trends, too, are not considered reversible at present.

The links between desertification and biodiversity are indirect and draw mostly on their shared determinants that were outlined above. Detsis and Briassoulis (2004) have suggested that efforts to achieve synergies among biodiversity and desertification-related policies in the Northern Mediterranean should be based on addressing these shared determinants of biodiversity decline and desertification. Before examining if and how this suggestion can be realized, it is useful to highlight some more general similarities between biodiversity policies and desertification-control related measures. Although science serves to legitimatize measures for addressing both issues, the translation of science into policy is often poorly suited to concrete local situations and may contradict the understanding of the environment held by locals. This is why conservation measures are often seen as being imposed on local communities from outside (Adger et al., 2001; Peuhkuri and Jokinen, 1999).

Mitigation measures addressing both issues are unpopular to those that have to adopt and implement them. Biodiversity conservation measures were and are often met with hostility by the affected populations (Theodossopoulos, 2003; Hiedanpää, 2002). The long-term interest of mankind, which forms the rationale of biodiversity conservation, conflicts with the concrete, short-term, interests of those bearing the costs and who are rural people with vested interests in areas where the necessary management interventions occur (Peuhkuri and Jokinen, 1999; Alphandéry and Fortier, 2001). Biodiversity conservation usually brings no immediate benefits.[6] Very often, biodiversity and the consequences of its decline are not perceived or felt at all by ordinary people. Similarly, the costs of desertification mitigation measures in agriculture very often exceed the benefits, so the rational farmer will not adopt them (Dregne, 2002). The same conflict thus emerges: mitigation of desertification may be beneficial for society but not for the individuals who have to bear the costs of its control.

Finally, both issues receive low political priority. Biodiversity policies are implemented half-heartedly (Fairbrass and Jordan, 2001) or they are under-resourced to carry out the prescribed tasks (Alphandéry and Fortier, 2001). The EU does not have desertification-specific policies[7] while at the MS level the issue is loosely framed by national action plans.[8] Even worse, the activities that account for both phenomena are driven by deeper forces, as discussed above, that lie beyond the reach of biodiversity or desertification policies.

Policy Integration-Theoretical and Methodological Issues

The ultimate task of PI should be to contribute to sustainable development by resolving the conflict between the environmental and the socio-economic aspects of development problems.[9] Although this may be extremely difficult if the prevailing general trends identified previously prove to be correct, an appropriate[10] form of PI could assist in mitigating at least some problems related to biodiversity decline and desertification.

An obvious precondition for analysing and attempting to achieve the integration of policies is that they concern or affect, intentionally or not, a common object. The changes in land use patterns discussed previously could be considered

---

[6] Eco-tourism might be an exception.

[7] However, the 6th EAP acknowledges the existence of desertification as a problem.

[8] Countries that are signatories to the CCD are obliged to prepare National Action Plans to Combat Desertification, considered to be the means to implement the Convention at the national and sub-national levels.

[9] See, Chapter 1, this volume, for a broader discussion of PI.

[10] Because some integration tools may be predisposed to work in favour of economic interests and against the environment (Scrase and Sheate, 2002).

as common policy objects linking together biodiversity, desertification and other related environmental issues. Another crucial requirement is that policies share a common understanding of policy objects as well as compatible, shared or complementary broad goals. A clear goal setting is needed for every policy, which can be combined with that of other policies. In the case of mitigating biodiversity decline and desertification, a shared understanding of the rural environment by the policies that affect it is required, building on which some agreement on the desired state and dynamics of the coupled socio-environmental systems is developed that can then take the form of concrete goals.

In the Mediterranean, hardly any ecosystem exists that does not have a very long history of human influence (Grove and Rackham, 2001). As society changes so does its influence on ecosystems. It is meaningless to define goals solely in environmental terms without considering human activities as part of the ecosystem processes to be managed (Christensen et al., 1996). In this sense, not only the agricultural policy, for example, should include goals related to the mitigation of biodiversity decline and/or of desertification but policies concerning these two phenomena should, at least, take into account the socio-economic conditions of the areas to which they apply.

The determinants of biodiversity decline and desertification as well as the relationship between biodiversity and desertification are context-specific depending on the peculiarities and history of the interaction between a local society and its natural environment. Therefore, flexible approaches should be adopted facilitating the development of appropriate policy mixes for every individual situation at sub-national levels.[11] In this sense, goals should be defined for distinct, appropriate spatial units that should be shared by the policies whose integration is contemplated. These spatial units must be integral in terms of the resources and activities they include. This sounds easier in theory than can be achieved in practice. For example, administrative subdivisions rarely coincide with ecological units. Besides administrative and ecological subdivisions, other spatial subdivisions according to, for example, land ownership and tenure[12] and land use types[13] exist associated with different institutional regimes. The use of administrative subdivisions often leads to a dual problem; an administrative unit may be competent for several types of resources[14] within its jurisdiction, whereas resources have different spatial boundaries or, even worse, a single resource may fall under the competence of more than one administrative unit concerned with distinct, and not always compatible, uses of this resource (Cortner et al., 1998).

Procedures have to be developed to reconcile fragmented administrative subdivisions and sectoral areas of competence and their associated perspectives

---

[11] This approach is not uniquely suited to the Mediterranean environment but it gains in importance in this region due to the relatively large variability encountered within limited space (Grove and Rackham, 2001)

[12] Public, communal, private, to mention the more common types (Cortner et al., 1998).

[13] Forest, agricultural, residential, recreational, etc.

[14] Focusing, however, on particular aspects; e.g. use of water for agriculture.

(Cortner et al., 1998) as well as to establish common, long-term objectives and priorities for action. This will provide an enabling environment for the proper combination of instruments in the appropriate geographic areas and at the proper spatial scales to produce multiple benefits and avoid costs. Thus, policy fields with different understandings as well as use of concepts have to initiate and sustain communication and discussion. This is especially important in the present case that concerns the interaction of policy areas with widely contrasting ways of thinking; conservation on the one hand and production on the other.

Drawing on the above considerations and following the methodological guidelines of Chapter 2 of this volume, the integration of the EU biodiversity with the EU rural development policy is analyzed using criteria that concern the relationships among their objects, goals and objectives, actors and actor networks, structures and procedures, and instruments. Their present interpretation is detailed below.

Relationships among policy objects. In the present context, the integration of policies refers to rural areas; in particular, to their geographic location, human activities and uses of land. The issue of land management is particularly important in this case.

An important consideration is whether the definitions of concepts central to the common policy object (rural space, environmental problems, etc.) are identical, compatible, congruent or complementary among policies, reflecting a shared understanding of the common object. The answer partly depends on whether concepts are defined and used concretely and consistently. Vague and inconsistently used concepts might fit well or not to another policy depending on who interprets them.

Relationships among policy goals and objectives. Political commitment for integration with other policies should be included in the goals of a policy, if PI is to be pursued. A crucial requirement is that policy goals are compatible, complementary or, the goals of one policy embody concerns of other policies. Then it may be possible to streamline actions to produce a combined outcome that cannot be achieved by means of a single policy. This should be translated in targets that refer to this outcome.

Relationships among policy actors and actor networks. If policies share common, formal or informal, actors, then there is a good chance that these might act as integration agents. Here the term 'actors' excludes policy recipients as these simply coincide, being the inhabitants of rural areas.

Relationships among policy structures and procedures. To effectively pursue PI, the structures and procedures associated with the policies of interest should be coordinated or combined. Ideally, joint decision making structures and procedures should be established for both policy formulation and implementation.

Relationships among policy instruments. Relevant questions are (a) whether instruments of one policy might be used to achieve the goals of another policy;[15] (b) whether instruments of one policy may restrict the use of otherwise legally acceptable instruments of another policy;[16] (c) whether policy instruments can be combined to produce outcomes promoting the goals of each or all policies and beneficial to their common object (e.g. sustainable rural development).

The EU Biodiversity and Rural Development Policies

The Biodiversity Protection Policy

The EU biodiversity protection policy mainly comprises the Birds and the Habitats directives, and the EU Biodiversity Strategy (EUBS) and the associated Biodiversity Action Plans (BAPs). This section discusses also the crucial issue of financing these policies.

The EU Habitats and Birds directives   The Birds Directive (EC Directive 79/409/EEC) aims at the maintenance of populations of naturally occurring wild birds at levels corresponding to ecological requirements, regulating trade of birds, limiting hunting of species able to sustain exploitation, and prohibiting certain methods of capture and killing. It also asks the MS to take special measures to conserve the habitat of listed threatened species through the designation of Special Protection Areas.

The goal of the Habitats Directive (Directive 92/43/EEC) is to conserve the fauna, flora and natural habitats of EU importance. Its major target is to establish a network of protected areas throughout the Community designed to maintain the distribution and abundance of threatened species and habitats. This network, called Natura 2000, comprises the Special Areas of Conservation, designated pursuant to this Directive, and the Special Protection Areas, designated pursuant to the Birds Directive. The Habitats Directive includes also provisions for the protection of threatened species outside protected areas as well as other measures such as the creation or maintenance of corridors or stepping-stones between the network sites.

The implementation of these directives has encountered a number of problems,[17] which the integration of biodiversity with other policies is expected to mitigate. The emphasis in the following is on the Natura 2000 network since, for the time being, the implementation of other provisions is lagging behind.[18]

---

[15] This is the case of a complementary, positive relationship among instruments of different policies.

[16] In this case, one policy acts as a boundary condition for another; such as the EIA requirement required of regional policy or transport policy-funded projects.

[17] Reflecting perhaps what Jordan (2002) calls a 'built-in' implementation deficit in EU environmental policy (p.321).

[18] This fact alone is a cause of concern.

When the MS started compiling the lists of proposed Natura 2000 sites, local tensions arose in many cases (Ledoux et al., 2000). Although the principal reasons for local opposition to the establishment of protected areas were the anticipated constraints on economic activities (Hiedanpää, 2002), these were variously coupled with cultural reasons (Alphandéry and Fortier, 2001).[19] According to the legislative provisions, the compilation of site lists should be based on scientific criteria (concerning species and habitats) only, whereas public involvement was contemplated for the next phase of the elaboration of the management plans of the selected sites. The European Economic and Social Council (ESC, 2001) noted that this procedure "... has been seen by the parties concerned to show that decisions are being taken over the heads of the owners and users of the land concerned" and that this way "... opposition grows and nature conservation acquires a negative public image" (section 6). The general feeling is that the directives do limit the economic development of the proposed areas (Fairbrass and Jordan, 2001), creating a feeling of injustice concerning the divide between land owners and users within and outside protected areas (Krott et al., 2000).[20] Consequently, the question of compensation for the anticipated restrictions taking into account not only the current land use but also its potential is another recurring critical issue (Krott et al., 2000; ESC, 2001). Not surprisingly, the Habitats directive is one of the most litigated EU environmental instruments (Lasén Diaz, 2001).

The EU Biodiversity Strategy and Action Plans While the Habitats and the Birds directives are regulatory, legally binding, measures, the EUBS and especially the associated BAPs seek to promote biodiversity conservation through integration of biodiversity concerns into sectoral policies. An important feature of these policy texts is that they are not legally binding.[21] So far, the negative impacts of sectoral policies on biodiversity are considered to override the positive ones and progress towards sectoral integration of environmental concerns is limited in general (EC, 2000).

According to the EUBS, "Action Plans should be practical tools to achieve the integration of biodiversity into sectoral and cross-sectoral policy areas and instruments relevant to the conservation and sustainable use of biodiversity within the Community" (p. 20). The steps taken in the direction of integrating biodiversity concerns into sectoral policies are quite recent. The first BAPs (Conservation of Natural Resources, Agriculture, Fisheries, and Development and Economic Co-operation) were adopted in 2001. Their development has been led by the services

---

[19] Similar opposition had already occurred in areas protected through national legislation predating the Habitats directive (Theodossopoulos, 2003).
[20] This reaction was encountered also in areas protected through national legislation (Theodossopoulos, 2003).
[21] It has been argued that the wording of Article 6 of the Amsterdam Treaty that provides the legal basis for the integration of environmental concerns into sectoral policies is itself non-binding (Grimeaud, 2000).

of the respective competent DGs, in cooperation with each other and the DG-Environment, as well as with the EEA and MS experts.

The European Parliament considered the EUBS and the BAPs a welcome first step but also considered the measures announced insufficient and in need of being complemented and enhanced (EP, 2002). In some cases the EUBS and the BAPs lack concrete targets or concrete time frames. However, targets and time frames serve only the purpose of benchmarking or assessing the effectiveness of the relevant provisions. Several policy measures, not developed within the frame of biodiversity policy, are identified in the BAPs and their use is proposed to achieve biodiversity-related goals, but no sanctions are provided (and, indeed, could not be provided) in case these measures are not utilized.

Financing biodiversity conservation policies   The major financial instrument of the EU in the field of biodiversity conservation is the LIFE-Nature fund. Because the level of funding provided is insufficient, various policy texts often stress the possibility of using financial instruments provided through other policies (e.g. Reg. 1257/99) to achieve biodiversity conservation targets. However, the effectiveness of this approach has been questioned (Sutherland, 2002). As regards the Natura 2000 network specifically, the existing EU co-financing arrangements are reported to be plagued with problems. There are numerous potential funding sources, but each one of them involves different application procedures and is designed to achieve its own distinct objectives, rather than those of Natura 2000. The generally weak references to environmental and nature conservation in the funding rules complicates the situation further (Markland et al., 2002).

The EU Rural Development Policy

Several EU policies target rural areas either directly, such as the CAP, or indirectly, such as the regional,[22] transport and other policies. Rural development policy is a recent EU policy field, comprising the second pillar of the CAP, which is in part linked to EU regional policy. Its design and application are largely governed by Council Regulation 1257/1999, the Rural Development Regulation (RDR), as well as by Commission Regulation (EC) 1750/1999. The instruments and provisions of the RDR, which is the most important piece of EU's rural policy, that are most relevant in the present context are presented in the following.

According to Article 1, the RDR "establishes the framework for Community support for sustainable rural development" beginning January 1st 2000, with a view that it "shall accompany and complement other instruments of the Common Agricultural Policy". The preamble of the RDR declares its objective to introduce a sustainable and integrated rural development policy, to ensure better coherence

---

[22] E.g. the INTERREG or programs implemented in Objective I regions through the Community Support Framework (CSF) of each country and concerning quite diverse issues like infrastructure, tourism, education, etc.

between rural development and the prices and market policy of the CAP and to promote rural development with a view at "restoring and enhancing the competitiveness of rural areas" in the face of the new realities concerning market evolution, market policy and trade rules, consumer demand and preferences as well as the Union's next enlargement.

The rural development measures eligible under the RDR fall into two groups. The first group includes the accompanying measures of the 1992 CAP reform, early retirement, agri-environmental and afforestation, complemented by the Less Favoured Areas (LFAs) and areas under environmental restrictions schemes The second group includes measures to modernize and diversify agricultural holdings, farm investment, setting-up of young farmers, training, investment aid for processing and marketing facilities, additional assistance for forestry, adaptation and development of rural areas. Minimum environmental standards are required for activities to be eligible for funding. The Commission explicitly requires the operation of the SFs in general, to which the RDR belongs, to be consistent with other Community policies, including environmental policies.

Exploring the Linkages of the EU Biodiversity with the EU Rural Development Policy

Introduction

This section explores the integration of the EU biodiversity with the EU rural development policy, focusing on the RDR. Although the EU requires the integration of environmental concerns into sectoral policies only, this section examines the integration between the two policies. More specifically, it asks whether these policies are appropriate candidates for this task, whether PI is being, intentionally or unintentionally, pursued and what form it does take. The ensuing analysis is structured according to the criteria for PI presented previously.

Relationships among Policy Objects

The EU biodiversity policy affects directly all rural areas since this is where biodiversity is found. A large proportion of the areas proposed for inclusion in the Natura 2000 network consist of farmland as well as forests managed for production purposes. Certain provisions of the policy, such as the maintenance or creation of corridors and stepping-stones between protected areas and the management of priority species outside protected areas, affect rural areas not included in the protected areas network. On its part, the RDR concerns all rural areas of the EU focusing on agricultural activities mostly and less so on forestry.

In terms of content, the RDR targets, to a considerable extent, agricultural and forestry activities whose management is necessary for the conservation of biodiversity. It is widely recognized by now that the CAP has encouraged the adoption of intensive production methods, which have had significant adverse impacts on biodiversity.[23] Up to now the link between 'nature' legislation (Habitats and Birds Directives) and the CAP has been considered inadequate (EC, 1998b). The RDR emerged as the second pillar of the CAP aiming to alleviate some of the negative features of the first pillar. Although there is no explicit reference to biodiversity (except agricultural biodiversity), the implementation rules (Reg. 1750/99) state that it should be taken into account. The RDR exhibits a strong focus on environmental issues in general and is considered by the BAP for agriculture as the key instrument for introducing biodiversity concerns in the CAP). For its mere activation in any MS, the Commission has required the completion of the lists of sites proposed for inclusion in the Natura 2000 network (Baldock et al., 2002).[24]

The objects of the two policies are thus linked through their reference to the same geographic entities mostly and, to some extent, to the same activities (agriculture and forestry). The conceptual and theoretical commonalities between the two policies are limited, however, to the use of the terms 'sustainability' and 'environment' which are loose enough to accommodate any priorities and that are usually used with varying meaning and content depending on the context. This has important implications on the framing of concrete goals as discussed next.

Relationships among Policy Goals and Objectives

The purpose of the RDR is the achievement of sustainable development and of environmental integration. However, since both concepts are used very loosely and are open to any convenient interpretation, the issue requires further consideration. It seems that the goals of the RDR are threefold: (a) to enhance the competitiveness of agricultural holdings and areas (mainly through investment support and adaptation measures), (b) to provide some income support decoupled from production outputs (e.g. LFA allowances and AEMs) and (c) to promote environmental protection (through, e.g., agri-environmental and some forestry measures). These three broad goals are not necessarily compatible with one another let alone with the goals of other policies.

The environment has been one of the many, not necessarily compatible between them, priorities of the CAP reform leading to the emergence of its 'second pillar' (Barnes and Barnes, 2001). Therefore, an inherent inconsistency seems to exist in the goals of the RDR. On the one hand, it promotes competitiveness

---

[23] To be fair, in some marginal farming landscapes, support provided to farmers under the CAP has enabled the continuation of traditional farming practices that are of critical importance in maintaining biodiversity (EC, 1998b). This support may be far more substantial than the agri-environmental payments (van Dijk, 2001). Note also that frequently the effects of the CAP are difficult to separate from those of other drivers of change, like technology and global market price changes (Baldock et al., 2002).

[24] Fairbrass and Jordan (2003) described the move was as "informal greening" of the policy (p.102).

through investment support that may contribute[25] to production intensification in areas where investments are worth making (Baldock et al., 2002), widening the gap between productive and marginal areas and known to have negative environmental effects. On the other hand, it promotes non-competitive practices, through the AEMs, or contributes to sustaining production in marginal areas, through the LFA allowances.

Of the nine groups of policy measures that can be promoted through support provided by the RDR, "the preservation and promotion of a high nature value and a sustainable agriculture respecting environmental requirements" is more or less directly related to biodiversity while the "sustainable forest development" and "the maintenance and promotion of low input farming systems" indirectly. The other six are directed to distinctly different issues. The integration of biodiversity concerns may be directly pursued through funding made available in the frame of the biodiversity-related measures. Nevertheless, it is entirely plausible that funding can be directed to other environmental activities without any effort to integrate biodiversity concerns other than those arising from the provisions of the Birds and the Habitats directives. While RDR programming must be consistent with other Community policies, including conservation of biodiversity, in practice this translates into respecting the legal provisions only. Mid-term and ex-post evaluations of the RDR-financed programs must review effects on the environment. But, excluding measures directly targeting environmental issues, the cost of a real evaluation of other options (e.g. investment support) in environmental terms would become prohibitive, especially if environmental aspects, such as biodiversity, are detailed.[26] So this requirement can only be interpreted and applied in quite loose and broad terms.

Turning to the biodiversity policy, its goal is to reverse present trends in biodiversity loss, to place species and ecosystems at a satisfactory conservation status and to "anticipate, prevent and attack the causes of significant reduction or loss of biological diversity at the source" (EC, 1998a, p.3). In principle, the goals of this policy do not include any explicit reference to rural development. The wording of Article 2.3 of the Habitats directive "… measures taken pursuant to this Directive shall take account of economic, social and cultural requirements and regional and local characteristics" implies that these requirements and characteristics are viewed more as inevitable constraints or, in the best case, as parameters to be considered rather than objects of intervention on the side of biodiversity policy. This may be due to the fact that at the time of its formulation, the consequences of this policy for rural areas were not widely felt. It seems that only with the implementation of the Habitats directive the effects of biodiversity policy on rural development were widely realized for the first time.

---

[25] And similar instruments in the past have indeed contributed.
[26] The lack of a comprehensive evaluation of the effects that even AEMs have had on biodiversity up to now is worth noting.

Although rural development is not included in the goals of biodiversity policy, the latter relies on the RDR, among others, for funding certain activities, since resources earmarked for its implementation are largely insufficient to accomplish its goals. Especially the implementation of the BAP for agriculture relies heavily on the RDR, but the latter has its own priorities and is not legally linked to the EUBS and the BAPs. Therefore, the RDR is constrained by the regulatory but not by the 'integrative' part of the EU biodiversity policy. By the time the BAP entered into force, the implementation process of the RDR had already taken its course. It is difficult to conjecture what effect a non-compulsory instrument, such as the BAP, might have on a compulsory one, such as the RDR, that preceded it.

Relationships among Policy Actors and Actor Networks

DG-Agriculture is responsible for the CAP and its second pillar that currently consists largely of the RDR, while DG-Environment is responsible for biodiversity policy. Groups critical to the biodiversity policy, namely, the Forum Natura 2000 platform, have formed a coalition with DG-Agriculture, which is, however, less firm and influential than the coalition formed between DG-Environment and environmental NGOs (Weber and Christophersen, 2002). The latter have a strong influence on biodiversity policy, though its strength declines as one moves down from the EU to the national levels (Fairbrass and Jordan, 2003). Informal actors with a major say in agricultural affairs in the EU have a too strong preoccupation with price support matters to be concerned with environmental issues (Barnes and Barnes, 2001). The same pattern of distinct formal and informal actors in the two policy fields can be found throughout the implementation chain. In all southern MS (i.e. Greece, Italy, Portugal and Spain), agricultural authorities are responsible for rural development while environmental authorities for biodiversity protection, with a varied emphasis between central and regional level actors. Agricultural authorities are also responsible for elaborating the Good Agricultural Practice Codes that farmers have to respect to be eligible for support. These codes are also negotiated with the Community services prior to final approval.

Although the relevant information is rather limited, it appears that agricultural authorities are inclined to prioritize production over the environment, which makes sense since this is their responsibility (Fairbrass and Jordan, 2001). Together with agriculture-related informal actors, they are dissatisfied with biodiversity legislation and the fact that environmental concerns threaten to affect the way agricultural policy is developed and implemented.[27] Therefore, it seems that not only the administrative structures of the two policies operate in parallel but also

---

[27] The process is sometimes called 'environmentalization' of rural land (Alphandéry and Fortier, 2001, p.316).

that their priorities and the mind-sets of the respective actors and actor networks run largely in parallel without contact points.[28]

Relationships among Structures and Procedures

Apart from the 'environmental integration correspondents',[29] there does not seem to be any formal structure allowing interaction between the competent DGs with regard to PI. In general, there exists still much room for better coordination between the various Commission services (Mazey and Richardson, 2002). The RDR provides for vertical interaction between the competent authorities of the MS (national or regional agricultural authorities) and the relevant Commission services, but it does not explicitly provide for any kind of horizontal interaction at any level, apart from very general statements that do not seem to have any effect in practice.

On the whole, there is no evidence of structures and procedures favouring the integration of the two policies. There seem to be no procedures for formal interaction between the respective policy actors, no joint decision making procedures, no administrative structures responsible for promoting PI, no concrete provisions for consultation with respect to other policy goals and so on. The only promising procedure was the formulation of the BAP for agriculture, where a number of actors got involved and interacted. However, this was a one-off assignment that ended with the approval of the BAP. Nevertheless, the whole process raises questions about the commitment to the integration task. The BAP relies on the RDR for funding but it entered into force after the RDR. The integration of biodiversity concerns are not built in the design of the RDR; rather the BAP seeks to promote integration ex post by trying to exploit whatever possibilities are independently offered by the RDR.

The formulation of biodiversity legislation was also characterized by the lack of horizontal consultation (Fairbrass and Jordan, 2003). The rulings of the ECJ that exclusion of sites from the Natura 2000 network cannot be granted on economic and social grounds and the prevailing lack of consultation in the process of assembling national lists of proposed sites have strengthened the image of a policy unwilling to interact with other policy fields. A clear provision for developing procedures to facilitate horizontal consultation exists with respect to the elaboration of management plans for protected areas that is essentially a locally-oriented process. As already mentioned, the BAPs were an exception but this was a one-off procedure targeting the integration of biodiversity concerns in sectoral policies and not the opposite.[30] Once again, and although insufficient information

---

[28] Exceptions to this general picture may be, e.g., agri-environmental programs specifically targeting biodiversity implying some limited of interaction between the relevant actors.

[29] Officials designated by each DG to liaise with the DG-Environment and ensure that environmental concerns are given proper account (Lenchow, 2002).

[30] Note, however, that biodiversity policy is in a defensive position. It may be legally strong but essentially its object is negatively affected by several other socio-economic factors

is available, the two policy fields do not appear to possess any effective means for horizontal interaction.

Relationships among Policy Instruments

Reflecting the variety of its goals, the RDR includes a number of quite diverse financial instruments that are grouped into two categories for the present purposes: those mainly promoting competitiveness of agricultural holdings and areas and those mainly promoting environmental concerns. A third category may be distinguished including instruments promoting very specific concerns such as the LFA allowances and certain forestry measures. In the first (and the third) category, environmental concerns are placed as boundary conditions whereas in the second category support is granted on the basis of actions favourable to the environment, above those that are compulsory.

Instruments of the first category, e.g. investments in agricultural holdings, are designed to deliver their own objectives without sharing any concerns with biodiversity policy. These are linked to biodiversity conservation through legal provisions, for example the EIA requirement in certain projects. Apart from the more general effect these measures might have on the coupled processes of production intensification and land abandonment, their effect on biodiversity is context- and site-specific. In limited cases they might be coupled with other biodiversity favourable instruments; for example, investments for the uptake of organic farming practices by a farmer may receive priority relative to others. This effect cannot be generalized across quality products that are particularly favoured by the RDR. For example, conditions for producing Protected Denomination of Origin products do not usually contain environmental requirements beyond the legally compulsory.

On the contrary, instruments of the second category, e.g. AEMs, may be, and are indeed, used to achieve certain biodiversity-related goals. These measures have a prominent position in the discussion of how to reconcile modern farming with environmental concerns in general. They are considered as a way of correcting market failures, as society pays for common goods that the market fails to provide. They are also means of agricultural income support, which is accepted as non trade-distorting and, thus, less prone to be challenged in the WTO talks (Potter and Goodwin, 1998).[31] However, these are not principally designed to favour biodiversity, maybe reflecting the fact that agricultural authorities have different priorities. Currently, about 20 per cent of EU's farmland is under agri-

including sectoral policies that render the policy's outcome ineffective. In this sense, accounting for concerns of other policies might seem an undue luxury. Even within protected areas the effects of other policies can be stronger than those of the biodiversity policy itself (Beaufoy, 1998).

[31] However, policy measures considered positive for the conservation of biodiversity such as the AEMs and the LFA payments have sometimes come under fire in past WTO talks (IUCN ESUSG, 1999; van Dijk, 2001; ELO Policy Group, 2001).

environmental agreements, but the share of agreements targeting biodiversity protection has not been calculated. It seems that an insufficient area is under such agreements. This is certainly true for the Mediterranean countries, where large extensively farmed areas important for biodiversity conservation still exist (van Dijk, 2001).

A review of published assessments (Kleijn and Sutherland, 2003) concludes that no definitive evaluation of the effectiveness of AEMs can be made as yet with respect to their effect on farm biodiversity.[32] Most of the measures assessed were not designed to promote biodiversity. This review revealed that successful measures were embedded in a wider effort that included additional management measures. Where a favourable effect on biodiversity was found, the relevant issue did not really concern biodiversity policy; namely, the number of common species had risen but no positive effect was reported concerning rare species, which are the object of biodiversity policy. This review indicated that there is no general positive outcome of AEMs with respect to biodiversity goals. Naturally, not every measure could be designed to have an effect on rare species but these findings may as well suggest a more narrow agriculture focus of these measures than could have been the case. As some AEMs do target rare species specifically, it may be concluded that at least some measures must have had an effect on these (Baldock et al., 2002). In general, AEMs can be used to achieve biodiversity policy goals. They may also act complementary to projects funded by the LIFE-Nature financial instrument. These are, however, open possibilities and not explicit requirements of the RDR.

LFA allowances may favour continuation of extensive farming and grazing in marginal lands threatened by abandonment that is often vital for biodiversity conservation. Within LFAs eligibility is independent of the natural environment setting and the agricultural systems, since this measure targets a social rather than an environmental problem. Although LFA allowances will probably be beneficial for biodiversity conservation, there is no point in arguing about integration; biodiversity does not seem to play any role in the design of the measure. LFA allowances may favour continuation of agriculture in marginal areas and large tracts of high nature value farmland do exist in these areas, but this is a general statement not implying that all farming activities supported by LFA allowances are environmentally friendly. On the opposite side, allowances in areas with environmental restrictions can be used to compensate farmers in protected areas for income forgone due to restrictions arising from biodiversity conservation measures. This option can be directly used to achieve biodiversity policy goals.

In the field of forestry, a variety of financing options are provided, including those related mainly to production, processing and marketing of forest products as well as afforestation and aid for the establishment of forest-related associations. Apart from afforestation, these measures are little related to biodiversity policies. Support is subject to the constraints arising from biodiversity legislation, this being

---

[32] The authors note a complete lack of relevant studies in Mediterranean countries.

the sole point of contact between the two policies. The afforestation of agricultural land is discussed in more detail in the next section.

Regarding the allocation of funds among the various measures of the RDR, measures with potentially positive effects on biodiversity[33] receive around 50 per cent of the budget. These figures are lower in the southern MS: around 20 per cent for Greece, 39 per cent for Italy and Portugal and 33 per cent for Spain.[34] A similar differentiation is exhibited in Italy between non-Objective I and Objective I regions; the same measures account for about 60 per cent of EU spending in the former and about 26 per cent in the latter. Poorer MS and regions appear to place more emphasis on investments than on potentially environmentally-friendly measures. These figures do not imply that the relevant amounts are indeed used for environmental purposes, let alone biodiversity-related ones. As mentioned before, LFA allowances and forestry projects are not necessarily environmentally friendly. The figures are only indicative of the proportion that might have a more or less positive effect.

Compared to the first pillar of the CAP, the RDR exhibits a bias against poorer players, countries (or regions) and agricultural holdings alike. Direct payments under the first pillar are entirely EU-financed, while RDR measures must be co-financed by the MS, with the poorer ones finding it more difficult to mobilize the necessary funds. Investment support also requires a contribution from the beneficiary. The poorer players face the greatest difficulties to find capital for their own contribution and thus are not eligible for support, although they are those in most need of it. This might mean that less support will be available for those players, thus forcing them to be less competitive and promoting agricultural abandonment (Baldock et al., 2002).

Summing up, RDR instruments unrelated to biodiversity are linked to it through legal restrictions while their use might as well conflict the targets of biodiversity policy. Other instruments, developed for different purposes, can be favourable for biodiversity conservation in general while their target setting and indiscriminate use in environmental terms precludes any further utilization (LFA allowances). Instruments specifically targeting environmental issues, which can be used to directly pursue biodiversity targets, and certain instruments related to forestry may be used in favour of biodiversity or environmental protection in general. Seeking PI through these latter instruments is an open question: there is no legal requirement to do so, there are no structures or procedures (as explained previously) in place to facilitate it and the RDR does not require it. There are no common assessment criteria, institutional structures, shared resources or anything

---

[33] LFA allowances, including areas with environmental restrictions, agri-environmental and forestry.

[34] These figures are calculated from data taken from various fact sheets and other rural development documents available through http://www.europa.eu.int/comm/agriculture/index_en.htm. They should be considered as indicative since in some cases it was difficult to differentiate between the various measures presented in sum by some regions.

pointing to any formal way of seeking PI. In this sense, any integration between biodiversity policy and the RDR essentially concerns a segment of the instruments provided by the RDR and does not draw on the design of the RDR but rather from the will of lower (national or regional) levels to do so.

As regards the potential of the biodiversity policy instruments to be used to achieve the goals of the RDR, the answer is generally negative, letting alone the fact that the legal provisions of the Birds and Habitats directives are taken as boundary conditions for the implementation of the RDR. Some exceptions might occur, however. The goal of improving the competitiveness of agricultural holdings and regions is certainly not served by biodiversity policy. On the contrary, much of the local dissatisfaction with the biodiversity legislation arose from the fact that its restrictions would worsen the competitive situation of holdings in protected areas. In limited specific cases, biodiversity policies may help pursue the goal of the RDR to diversify activities in rural areas through eco-tourism. Protected areas, as well as infrastructure financed by conservation-related instruments, could promote the development of eco-tourism. Complementary activities can be undertaken in this respect by projects financed by the RDR and biodiversity-related instruments.

Summary

Taking into account that rural development is still receiving around 10 per cent of the total CAP budget and that the RDR is oriented towards increasing competitiveness, the environmental dimension seems to be underrepresented in the EU rural policy. Increased competitiveness usually requires intensification while environmental restrictions, such as those resulting from the biodiversity policy, usually reduce competitiveness. Hence, whereas some aspects of the RDR might become integrated with the biodiversity policy, the RDR cannot be considered as suitable for being integrated with the biodiversity policy on the whole due to its distinct priorities that are unrelated, or even contradict, biodiversity conservation.

If the RDR is placed in a wider policy context, this contradiction becomes even more apparent. Trade liberalization, to which the EU is committed, due to distinct other priorities, is acknowledged to induce rural restructuring that will, among others, spur further production intensification and land abandonment and create general environmental and biodiversity-related problems (EC, 2001; Baldock et al., 2002). This commitment actually undermines the PI discussion, since there appears to be a case of conflicting priorities. Considering that the initiation of the CAP reform, in the frame of which the RDR was produced, was a closed process triggered by the stagnation of the Uruguay round (Roederer-Rynning, 2003), it is not surprising that the RDR is not designed to promote integration with biodiversity or other policies. It seems to respond to challenges posed by international trade negotiations either by helping the 'winning horses' to become more competitive or by providing some support to the losers (LFA allowances) and some support in the name of the environment but decoupled from production and also helping reduce agricultural surpluses (AEMs).

Where one goes from here? The RDR includes a number of instruments, some of which are unsuitable for being integrated with the EU biodiversity policy while some others are indeed suitable. But the possibility of integration is an open option rather than an obligation. The environmental measures of the RDR seem to be promoted in the name, rather than for the sake, of the environment. This does not mean, however, that they cannot be effectively used to address environmental issues. Indeed, they should be used that way, becoming integrated with biodiversity policy to make the most out of the limited resources provided. However, the question remains: what is the chance of PI if these aspects are not primarily developed to tackle environmental problems and, therefore, the incentive to effectively use them to this end is low? To answer this question, the environmental measures of the RDR should be examined more thoroughly. The next section offers an example by examining the "afforestation of agricultural land" measure of the RDR.

Exploring the linkages of the EU Biodiversity Policy with the "Afforestation of Agricultural Land" Measure of the RDR

Introduction

A complete assessment and interpretation of the use of the open options for integration mentioned before would require a case-by-case examination of the environmental measures of the RDR in every country or region, a task which is beyond the present scope. Therefore, one measure, "afforestation of agricultural land", is discussed, drawing on the Greek experience, to illustrate the type of examination required and to arrive at some more general conclusions. This measure is chosen because it is explicitly considered as a measure that should be used in promoting biodiversity as well as other environmental objectives. The EU legislation and policy texts, the implementation provisions for Greece[35] and available experience with its implementation are analyzed to judge whether its conception and implementation is in the direction of achieving integration with biodiversity policies. The evaluation criteria used are those employed before for the general case of the RDR.

Afforestation of agricultural land was promoted through Reg. 2080/92 and currently through the RDR whose preamble states that "the afforestation of agricultural land is especially important from the point of view of soil use and the environment and as a contribution to increasing supplies for certain forestry products". Article 31 states that "Support shall be granted for the afforestation of

---

[35] Rural Development Programming Document and Common Ministerial Decree 312268/9023/507/3-12-2002.

agricultural land provided that such planting is adapted to local conditions and is compatible with the environment". In Greece, the support granted covers the costs of establishing plantations and the management costs for the first five years and is complemented by an annual subsidy for the agricultural income forgone until trees are felled for a maximum of 20 years.

According to the Greek Rural Development Programming Document, this measure aims at the development of forest resources, the enhancement of biodiversity, the protection and enhancement of the environment, the protection of soil resources and the conservation of the landscape and of the countryside in general. The concrete targets defined in the case of Greece do not seem to be consistent with these aims, however. These are:

- Offering supplementary income to the population of mountainous and geographically marginal areas;
- Mitigation of population desertization in LFAs;
- Alternative use of agricultural land in favour of the natural environment and the enhancement of the quality of life of the inhabitants;
- Contribution to the balanced and sustainable development of rural areas.

Relationships among Policy Objects

The measure shares a common object with the biodiversity policy both in physical terms (farmland) and content (land management practices). It belongs to the group of measures that may be readily applied to promote environmental objectives and become integrated with environmental policies in general or biodiversity policies in particular. However, this option is not obligatory due to the lack of conceptual clarity. The RDR declares afforestation as important from the point of view of soil use and the environment. Apart from the fact that soil is also part of the environment,[36] the environment as a term is loose enough to accommodate any priority. Which aspect of the environment is supposed to benefit from afforestation and in which respect plantations should be compatible with environmental protection is left open to any interpretation. It may be argued that this loose formulation is consistent with the subsidiarity principle, but this is oversimplified. The RDR also requires that afforestation must be adapted to local conditions; this might mean suitable for timber production under the particular local conditions, or favourable to biodiversity under these same conditions, or anything else.[37] Hence, it seems that this measure is designed, like the whole RDR, to accommodate any number of (even) conflicting priorities.

---

[36] However, not every afforestation is positive for soil protection (Romero-Díaz et al., 1999).

[37] Sustainability of a timber producing forest stand in economic terms might not be compatible with the conservation of biodiversity (Glück, 2000).

Relationships among Policy Goals and Objectives

Although it is not specified in what sense afforestation of agricultural land is important for the environment, it is implied that it will have a positive impact on some environmental aspect(s). Moreover, this measure implicitly assumes that agricultural land contraction in favour of forests can be tolerated or it is even desirable. Otherwise, it would be difficult to explain why it provides subsidies for afforesting agricultural land in times where agricultural land is being abandoned and taken up by natural vegetation, including forests, anyway. This assumption is particularly important in cases where extensively managed agricultural land is vital for biodiversity conservation, which can be mainly found in marginal areas, already under pressure from agricultural abandonment. The use of this measure in these areas would directly conflict the goals of the biodiversity policy. All in all, the conception of this particular measure does not seem to take into account the conditions under which afforestation of agricultural land could be favourable for the environment in general or for biodiversity in particular, both in terms of what kind of afforestation might prove beneficial or in what kind of areas these conditions might be met. So there is nowhere an explicit obligation of taking biodiversity into account apart from legal restrictions arising from conservation policies, and no concrete targets are set. It can be concluded then that the goals of this measure are restricted to its own domain; it may be used at will to achieve biodiversity goals or, it may even conflict biodiversity policies by encouraging replacement of farmland by forests where the existence of the former may be irreplaceable in conservation terms.

From the biodiversity policy point of view, the BAP for natural resources emphatically encourages "nature-oriented forest management techniques", among which the use of native species figures prominently to "ensure the conservation and sustainable use of biodiversity in European forests"; it also suggests that "new forests should not negatively affect ecologically interesting or noteworthy sites and landscapes" (p.15). A section referring to the Kyoto protocol emphasizes that afforestation actions, in the frame of this protocol, should recognize biodiversity conservation as a precondition.[38] The EUBS that predated the RDR states that Reg. 2080/92 should be further developed "to enhance its benefits to biodiversity" (p.16). However, these suggestions remained non-binding. The BAP came into force after the RDR but it seems that no effort has been made since then to bring in line the provisions of the Regulation with the BAP.[39]

It follows that there is a one-way view of integration between this measure and the biodiversity policy. The latter views it as a means to achieve its objectives but

---

[38] Jones (2003) also proposed the use of greenhouse gas emissions trading to promote various other environmental issues including biodiversity. Afforestation positive for biodiversity could receive additional credits in this frame.

[39] However, the recognition that insensitive afforestation has the potential to reduce biodiversity predated the RDR (EC, 1998b).

no such obligation exists in the frame of the RDR where a general reference to the environment seems insufficient to actually safeguard not only the intention of integration with biodiversity policies but also to prevent this measure from being implemented in biodiversity-harmful ways. The lack of concrete targets in this respect as well as of clear directions included in the formulation of the measure reaffirms the conclusion that its integration with the biodiversity policy is an open option rather than an obligation for development programming at lower levels.

The Greek experience seems to confirm this conclusion. When it comes to concrete targets, the relevant implementation provisions exhibit a shift from the environmental focus of the aims of the measure to a predominantly social one. As in the RDR, the same loose references to the environment prevail, accounting for the lack of concrete environmental provisions. The implementation of the measure is planned to prioritize mountainous and semi-mountainous areas, where agricultural land abandonment and forest encroachment are taking place anyway (a process that is generally a cause of concern in the frame of biodiversity policies). Thus, it can be conjectured that the measure is considered as an income subsidy for populations living in these areas, a point strengthened by the social focus of the targets set, the most important among which is to provide some compensation to the losers from agricultural restructuring rather than to promote environmental goals. The former may be a legitimate and justified concern, but it is pursued inappropriately. Instead of finding ways for agriculture to continue providing a decent income, this measure will probably accelerate agricultural land abandonment. Again there is a lack of concrete targets relating to biodiversity as well as of concrete commitment to integration with the relevant policies. Thus, the goal setting of the implementation of this measure in Greece appears to be potentially in conflict with biodiversity policies.

Relationships among Actors and Actor Networks

The general discussion on RDR applies here also. In Greece, the Forestry Service is the sole institution responsible for setting funding priorities for administrative subdivisions, examining plans and overseeing their proper implementation. No provisions are made for coordination with actors of other policies neither evidence of any informal interactions occurring exists.

Relationships among Structures and Procedures

The general discussion on the RDR is relevant here, too. The one-off procedure of compiling the BAP for natural resources that examined the possibilities of using this measure to achieve biodiversity-related targets led, as indicated in the "goals and objectives" subsection, to a one-sided result: the biodiversity policy considers the measure as a potential tool but this consideration does not seem to have had any effect on the RDR implementation process.

In Greece, the same picture is encountered at lower levels; there is no structure or procedure allowing interaction among the actors associated with this measure and those of biodiversity policy. This is best exemplified in the case of protected areas. A provision of the national implementation rules states that in Natura 2000 sites the physiognomy of the site must be preserved through the choice of suitable species to avoid the alteration of the character of the natural forests occurring in the site. However, in the few cases of protected areas where management authorities exist, they do not have any formal means of participating in decision making with respect to the above requirement. Compliance with this condition is judged only by the local forest authority.

Even without a formalized procedure allowing interactions, it might be argued that a local forest officer can reject an application on the basis that the submitted plan is not suited for the area in environmental terms, including its conflict with biodiversity-related goals. Experience, however, suggests that such a procedural possibility is merely theoretical. An application cannot be rejected by the administration unless the rejection is founded on the violation of explicitly stated implementation rules in the absence of which there is no room for a local forester to take into account the particularities of his/her own territory when approving or rejecting an application. Therefore, the formulation of the implementation provisions discourages even the use of the existing one-player procedures for avoiding conflicts with biodiversity policies, except, perhaps, in some cases within protected areas.

Relationships among Policy Instruments

In Greece, the implementation rules include a number of technical provisions to be followed by an applicant to be eligible for support. A list of 43 taxa that can be used is defined; 13 of them are exotic (Georgiadis, 2003). Although some of the promoted species, both native and exotic, are highly flammable, the only environmental parameter to be clearly taken into account outside Natura 2000 areas is fire risk, which is a major issue in forest policies. Any afforestation plan must be examined prior to approval for not increasing fire risk at the landscape level. Concerning the list of species promoted, no restrictions exist so as to avoid hybridization among cultivars and wild populations of the same species or among cultivated species and wild occurring relatives. Thus, the implementation of this measure may threaten the genetic diversity of certain species in direct conflict with conservation policies. In sum, there are no common resources, instruments or any framework bringing together biodiversity policies with this measure. Furthermore, the implementation provisions of this measure not only do not take into account biodiversity policy goals but also contain provisions that are in conflict with conservation goals.

Implementation Experience

The actual implementation of the measure is examined looking to confirm the points made while examining its design at the EU and at the MS (Greek) level. 43 per cent of all plantations established in Greece between 1992 and 2002 (mostly in the frame of Reg. 2080/92) consisted of black locust (*Robinia pseudoacacia*) (Dini-Papanastasi, 2004), an exotic species.[40] A case study of the use of this species in the Evros prefecture (Georgiadis, 2003) offers some hints as to what might be happening in other areas also. In this prefecture, the vast majority of black locust plantations are located in marginal fields unsuitable for timber production. Thus, the main incentive for afforestation seems to be the subsidy as such and not income earned from timber production. The measure appears to affect more marginal, extensively managed land and less intensively managed land where more benefits in biodiversity terms could be gained. In this area, a further paradox is observed. The Dadia-Leukimi-Soufli national park exists there, which is proposed to be included in the Natura 2000 network. Within the borders of the protected area, a LIFE project is being implemented which, among others, targets the clearing of forest openings not used any more for cultivation or grazing and in danger of becoming lost as the forest encroaches. Just outside the borders of the area, the RDR finances afforestation of agricultural land leading to forest taking over previously open spaces. Considering that clearing of forest openings is a management measure for raptors whose mobility does not confine them within the borders of the protected area, this is clearly a case of actions running counter to one another (Liarikos, personal communication).

The conclusion is that the call of the BAP to use native species for afforestation purposes has been disregarded, the geographic areas where afforestation has taken place are not those where this kind of management might prove beneficial for biodiversity and that even the purpose of increasing timber production is not generally served. There is also evidence that the use of this measure is in conflict with actions initiated through biodiversity policies. On the other hand, the purpose of delivering income support to people in marginally productive areas[41] seems to have been served. The available evidence suggests that the lack of an integration focus in the development of this measure at the EU level has indeed led to the lack of integration on the ground.

Summary

The aim of the "afforestation of agricultural land" measure is to promote the contraction of agricultural land in favour of forestry. Unlike other RDR measures,

---

[40] This species has become invasive in other parts of the world but this does not seem to be a major problem under Greek conditions (Dini-Papanastasi, 2004; Georgiadis, 2003).

[41] How effective, in social terms, this support has been or still is, is a separate issue beyond the purposes of the present analysis.

this aim does not a priori contradict biodiversity policies. But the formulation of the RDR does not explicitly require integration with other policies. No explicit provisions were made in terms of the legal, institutional and procedural infrastructure to facilitate this integration, which remains an open option to be considered at the national or the regional level. In the case of Greece, the measure is obviously being implemented without intention to achieve integration with biodiversity policy. Since all programs are approved by the Commission, it is also clear that the observed absence of integration is an option which is not considered inconsistent with the EU policy so as to lead to rejection of the program.

The measure could prove beneficial for biodiversity if certain constraints applied and certain considerations were explicitly taken into account. From the original legal text down to its implementation on the ground, at least in the case of Greece, the only case where some room was made for biodiversity concerns is in Natura 2000 sites. It is emphasized that this case results from a regulatory measure, the Habitats directive, and not from the non-compulsory BAPs. As it stands now, the measure may produce a number of negative effects on biodiversity from hybridization of wild populations to loss of valuable, in terms of biodiversity, agricultural land.

The measure seems to target other than environmental problems in which case biodiversity concerns could be left in the background since no one has a real incentive to promote them. To the extent that biodiversity concerns and support to a certain farmer group could be combined, the option for PI is left open. If it was possible to combine the concerns served by this measure with environmentally- (or specifically, biodiversity) friendly afforestation in areas where this could be beneficial, then this might be a success story. But this is not the case yet.

Concluding Remarks

It is high time to return to the question set in the introduction. The processes of biodiversity decline and desertification take place in a complex world. A number of common direct causes underlie both processes that are not uniform,[42] operating in unique ways in each and every case and in variable combinations. Policies affecting, intentionally or unintentionally, the development of desertification-sensitive rural areas, a policy object relevant to both biodiversity decline and desertification, yield variable results depending on the particular socio-environmental conditions under which they are implemented, the site, or even the individual policy recipient. To reverse current trends in biodiversity decline and desertification and manage their complex relationships productively, the task ahead is to try to properly integrate relevant policies, including in their scope the task of halting these processes. This sounds too good to be true and indeed it is.

---

[42] Although they may be attributed to a single driving force.

The analysis of the current state of integration of the EU biodiversity and rural policy revealed a considerable departure from a theoretically desired PI. As noted already, the RDR is internally inconsistent by promoting both competitiveness-inducing and non-competitive practices. However, the driving forces behind the RDR formulation are not inconsistent. Agricultural restructuring produces winners and losers. The winners are aided in order to retain or improve their competitive position in the global market while the losers receive some income support. AEMs may apply to both. In this case, support is decoupled from production and these measures also help reduce agricultural surpluses, since by definition, they compensate farmers for lost income due to reduced production.

This 'double identity' underlies the potential of the RDR to become integrated with biodiversity policy. In the cases that it promotes increased competitiveness, the two policies have no points of contact in their goals and they might even be in conflict, as competitiveness usually requires production intensification. This conflict exists also the other way around: restrictions arising from the biodiversity policy may hinder the exploitation of the full potential of a holding or of an area to increase its competitiveness. In this case, the RDR de facto views biodiversity conservation as a constraint rather than something to be pursued. Measures promoting competitiveness cannot become integrated with biodiversity policies due to their divergent priorities.[43]

On the other hand, promotion of non-competitive practices may work well together with biodiversity policies. It is generally agreed that in the fields of agriculture and forestry, concern for biodiversity would limit the economic performance of the relevant holdings and thus it has to be compensated[44] (Potter and Goodwin, 1998; Glück, 2000). The AEMs are the only RDR instruments that have to produce results in environmental terms, though not necessarily in biodiversity terms. Other options of the RDR, not pursuing the goal of increasing competitiveness, might be used to achieve environmental goals as well.

Are things moving in that direction? Although the picture is too complicated, the example of the 'afforestation of agricultural land' measure indicates that the answer might be 'no'. If it is assumed that the rationale behind this measure was to contribute to a reduction of agricultural production while providing some income support to farmers, then it seems to have achieved its target. If it was desired to contribute also to biodiversity conservation then it seems that this has been a weak desire. The measure was loosely formulated from the beginning, leaving all options open. The implementation rules for Greece made clear that the services of this measure to the environment would be questionable but they were still approved by the Commission. The limited information on its implementation indeed appears to confirm these fears. The findings of this case may not be generalizable but the

---

[43] Even integration of concerns of one policy into another might fail because the two policies strive for incompatible goals (Hudson, 1999).

[44] In the case studies briefly discussed in footnote 5, biodiversity conservation relied on public spending to become effective.

example is nevertheless indicative. A clear goal setting requirement that these measures bring environmental benefits by becoming integrated with, for example, biodiversity policies, is missing. The option was simply left open for lower level programming. In the example discussed here this option was not utilized.

As regards the implications of the state of integration of the EU biodiversity with the rural policy for combating desertification, it is conjectured that, because desertification and biodiversity decline can be attributed to common, direct causes, effective PI in terms of biodiversity conservation should generally favour mitigation of desertification. Experience from the implementation of a specific biodiversity instrument, the LIFE-Nature fund, seems to confirm this point; a large number of projects funded by LIFE-Nature in southern Europe include actions for erosion control or restoration of vegetation cover, that are critical for combating desertification. Similarly, actions taken by other policies in favour of biodiversity goals should be generally, though not always, favourable for combating desertification since they will usually address issues that concern both phenomena.[45]

The issue is not that simple, however; as already suggested, biodiversity decline and desertification are not linked to one another in a straightforward manner but rather in context-specific ways. In this sense, PI for biodiversity conservation might be usually, but not always, favourable for combating desertification. Furthermore, results achieved this way will be sub-optimal. It is not sufficient to rely on combating desertification resulting as a by-product of biodiversity conservation; desertification concerns must enter the PI equation on their own so as to achieve results by design and not by chance. However, if a well-established environmental policy domain like biodiversity fares so badly in terms of PI, then the chances of a 'shadow' policy domain like desertification seem quite meagre.

So why did the integration of biodiversity and rural development policy fail and to what extent was the failure inevitable? If the CAP reform was driven by international trade negotiations rather than by environmental concerns then it is reasonable that this reform is far from leading to an environmentally friendly agriculture (Barnes and Barnes, 2001).[46] To a certain extent this was inevitable. The CAP is a predominantly market policy and the market does not provide a value for environmental goods arising from biodiversity conservation; it does provide for a difference in value of products produced by resource-consuming intensive agriculture as opposed to products produced by environmentally-friendly, extensive agriculture.[47] As both biodiversity decline and desertification are driven by deeper socio-economic processes, such as changing production and consumption patterns and global market competition, prioritizing the

---

[45] For example, overexploitation of freshwater resources and conservation of wetlands.

[46] Although in this case the pressure to utilize the 'green box' option of the WTO (the group of agricultural subsidies not distorting free trade) may produce environmental positive results as a by-product.

[47] The reverse may be true when some product labeling schemes are considered but these do not alter importantly the general picture.

competitiveness of agriculture necessarily leads to respecting the processes facilitating it and, thus, is largely insensitive to conservation considerations.

If the rationale for environmentally-oriented spending was to provide an alternative of keeping some level of income support while not distorting free trade rules, then environmental issues in the case of the CAP reform just happened to fit the policy planning purposes that arose from non-environmental concerns. In this case, the incentive to build in the RDR the obligation of producing concrete positive environmental results, at least through some measures, must have been low. On the contrary, the incentive to let implementation be as unconstrained as possible to permit the policy develop and produce the desired socio-economic results regardless of concrete environmental outputs must have been powerful. If the experience of the "afforestation of agricultural land" measure can be generalized, attaining PI with some of the measures provided by the RDR failed, although it was technically achievable. The most plausible explanation is the lack of political and societal commitment at all levels.

Examples of policies that have performed better or worse in terms of PI than those examined here might be probably found. But the position outlined in the beginning seems to hold; it is not reasonable to consider current trends in biodiversity decline and desertification reversible in the short to medium term. Grove and Rackham (2001) are right when they suggest that "many of the problems [....] can be mitigated by improved management, but not solved: they will not go away" (p.17). It cannot be reasonably assumed that the changing equilibrium between society and nature will be seriously affected by conservation policies when their targets run counter to deeper socio-economic tendencies.

If this is so, the maximum that can be reasonably achieved in the short to medium term is the integration of environment-related policies. Although the overall trends cannot be reversed, opportunities exist to develop certain policy mixes that will help improve the situation in some cases, but these are currently underutilized. If the necessary commitment develops and if the necessary scientific and administrative resources are mobilized, it is technically feasible to achieve more out of the limited policy opportunities available. The RDR measures with a potential to be used in this way account for around a modest 5 per cent of the total CAP spending. It is an undue luxury not to exhaust their potential. The effectiveness of mitigation efforts could be enhanced if these measures were properly coordinated with an array of policies that are related to the rural environment and either address directly environmental protection (e.g. the Habitats and Birds directives) or include such aspects in their conception (e.g. the Water Framework Directive). This coordination necessitates certain actions. At the EU level, a clear obligation must exist to integrate these aspects of policies that are suitable for the task with other policies such as the biodiversity policy. The desertification issue must also attain a formal status in the EU, even if it concerns only its southern MS, and must be introduced as an obligatory component of PI. If this kind of obligations at the EU level is absent, there are still possibilities to achieve some degree of PI at lower programming levels if the requisite political will exists.

The implications of these actions for combating desertification draw from the fact that PI concerning biodiversity must necessarily touch upon the more general subject of conservation of land resources. The outcome should then be generally positive in terms of desertification. However, what is more important is the creation of a proper political culture. If the process of integrating biodiversity with other policies is set in motion, then, it will require just one more little step to introduce the desertification parameter in the equation. In one sense, it may be easier to introduce desertification concerns in the PI discussion than is to introduce biodiversity concerns; desertification, unlike biodiversity decline, may have economic implications and, therefore, it may be more suitable for integration with production-related policies. This 'advantage', however, cannot be efficiently utilized due to the lack of a formal status of desertification as a policy problem.

In closing this chapter, it is acknowledged that the discussion and the suggestions offered are largely based on insufficient information. For example, the view of LFA allowances as generally, though not universally, positive for biodiversity conservation cited is derived through a logical thought process; most high-nature-value farmland is found in LFAs and allowances are supposed to help this kind of farming survive. Nevertheless, this view has not been empirically assessed and, perhaps more importantly, the complexity of the issue remains untouched: under which conditions might these allowances prove positive or otherwise? Clarification of such issues would greatly benefit the PI discourse.

Another point worth clarifying concerns the unpopularity of both biodiversity policy and the CAP reform, from which the RDR measures emerged. The drafting of two key steps in the respective policy processes, the McSharry reform of the CAP and the formulation of the Habitats directive, seems to have been extremely closed processes (Roederer-Rynning, 2003; Fairbrass and Jordan, 2003). The precise effects of the closed character of the processes on incorporating concerns of other policies during the crucial drafting phase are insufficiently known.

A point that particularly needs further elaboration is the relationship between local circumstances and the EU level policy provisions. It is known, for example, that CAP provisions have had a different effect in the same sector in different geographic areas. How to accommodate the variability encountered in the different environmental and social settings in the EU within EU policy making should be considered an important open question. The principle of subsidiarity is only part of the answer. Reference to this principle may result in lack of concrete obligations, thus creating incentives for avoiding PI.

The findings of this work may have wider implications exceeding the issues of biodiversity, desertification and rural development. The fact that two or more policies affect or concern a common object is a necessary though not sufficient condition to permit integration between them. PI might fail due to diverging goals pursued by the separate policies for the same object. This need not be a 'technical' failure: if the rationale underlying policy-making is examined it may come out that the drivers that lead to the specific formulation of a policy have pushed its development in directions that are a priori incompatible with the directions of other policies. In this case the question of integration becomes no more than a merely theoretical quest.

References

Adger, W.N., Benjaminsen, T.A., Brown, K., and Svarstad, A. (2001), 'Advancing a Political ecology of Global Environmental Discourses', *Development and Change*, Vol. 32, pp. 681-715.

Alphandéry, P. and Fortier, A. (2001), 'Can a Territorial Policy be Based on Science Alone? The System for Creating the Natura 2000 Network in France', *Sociologia Ruralis*, Vol. 41, pp. 311-28.

Baldock, D., Dwyer, J. and Sumpsi Vinas, J.M. (2002), *Environmental Integration and the CAP*, A report to the European Commission, DG Agriculture, IEEP.

Barnes, P.M. and Barnes, I.G. (2001), 'Understanding the costs of an environmentally friendly Common Agricultural policy for the European Union', *European Environment*, Vol. 11, pp. 27-36.

Beaufoy, G. (1998), 'The EU Habitats Directive in Spain: can it contribute effectively to the conservation of extensive agro-ecosystems?', *Journal of Applied Ecology*, Vol 35, pp. 974-8.

Bongaerts, J. (2000), 'Agenda 2000 and the Common Agricultural Policy', *European Environmental Law Review*, Vol. 9, pp. 243-45.

Brand, U. and Görg, C. (2003), 'The state and the regulation of biodiversity International biopolitics and the case of Mexico', *Geoforum*, Vol. 34, pp. 221-33.

Christensen, N.L., Bartuska, A.M., Brown, J.H., Carpenter, S., D' Antonio, C., Francis, R., Franklin, J.F., MacMahon, J.A., Noss, R.F., Parsons, D.J., Peterson, C.H., Turner M.G. and Woodmansee, R.G. (1996), 'The report of the Ecological Society of America Committee on the scientific basis for Ecosystem Management', *Ecological Applications*, Vol. 3, pp. 665-91.

Convention on Biological Diversity (1998), *The relationship of the Convention on Biological Diversity with the Commission on Sustainable Development and biodiversity-related conventions, other international agreements, institutions and processes of relevance*, COP decision IV/15.

Convention on Biological Diversity (2000a), *Consideration of options for conservation and sustainable use of biological diversity in dryland, Mediterranean, arid, semi-arid, grassland and savannah ecosystems. Possible elements for a joint work programme between the Secretariat of the Convention on Biological Diversity and the Secretariat of the Convention to Combat Desertification on the biological diversity of dry and sub-humid lands*, Note by the Executive Secretary, UNEP/CBD/COP/5/INF/15.

Convention on Biological Diversity (2000b), *Consideration of options for conservation and sustainable use of biological diversity in dryland, Mediterranean, arid, semi-arid, grassland and savannah ecosystems*, COP decision V/23.

Cortner, H.J., Wallace, M.G., Burke, S. and Moote, M.A. (1998), 'Institutions matter: the need to address the institutional challenges of ecosystem management', *Landscape and Urban Planning*, Vol.40, pp. 159-66.

Detsis, V. and Briassoulis, H. (2004), 'Conserving biodiversity and combating desertification: achieving synergies', in M. Arianoutsou and V.P. Papanastasis (eds), *Ecology, Conservation and Management of Mediterranean Climate Ecosystems,* Proceedings of the 10th International Conference on Mediterranean Climate Ecosystems, April 25-May 1, 2004, Rhodes, Greece, Millpress, Rotterdam, (published on CD-ROM).

Dini-Papanastasi, O. (2004), '*Robinia pseudoacacia* L.: a dangerous invasive alien or a useful multi-purpose tree species in the Mediterranean environment?', in M. Arianoutsou and V.P. Papanastasis (eds), *Ecology, Conservation and Management of Mediterranean Climate Ecosystems,* Proceedings of the 10th International Conference on Mediterranean Climate Ecosystems, April 25-May 1, 2004, Rhodes, Greece, Millpress, Rotterdam, (published on CD-ROM).

Dregne, H.E. (2002), 'Land Degradation in the Drylands', *Arid Land Research and Management*, Vol. 16, pp. 99-132.

European Commission (1998a), *Communication of the European Commission to the Council and to the Parliament on a European Community Biodiversity Strategy*, COM (98)42.

European Commission (1998b), *First Report on the implementation of the Convention on Biological Diversity by the European Community.*

European Commission (2000), *Global assessment. Europe's Environment: What directions for the future?*, Office for Official Publications of the European Communities, Luxembourg.

European Commission (2001), *Communication from the Commission to the Council and to the European Parliament Biodiversity Action Plans in the areas of Conservation of Natural Resources, Agriculture, Fisheries, and Development and Economic Co-operation*, COM(2001)162 final.

European Economic and Social Committee (2001), *Opinion on "The situation of nature and nature conservation in Europe"* (NAT/081).

European Landowners Organization Policy Group (2001), WTO: the outcome of Doha, what next?, EPG 143/01.

European Parliament (2002), *European Parliament resolution on the communication from the Commission to the Council and the European Parliament on the biodiversity action plans in the areas of conservation of natural resources, agriculture, fisheries, and development and economic cooperation*, P5_TA(2002)0121.

Fairbrass, J. and Jordan, A. (2001), 'Protecting biodiversity in the European Union: national barriers and European opportunities?', *Journal of European Public Policy*, Vol. 8(4), pp. 499-518.

Fairbrass, J. and Jordan, A. (2003), 'The informal governance of EU environmental policy: the case of biodiversity protection', in T. Christiansesn and S. Piattoni (eds), *Informal Governance in the European Union*, Edward Elgar, Cheltenham, pp. 94-113.

Firmino, A. (1999), 'Agriculture and landscape in Portugal', *Landscape and Urban Planning*, Vol. 46, pp. 83-91.

Georgiadis, N.M. (2003), *Utilization of black locust in the implementation of the regulation on afforestation of agricultural land in the Evros prefecture: literature review, use, potential and distribution of the species*, unpublished study, WWF Greece, Athens (in Greek).

Glück, P. (2000), 'Theoretical perspectives for enhancing biological diversity in forest ecosystems in Europe', *Forest Policy and Economics*, Vol. 1, pp. 195-207.

Grimeaud, D. (2000), 'The integration of Environmental Concerns into EC Policies: A Genuine Policy Development?', *European Environmental Law Review*, Vol. 9, pp. 207-18.

Grove, A.T. and Rackham, O. (2001), *The nature of Mediterranean Europe: an ecological history*, Yale University Press, New Haven.

Haila, Y. (1999), 'Biodiversity and the divide between culture and nature', *Biodiversity and Conservation*, Vol. 8, pp. 165-81.

Harrison, C.M., Burgess, J. and Clark, J. (1998), 'Discounted knowledges: farmers' and residents' understandings of nature conservation goals and policies', *Journal of Environmental Management*, Vol. 54, pp. 305-20.

Hiedanpää, J., (2002), 'European-wide conservation versus local well-being: the reception of the Natura 2000 Reserve Network in Karvia, SW-Finland', *Landscape and Urban Planning*, Vol. 61, pp. 113-23.

Hudson, R. (1999), 'Putting policy into practice: policy implementation problems, with special reference to the European Mediterranean', in P. Balabanis, D. Peter, A. Ghazi and M. Tsogas (eds), *Mediterranean Desertification Research results and policy implications Proceedings of the International Conference 29 October to 1 November 1996 Crete, Greece*, European Commission Directorate-General Research.

IUCN European Sustainable Use Specialist Group (1999), *Biodiversity and Landscape Conservation in Pan-European Agriculture: The IUCN ESUSG/AWG Approach.*

Jones, D. (2003), 'Trading for climate without trading off on the environment An Australian perspective on integration between emissions trading and other environmental objectives and programs', *Climate Policy*, Volume 3 (Supplement 2), pp. 125-41.

Jordan, A. (2002), 'The implementation of EU environmental policy: A policy problem without a political solution?', in A. Jordan (ed.), *Environmental policy in the European Union*, Earthscan Publications Ltd., London, Sterling VA, pp. 301-28.

Kleijn, D. and Sutherland, W.J. (2003), 'How effective are European agri-environment schemes in conserving and promoting biodiversity?', *Journal of Applied Ecology*, Vol. 40, pp. 947-69.

Knickel, K. (2001), 'The Marketing of Rhöngold Milk: An Example of the Reconfiguration of Natural Relations with Agricultural Production and Consumption', *Journal of Environmental Policy and Planning*, Vol. 3, pp. 123-36.

Krott, M., Julien, B., Lammertz, M., Barbier, J.-M., Jen, S., Balesteros, M. and de Bovis, C. (2000), 'Voicing Interests and Concerns: Natura 2000: An ecological network in conflict with people', *Forest Policy and Economics*, Vol. 1, pp. 357-66.

Lasén Diaz, C. (2001), 'The EC Habitats Directive Approaches its Tenth Anniversary: An Overview', *RECIEL*, Vol. 10(3), pp. 287-95.

Ledoux, L., Crooks, S., Jordan, A. and Turner, R.K. (2000), 'Implementing EU biodiversity policy: UK experiences', *Land Use Policy*, Vol. 17, pp. 257-68.

Lenschow, A. (ed.), (2002), *Environmental Policy Integration: Greening sectoral Policies in Europe*, Earthscan, London.

Markland, J., Liebel, G., Alvarez, G., Lago, A., Chevin, C., Jensen, J., Norman, Å., Terstad, J., Rayment, M., Naveso, M.A., de Galembert, B., de l'Escaille, T. and Welge, A. (2002), *Final Report on Financing NATURA 2000*, Working Group on Article 8 of the Habitats Directive.

Mazey, S. and Richardson, J. (2002), 'Environmental groups and the EC: Challenges and opportunities', in A. Jordan (ed.), *Environmental policy in the European Union*, Earthscan Publications Ltd., London, Sterling VA, pp. 141-56.

Mills, J. (2002), 'More than Biodiversity: The Socio-economic Impact of Implementing Biodiversity Action Plans in the UK', *Journal of Environmental Planning and Management*, Vol. 45(4), pp. 533-47.

Moreira, F., Rego, F.C. and Ferreira, P.G. (2001), 'Temporal (1958–1995) pattern of change in a cultural landscape of northwestern Portugal: implications for fire occurrence', *Landscape Ecology*, Vol. 16, pp. 557-67.

Peuhkuri, T. and Jokinen, P. (1999), 'The role of knowledge and spatial contexts in biodiversity policies: a sociological perspective', *Biodiversity and Conservation*, Vol. 8, pp. 133-47.

Potter, C. and Goodwin, P. (1998), 'Agricultural Liberalization in the European Union: An Analysis of the Implications for Nature Conservation', *Journal of Rural Studies*, Vol. 14(3), pp. 287-98.

Rabbinge, R. and Van Diepen, C.A. (2000), 'Changes in agriculture and land use in Europe', *European Journal of Agronomy*, Vol. 13, pp. 85-100.

Roederer-Rynning, C. (2003), 'Informal Governance in the Common Agricultural Policy', in T. Christiansesn and S. Piattoni (eds), *Informal Governance in the European Union*, Edward Elgar, Cheltenham, pp. 173-88.

Romero-Díaz, A., Cammeraat, L.H., Vacca, A. and Kosmas, C. (1999), 'Soil erosion at three experimental sites in the Mediterranean', *Earth Surface Processes and Landforms*, Vol. 24, pp. 1243-56.

Scrase, J.I. and Sheate, W.R. (2002), 'Integration and Integrated Approaches to Assessment: What Do They Mean for the Environment?', *Journal of Environmental Policy and Planning*, Vol. 4, pp. 275-94.

Sutherland, W.J., (2002), 'Restoring a sustainable countryside', *Trends in Ecology and Evolution*, Vol. 17, pp. 148-50.

Theodossopulos, D. (2003), *Troubles with turtles. Cultural Understandings of the Environment on a Greek Island*, Berghahn Books, New York.

Van Der Leeuw, S. (1999), 'Degradation and Desertification: some lessons from the long-term perspective', in P. Balabanis, D. Peter, A. Ghazi and M. Tsogas (eds), *Mediterranean Desertification Research results and policy implications Proceedings of the International Conference 29 October to 1 November 1996 Crete, Greece*, European Commission Directorate-General Research.

Van Der Leeuw, S. (2004), 'Vegetation Dynamics and Land Use in Epirus', in S. Mazzoleni, G. di Pascuale, M. Mulligan, P. di Martino and F. Rego (eds), *Recent Dynamics of the Mediterranean Vegetation and Landscape*, John Wiley & Sons, Chichester, pp. 121-41.

Van Dijk, G. (2001), 'Biodiversity and multifunctionality in European agriculture: priorities, current initiatives and possible new directions', *Consultant background paper for the workshop: "Multifunctionality: Applying the OECD Analytical Framework. Guiding Policy Design"*, Paris, 2-3 July 2001.

Weber, N. and Christophersen, T. (2002), 'The influence of non-governmental organizations on the creation of Natura 2000 during the European Policy process', *Forest Policy and Economics*, Vol. 4, pp. 1-12.

Chapter 8

Sustainable Forest Management in the European Union: The Policy Integration Question

Aristotelis C. Papageorgiou, Georgios Mantakas and Helen Briassoulis

Introduction

Forests are complex, multi-level ecosystems comprising interrelated associations of flora and fauna as well as human communities. They produce valuable products and offer a wide array of services thus contributing to human welfare and quality of life. For this reason, forestry is a lot more than wood production concerning diverse human activities in forests aiming at the best possible satisfaction of human needs in the long term. This can be achieved only through integrated management that accounts for the socio-ecological complexity of forest systems.

The beneficial role of forests is not limited to the local or the national level; it extends to the mitigation of global environmental problems, such as climate change and desertification. The success of any strategy for combating desertification in particular, depends crucially on forests and their management. Healthy forest ecosystems slow down the process of soil erosion, regulate the water cycle, safeguard the biological diversity in forests and conserve the landscape to be enjoyed by citizens.

Forest management systems were first developed in Central Europe, aiming at the long-term production of forest products, mainly timber, without diminishing forest resources and the productive capacity of forests. These systems were based on maximum production and a classification of forests according to age classes and rotation periods. In the last decades, recognising the complexity of forests, forest management systems have adopted a more 'nature-oriented' style that gives as much weight to ecological and social functions as to wood production. Since 1990, an extensive scientific and political debate has centred on the definition of Sustainable Forest Management (SFM) in various fora such as the United Nations Forum on Forests, the European Union, the Ministerial Conference on the Protection of Forests in Europe, the 'G8' and meetings of the signatories to the Convention on Biological Diversity, the Convention to Combat Desertification and the Framework Convention on Climate Change. During the 1992 'Earth Summit' in Rio de Janeiro, the United Nations Commission on Sustainable Development defined SFM as: "...management to meet the social, economic, ecological, cultural and spiritual needs of present and future generations" (UN/DESA, 1992, p.1). It is widely recognized by now that SFM has an ecological, a social and an economic pillar, all of which are of equal importance for the maintenance of forest ecosystems and their productivity.

Forest policies, management systems, as well as products and services marketing mechanisms, have existed since the first organized attempts to exploit forests. In the last three decades, forest policy, defined as "the principles that govern the actions of people with respect to forest resources" (Worrel, 1970), has evolved into an independent scientific field (Merlo and Paveri, 1997) whose formation has been influenced by the recognition of the complexities ruling SFM. National and international forest policies reflect, at least on paper, the interests of non-forest sectors, paying attention to non-forest policies (agricultural, environmental, industrial, fiscal, commercial, etc.) with significant effects on forestry as well (Repetto, 1988). Since the object of forestry has shifted from forest products to forest ecosystems, forest policy has assumed a strong spatial character, concerned with the actions of the complex set of actors and stakeholders and their effects within a specific spatial unit, defined by the forest land use. The current trend favours public participation in both policy formation and implementation at all levels.

Despite its significant richness in forest resources and its well-developed forest sector, the EU does not have a common forest policy (European Commission, 2003a) and it seems that almost not one single member state favours its creation for different reasons. Countries with a significant forest sector do not want any restrictions or barriers on timber markets and trade; countries with a less strong forest sector do not want to enter a common 'system', that will be defined by partners with larger impact; other parties that place emphasis on the non-marketed services and social role of forests fear that a common EU forest policy would be defined by countries that give priority to timber production (Papageorgiou and Mantakas, 2004a).

At the same time, numerous provisions in EU policy areas such as agricultural, regional, environmental and energy policy and monopoly legislation, indirectly affect the forest sector (Hogl, 2000). In fact, because of the spatial and multidimensional character of forests, almost all policies relate somehow to forests and forestry and have some kind of forest policy component (Papageorgiou and Mantakas, 2004a). It follows that to achieve SFM, forest and non-forest policies should be coordinated and integrated. However, policy integration (PI) cannot be limited to simply incorporating forest-related concerns into non-forest policies (henceforth, IFC, for brevity), an equivalent to the notion of Environmental Policy Integration (EPI) (see, Chapters 1 and 2, this volume). Instead, it should strive for more meaningful and essential links among the theoretical and applied aspects of forest and non-forest policies (henceforth, FPI, forest policy integration, for brevity).

This chapter aims to promote the discussion on policy integration at the EU level towards the achievement of SFM. More specifically, it explores the following questions:

a) How well are forest-related concerns integrated in selected EU policies and how does IFC contribute to SFM in Europe?
b) Is it possible to synthesize and coordinate existing EU policies into an integrated policy complex to achieve SFM or is it preferable to design an integrated EU Common Forest Policy (EUCFP)?

The chapter comprises six sections. The second section examines the role of integration in forest policies at the international, EU and national levels. The third section discusses conceptual and theoretical issues in SFM and the fourth presents the methodological approach adopted in this chapter. The fifth section is devoted to the analysis of IFC in selected EU policies. The last section negotiates the possibility of FPI or the formulation of a EUCFP.

The Role of Integration in Forest Policies

The development of forest policies at the international and sub-global levels has a long history. This section provides a brief overview of how the integration of forest with other policies has been addressed at different levels.

The International Forest Policy Dialogue

Following on from the Brundtland Report in 1987 (WCED, 1987), the 1992 Earth Summit was the first major international conference to produce a substantive agreement on global forest management. Chapter 11 of Agenda 21 and the Non-Legally-Binding Authoritative Statement of Principles for a Global Consensus on the Management, Conservation and Sustainable Development of All Types of Forests (or, Forest Principles) represent a first attempt by governments to establish a global framework for the management, conservation and sustainable development of forests (UN, 1992).

Chapter 11 of Agenda 21 places emphasis on its first programme area entitled 'sustaining the multiple roles and functions of all types of forests, forest lands and woodlands', where the harmonization of 'policy formulation, planning and programming' is the main prerequisite to achieve 'a rational and holistic approach to the sustainable and environmentally sound development of forests' (UN, 1992). The text recognizes a failure of current policies in this respect. While the need for policy integration is clearly stated in the preamble, the focus is mainly on participation of several players in management decisions and the creation of national bodies to facilitate this participation. In the same spirit, the Forest Principles refer to the need that "...all aspects of environmental protection and social and economic development as they relate to forests and forest lands should be integrated and comprehensive..." (UN/DESA, 1992, p.2) and call the signatory countries to develop national programmes with the participation of all related actors.

In Rio, many countries pressed for a legally binding international agreement to provide a framework for the sustainable management of the world's forests. The objective was to have a Convention like those on Biological Diversity, Climate Change and Desertification. However, several countries opposed such a Convention, as they believed that it would conflict with their rights to use their natural resources to meet national policy objectives. The Conference adopted instead the above-mentioned non-legally binding Statement of Forest Principles. This failure to launch a binding forest policy text at the international level has marked future forest policy dialogues and has resulted in a gradual decrease of forest-related issues in many national and international policy agendas, as the implementation of the commitments associated with the three UN-Conventions has received priority.

The dialogue remained in this soft state with the meetings following up Rio establishing fora such as the Intergovernmental Panel of Forests (IPF), the Intergovernmental Forum of Forests (IFF) and finally the United Nations Forestry Forum (UNFF), that produced texts of more or less similar content, as far as PI is concerned. At the last UNNF meeting in Geneva (2004), several countries started considering the possibility to produce a more specific and binding Forest Convention (UN/ECOSOC, 2004).

The Ministerial Conferences for the Protection of Forests in Europe (MCPFE)

Inspired by the international forest policy dialogue, European countries have established a series of high-level political meetings, where the ministers responsible for forests meet and sign commonly agreed declarations and resolutions concerning the protection and sustainable management of forests in the European continent. So far, this regional process has produced several non-binding general and thematic texts. A number of technical documents have also been published, where new ideas concerning SFM have been introduced and new concepts have been promoted, such as a common European definition of SFM reflecting the global sustainable development discussion (Mayer, 2000).

At the Third Ministerial Conference held in Lisbon (1998), several PI parameters have been included in the resolution concerning the socio-economic aspects of SFM (MCPFE, 1998a). Emphasis was placed on rural development and public participation in decision-making. Finally, the last Ministerial Conference held in Vienna (2003) included a declaration prioritizing cross-sectoral issues (MCPFE, 2003a) and a separate resolution entitled 'Strengthen synergies for sustainable forest management in Europe through cross-sectoral co-operation and national forest programmes' (MCPFE, 2003b). Similar to the international processes, the texts signed in these conferences recognize the need to broaden the object of forest policies and promote it through participation of all related national level actors.

The EU Forestry Strategy

Following the principles and commitments adopted at the 1992 Earth Summit and the MCPFE, the European Commission (1998a) submitted to the Council a Communication proposing an EU forestry strategy (EUFS) in late 1998. A Council Resolution followed in December 1998 that officially declared the importance of the forest sector and established the framework conditions for the implementation of the EUFS (European Council, 1998). The overall objective of the EUFS is to strengthen sustainable forest development and management. Like the international and MCPFE procedures, the EUFS is a green paper without an implementation mechanism. The need for PI at the national level is emphasized, calling the Member States (MS) to implement the EUFS through national and regional forest programmes or equivalent instruments, in accordance with the principle of subsidiarity. The EUFS makes an attempt to organize existing forest-related regulations scattered in different policy texts. The Commission is invited to report to the Council on the implementation of the EUFS within five years. This report is currently under preparation.

The EUFS highlights the role of forests for the conservation and enhancement of biodiversity, the combat against climate change and the contribution of forestry and forest-based industries to income, employment and other quality-of-life enhancing factors. Forest management is understood as a tool to achieve the goals of other EU policies. On the other hand, the EUFS suggests the 'effective coordination between different policy sectors which have an influence on forestry and of coordination at Community level' using specific EU committees, working groups and *ad hoc* consultation fora (European Council, 1998). Policy integration is understood as a joint consultation effort to integrate 'measures in the forest programmes or equivalent instruments of the Member States'.

An important step towards PI on forest matters was taken in 2001, when the Commission established the Inter-Service Group on Forestry to increase the joint organizational capacity of the Directorates-General of the Commission and strengthen the co-ordination of forestry issues among the different Community policies (Van de Veelde, 2003).

Another body aiming at facilitating forest-related synergies among EU policies is the Standing Forestry Committee (European Council, 1989), an *ad-hoc* consultation forum on cross-sectoral issues such as rural development, Natura 2000 and forests, research and forest certification, etc. However, considering the overall actions taken during the last six years, it appears that the existing institutional framework and formal competences in relation to forestry matters in the EU have not changed since 1998, when the EUFS was launched.

National Forest Programmes

The concept of the National Forest Programmes (NFPs) was defined by the Intergovernmental Panel on Forests (IPF) and was considered by its successors, the Intergovernmental Forum on Forests (IFF) and the United Nations Forum on Forests (UNFF), as a vehicle to implement forest-related decisions taken within these international fora. NFPs have also become a major topic at the MCPFE and of the EUFS (Hogl, 2002). Considering that all processes mentioned transfer the responsibility for implementation to the national level, the NFPs are the only implementation mechanisms for international forest policy agreements.

The NFPs are national level participatory dialogue processes involving all forest-related sectors, aiming at the conservation, management and sustainable development of a country's forests, in accordance with local, regional and global needs and demands. They should result in agreed objectives, policies and strategies on SFM. NFPs adopt a holistic, inter-sectoral approach, thus becoming tools for achieving PI at the national level during both policy formulation and implementation.

Despite international support and guidance, the NFPs have found minimal application in European countries so far. Where the process has been put forward, mostly in countries where forestry has a large political and financial impact, several problems have arisen, such as: (a) total domination by national forest administrations and participation by directly forest-related actors, (b) lack of political commitment (no budget, no legally-binding procedures) and (c) where participation has broadened, it proved to be an impeding factor, due to the difficulties to co-ordinate heterogeneous groups (Pülzl, 2002). Evidently, under such conditions, the PI objective becomes questionable.

Summarizing the discussion so far, the following points are made:

- the need for PI in forestry has been recognized early due to the complexity of forests and of their exploitation;
- the multiple dimensions of forest sustainability have been addressed in policy formulation at all spatial/organizational levels;
- international level forest policies are green papers possessing no implementation mechanisms; there is a global hesitation in creating legally binding procedures (Convention, EU Forest Policy);
- sectors with binding Conventions and policies dominate the policy arena at all levels;
- although the importance of PI for forest development is recognized, in the lack of implementation mechanisms, the design of specific measures is transferred to individual countries, where the NFPs are seen as a panacea;
- the development of the NFPs has often faced serious problems, due to lack of experience of the actors involved and of national level political commitment.

Sustainable Forest Management – Conceptual and Theoretical Considerations

Sustainable Forest Management (SFM) is a tool to achieve the long-term maintenance of forest cover and the benefits deriving from forests. It is the principal concept of forestry and its development has been influenced by current trends in society, science and the economy. Since the beginning of forest exploitation, forestry has been trying to satisfy social and economic needs, using a natural biological resource. SFM is a unique concept requiring that different aspects are considered, combined and integrated in a balanced way. This section highlights the nature and object of SFM, especially in relation to PI, and presents evaluation criteria for its achievement.

Forest Ecosystems and their Management

Forest ecosystems are characterized by the predominance of trees and by the fauna, flora and ecological cycles (energy, water, carbon, genetic information, nutrients) with which they are closely associated. They are complex and dynamic in space and time. Forest dynamics, the ability of forest ecosystems and their parameters to change, is crucial for their adaptation to future environmental or human-induced changes (e.g. climate change and desertification) and a prerequisite for the long-term maintenance of forest cover.

Because of the dynamic and multi-faceted importance of forests for human communities, their management has to cope with institutional in addition to biological complexity. The large variety of marketed products and of environmental and social services they provide, implicates numerous economic and other interests in decision-making. Their differing perceptions and value systems with respect to forests often create confusion in planning and conflicts over priority setting (Rojas Briales, 2002).

Another important characteristic of forest ecosystems, from both a biological and a socio-economic point of view, is their long reference time. Forest trees grow slowly and reach their productive stage (rotation time) after many decades. As a result, all forest- related values have a long-term reference, a problematic feature when forest policies are compared with other sectoral policies in a globalizing economic environment where long-term investments may not be as attractive to actors as other more immediate profit-generating uses.

Balancing between conservation and production, nature and markets, forestry is called to provide for the production of goods and services in the long run. Forest management is one of the most well-documented environmental management instruments, because of a long-accumulated experience under the most varied natural, cultural and socio-economic conditions. To be effective and contribute to sustainability, forest management needs to take into account the dynamic character and complexity of forest ecosystems in both spatial and temporal terms.

Forests and Forestry in the EU

The forest area of the European Union (EU-25) is around 170 million hectares. The number of private forest holdings reaches 10-11 million. The proportion of private forests cannot be safely estimated, since much will depend on the ongoing restructuring process in Central and Eastern European countries (Arevalo, 2003). Yet, the current percentage of private forests is roughly 65 per cent, with the largest amount of public forests in the new MS that joined the EU in 2004 (MCPFE, 2003c).

Although the EU possesses only 3 per cent of the world's forest area, it is a leading producer of wood-based products and a leading world trader accounting for 38 per cent of exports and 37 per cent of imports. It has become the world's second largest paper and sawn lumber producer, its foremost importer of forest products, and third largest exporter of forest products. The EU forest-based industry has a value-added of around 300 billion Euros, representing 10 per cent of the total for all manufacturing, and employs some 2.2 million people in all parts of the Union. Sweden, Finland, Germany, France and Austria are among the world's top 10 forest product exporters (Papageorgiou and Mantakas, 2004a).

Most forests in Europe are managed on the basis of a formal or informal management plan. Management may concern wood production, biodiversity conservation, recreation or any other objective. Most MS have no 'virgin' forest or 'forest undisturbed by man'. Significant areas of undisturbed forests exist in the Nordic countries (over 6 million ha) 8.5 per cent of the total forest cover in the EU is 'plantations', meaning intensively managed stands of introduced or domestic species, usually with a primary goal of maximum wood production. The majority (85 per cent) of EU forests are considered 'semi-natural', i.e. neither 'undisturbed by man' nor 'plantation'. The semi-natural category contains a wide variety of forest types (UN/ECE/FAO, 2000). Many major wood producing countries have forests that derive from afforestations in this category.

Europe's forests are being damaged by wildlife, grazing, insects, diseases and pollution. In many countries, a significant percentage of trees are recorded as showing over 25 per cent defoliation, although the causes and significance of these figures are not fully clear (UN-ECOSOC and EC, 2000). In the south of the EU, forests suffer from frequent and extensive wildfires, especially during years with dry and hot summers. Wildfires, combined with unsustainable practices and conversion of forests to other uses (settlements, tourism, agriculture), result in large scale deforestation and eventually desertification in bioclimatically-sensitive Mediterranean regions (Pons and Quézel, 1985).

Centuries of non-organized use have reduced the forest area in Europe significantly. This has led to the development of production-oriented forestry models, based on maximum production and a classification of forests according to age classes and rotation periods. However, alternative forestry models, whose primary management objective was not wood production, had existed and developed over a long time also, especially in countries with mountainous forests, or in areas of high cultural and recreational value. During the last decades, a new trend towards 'nature-oriented forestry', giving as much weight to ecological and social functions as to wood production, has emerged owing to the development of scientific disciplines, such as conservation biology, and the rise of environmental movements worldwide.

The Concept of Sustainable Forest Management

The first attempt to establish SFM in the temperate forests of Central Europe aimed at regulating yield, so that the wood capital would not decrease in the long term. This management system focused on the assessment of growth and yield at the stand level. During the last decades, the concept of SFM has greatly evolved. In its broadest sense, SFM means more than just cutting forests at a level of sustainable yield. It is not limited to the preparation of a plan regulating activities in a forest stand, but considers the complexities of forest ecosystems, taking ecological, social and economic aspects into account.

The forestry and natural resources community has debated the concept of 'sustainability' for decades. A high level political commitment to promoting SFM was achieved at the 1992 Earth Summit. Beyond this, several international organizations, such as the IPF, the IFF, the UNCSD, the UNFF, etc., have taken over the dialogue on sustainability in forest management (Humphreys, 2001). Despite these efforts, SFM remains an unclear concept, understood differently by different actors.

In Europe, SFM was defined in 1993 at the MCPFE in Helsinki (MCPFE, 1993) as follows:

> The stewardship and use of forests and forest lands in a way, and at a rate, that maintains their biodiversity, productivity, regeneration capacity, vitality and their potential to fulfill, now and in the future, relevant ecological, economic and social functions, at local, national and global levels, and that does not cause damage to other ecosystems (MCPFE, 1993, p.1).

While this definition is in line with that decided in Rio (1992), it ascribes to a broad conception of sustainability and focuses clearly on the dynamic character of forest ecosystems as a prerequisite for the long term production of goods and services. It was clearly influenced by the Convention on Biological Diversity (CBD) and its concept of the ecosystem approach in managing natural resources. The MCPFE definition of SFM is adopted in this chapter. Its main characteristics are analysed below.

Resolution H1 of Helsinki refers to those characteristics that SFM should have in order to achieve its target. The most important are:

- stable and long-term land-use policies and regulations are needed, preventing land conversion of natural or semi-natural forests;

- SFM is based on periodically updated, scientifically sound plans or programmes at local, regional or national levels;
- SFM provides optimal combinations of goods and services to society;
- SFM considers areas of specific cultural, ecological and protective value;
- forest ecosystems must maintain, and if possible improve, their adaptability towards stresses, natural disasters, erosion, environmental changes, etc.;
- the species and races of plants used for the management of existing forests and the development of new ones must be chosen according to primarily adaptive characteristics;
- native species and local provenances should be preferred for forest management and afforestation purposes;
- public awareness and understanding of SFM, as well as training and human capacity building should be promoted.

Criteria and Indicators for SFM

While the discussion for defining and promoting SFM is still underway, some international forest policy fora have produced sets of Criteria and Indicators (C&I), determining the general objectives or values that must be maintained in SFM (Kneesaw et al., 2000). A Criterion is a category of conditions or processes by which SFM may be assessed. An Indicator is a quantitative or qualitative measure of an aspect of a Criterion showing current performance and its trends (Canadian Forest Service, 1997). Indicators are monitored periodically to assess change (McDonald and Lane, 2004).

Three main international processes have produced C&I for SFM: the 1995 Montreal Process that established C&I for temperate and boreal forests, the International Tropical Timber Organization (ITTO) and its Manual on C&I for Sustainable Management of Natural Tropical Forests (ITTO, 2002) and the MCPFE for Europe (MCPFE, 1998b). The MCPFE C&I, presented in Resolution L2, under the title 'Pan-European Criteria, Indicators and Operational Level Guidelines for Sustainable Forest Management', aim at the harmonization of economic, social and ecological interactions in forests across Europe (Glück, 1999). The signatory countries committed themselves to use the MCPFE C&I as a reference framework and to adjust them to the specific national conditions. The implementation of the MCPFE resolutions and the C&I were endorsed by the EUFS (Wulf, 2003). Most EU countries have prepared national C&I. The MCPFE C&I (Table 8.1) will be used in this chapter to evaluate the achievement of SFM.

Methodological Approach for the Analysis of Policy Integration towards SFM

Policy integration is necessary for the development and application of SFM on the ground, because, first, non-forest policies influence the components of SFM and the ways these relate to forestry; and, second, SFM requires a broad temporal and spatial scale, completely different from the classical forestry approach where forest management focuses on the stand level, thus, implicating the objects of other policies. Almost all EU policies influence forests and forestry in the MS.

Table 8.1   MCPFE C&I for SFM

Criteria	Indicator areas
Maintenance and appropriate enhancement of forest resources and their contribution to global carbon cycles	General capacity Land Use & Forest Area Growing Stock Carbon Balance
Maintenance of forest ecosystem health and vitality	Status of Forests
Maintenance and encouragement of productive functions of forest (wood and non-wood)	Wood Production Non-Wood Products
Maintenance, conservation and appropriate enhancement of biological diversity in forest ecosystems	General conditions Representative, Rare and Vulnerable Forest Ecosystems Threatened Species Biological Diversity in Production Forests
Maintenance and appropriate enhancement of protective functions in forest management (notably soil and water)	General Protection Soil Erosion Water Conservation in Forests
Maintenance of other socio-economic functions and conditions	Significance of the Forest Sector Recreational Services Provision of Employment Research and Professional Education Public Awareness Public Participation Cultural Values

Source: MCPFE (1998b)

For the present purposes, policies with the largest impact on forestry and relevant to the mitigation of desertification in Europe are singled out for examination. These are: rural development, biodiversity, water resources, and natural disasters policies.

Policy integration can be conceived in a vertical and in a horizontal sense. The vertical sense concerns the integration of forest-related concerns in extant sectoral policies (IFC). The horizontal sense implies the integration of policies with respect to their objects, goals, actors, structures and procedures, and instruments (FPI, Forest Policy Integration).

The methodology adopted to address the main questions of this chapter comprises the following steps. First, the degree of IFC in selected EU policies is identified and assessed, based on specific criteria. Then, using the C&I of Table 8.1, the contribution of the IFC achieved to SFM is evaluated. The results of the assessment and evaluation of the policies considered are synthesized and discussed to come up with a preliminary assessment of the possibility of FPI to achieve SFM and the prospects for formulating a EU Common Forest Policy (EUCFP).

The criteria used to assess the IFC in the selected EU policies refer to the policy object, goals and objectives, actors, structures and procedures, and instruments (see, Chapter 2, this volume, for details). More specifically, it is asked whether (a) the object of the policy examined takes into account, directly or indirectly, forest-related issues; the policy approach and underlying theory and concepts agree with the basic SFM principles, (b) the policy goals and objectives comply with those of SFM, (c) policy actors relate to, communicate, cooperate with forest-related ones, (d) policy structures and procedures accommodate the SFM requirements and (e) policy instruments address satisfactorily forest-related needs.

The IFC assessment and evaluation task presents several difficulties. The implementation of forest-related measures concerns 25 EU MS now and can be neither monitored nor evaluated here. Furthermore, differences in social, ecological and economic conditions and administrative structures do not permit a global implementation assessment. Therefore, what will be evaluated is not the direct effect of EU policies on specific conditions in individual countries, but rather the environment these policies create for the implementation of the forest-related measures, taking the diversity of MS into account.

The Integration of Forest-Related Concerns in EU Policies

This section assesses the degree of integration of forest-related concerns (IFC) in selected EU policies and evaluates its contribution to SFM. The findings of this section will be used to investigate whether it is possible to integrate existing EU policies (FPI) or whether it might be preferable to design an integrated EUCFP to achieve SFM.

The Integration of Forest-Related Concerns in Rural Development Policy

The EU rural development policy is part of the Common Agriculture Policy (CAP) which is administered by DG-Agriculture. Forest policy issues and forestry measures are included under the Rural Development Regulation. According to the

European Commission (European Commission, 1997), 80 per cent of the EU territory is considered 'rural'. The rural areas include a great variety of cultures, landscapes, nature and economic activities that shape specific rural identities. Rural areas undergo major transformations (Elands and Wiersum, 2001) which affect the role of forestry as well. Forest production is no longer the only function of forests, since emphasis is now given to their ecological and amenity services. Forest policy is considered to be an integral part of the future EU policy on rural development (Kennedy et al., 1998). Both the EU and various MS stress the importance of forestry in the diversification of the countryside (i.e. afforestation and woodland restoration) and the maintenance of the economic vitality, social attractiveness and ecological integrity of rural areas (Koch and Rasmussen, 1998).

Since the first Conference on Rural Development held in Cork, Ireland in 1996 and the Cork Declaration, where the forest land-use was mentioned in very general terms, the CAP has undergone several reforms with special significance for forestry in Europe (European Commission, 2003b). The accompanying measures of the 1992 CAP reform included Regulation 2080/92 that established a Community aid scheme for forestry measures in agriculture, providing mainly for financial aid to convert marginal agricultural land to forests (European Commission, 2003a).

In 1994, the European Commission decided to create the European Forestry Information and Communication System (EFICS). In 1999, with a view to establishing the EFICS, the Commission adopted a Work Programme for 1999-2002 aiming to compile existing information in a computerized system and to improve and harmonize data on forests, forestry products and trade (European Commission, 1999).

In 1999, the Council reached political agreement on the Agenda 2000 agricultural reform package that represented a deepening and an extension of the 1992 reform for market policy and the consolidation of rural development as the second pillar of the CAP (Papageorgiou and Mantakas, 2004a). The EU rural development policy seeks to establish a sustainable framework for the future of rural areas based on the following main principles:

- the multifunctionality of agriculture and forestry, i.e. their varied role, over and above the production of foodstuffs and raw materials, implying the recognition and encouragement of the range of services provided by farmers and foresters, compatible with the new understanding about forest ecosystems;
- a multisectoral and integrated approach to the rural economy to diversify activities, create new sources of income and employment, and protect the rural heritage;
- subsidiarity for MS to draw up their rural development programmes.

The most important measure included in Agenda 2000 is the Rural Development Regulation (RDR)[1] whose Chapter VIII (articles 29-32) is devoted to forestry measures (European Council, 1999). The RDR is meant to replace all past EU schemes relating to forestry and the CAP, thus becoming the major EU funding mechanism in forestry.

The main principles of the EUFS, such as multifunctionality and sustainability, are reflected in the EU rural development policy, which brings together economic, social and environmental objectives into a coherent package of voluntary measures, thus adding value to the implementation of forest programmes of the MS. At the same time, the EU forestry measures seek to contribute to global issues, such as climate change and biodiversity conservation. The integration of forestry aspects in rural development policy follows three pathways, especially for privately-owned and municipal forests (European Council, 1999):

- investments to improve the multifunctional role of forests (article 30);
- afforestation of agricultural land (article 31);
- improvement of forest protection values (article 32).

Emphasis is placed on linkages with other policy areas and land uses, as well as on specific socio-economic and ecological factors, in line with the following basic principles (European Council, 1999):

- interdependence of different sectoral and horizontal policy areas;
- regional diversity;
- bottom-up approach.

Rural development programmes (RDP) are based on plans prepared by the MS at the most appropriate geographical level for a seven-year period (2000-06). The MS are responsible for ensuring effective monitoring of the implementation of these programmes, by means of physical and financial indicators agreed with the Commission, setting up monitoring committees, where necessary, and reporting to the Commission annually. Evaluation is governed by the regulation on the financing of the CAP.

The implementation mechanisms of the EU rural development policy are rather complex and depend on the mechanisms of the EU Regional Policy. The main implementation instrument of the CAP is the European Agricultural Guidance and Guarantee Fund (EAGGF or FEOGA). The 'Guidance' section contributes to spending on the structural reform of agriculture and the development of rural areas. For 2000-2006, EAGGF Guarantee and Guidance support the rural development measures with 3.192 Mio Euro and 1.546 Mio Euro respectively. Of these, 2.385 Mio Euro are planned to support afforestation measures only (MEDACTION, 2004a).

---

[1] Regulation 1257/1999 on support of rural development.

The specific forestry measures of the RDR aim at maintaining and developing the economic, ecological and social functions of forests in rural areas. While the object of Regulation 1257/1999 is clear and multidimensional, there is an apparent confusion over the specific targets and objectives of the different rural development measures, since the MS have their own priorities (MEDACTION, 2004b). Furthermore, an agricultural bias is evident in the design and implementation of these measures, most of which do not correspond to rural development strategies.

As the RDPs are currently at different stages of implementation, it is not possible to present a consolidated overview of measures implemented in different countries. As far as article 30 is concerned, it seems that most countries have concentrated so far on silvicultural measures to enhance the overall quality of forest stands, investments to improve the ecological value of forests and to improve forestry operations, setting-up forest holders associations, protection against fire and restoring the forestry production potential damaged by natural disasters and fire (Papageorgiou and Mantakas, 2004a). It is worth noting that these measures have been planned for private forests only.

As far as Article 31 is concerned, the Commission presented a report in 2001 (European Commission, 2001a) that evaluates the economic, social and environmental impacts of the measures in the EU. The report indicates that all countries benefited from the favourable effects of diversification of agricultural activities and the development of activities connected with afforestation. It is estimated that 150.000 full-time equivalent jobs were temporarily created from afforestation operations. While the contribution of the measures to the reduction of agricultural production was rather limited, they facilitated the occupation of marginal agricultural land with lesser potential, thereby preventing its abandonment.

Article 32 is implemented in a limited number of countries. Several difficulties seem to hinder its implementation, including lack of tradition of contracts between forest owners and administrations, burdensome administrative procedures, and limited financial resources within the RDPs (European Commission 2003c).

Summing up, the RDR is an innovative tool with a broad menu of measures offering considerable potential to support sustainable rural development throughout the EU. However, despite the promising description of concepts and objectives, the RDPs focus on agriculture and do not use all available measures effectively. The integration of environmental objectives in forestry and agricultural measures in rural development should be improved as was stated at the Second European Conference on Rural Development (European Commission, 2003c). This conference pointed also to the large number of programmes, programming systems and rules seriously overloading the administrative mechanism of the MS and the Commission, reducing transparency and visibility, and to the need for significant simplification of the delivery system and strengthening partnerships between public and private organizations and the civil society (European Commission, 2003c).

Table 8.2 presents a rough assessment of the degree of integration of forest-related concerns (IFC) in Agenda 2000 and Regulation 1257/1999. The two pieces prioritize the concept of 'sustainable rural development' and connect it with SFM. However, their objectives and measures reveal a narrow understanding of 'sustainability' and its application. Sustainability refers to the economic aspects of agricultural production, with social and environmental benefits being positive externalities and not prerequisites. This conception of sustainable rural development resembles the original SFM definition of the early 20th Century. According to recent reports (European Commission, 2003c), both the Commission and the MS have recognized this problem and are moving towards more integrated approaches.

Table 8.2   Degree of IFC in the EU Rural Development Policy

Criteria	Description	Quantification
Policy object	'Sustainable rural development' is very similar to SFM. Certain concepts used (multifunctionality, multisectoral & integrated approach) are in line with forestry purposes.	***
Goals and objectives	The main objectives follow the pattern "ecological, economic, social", but focus mainly on agriculture.	**
Actors	DG Agriculture, Standing Forest Committee, European Parliament	**
Structures, and procedures	Member States design RDPs. Different priorities, administrative problems, agriculture bias, priority in afforestations as far as forestry measures are concerned. RDPs are agriculture-oriented. Weak implementation mechanisms due to the subsidiarity principle.	*
Instruments	Funding through EAGGF is complicated. Only 10% of funds are dedicated to forestry. More than half goes to afforestations.	*

(Integration level: * = weak, ** = moderate, *** = strong)

As a result of this conceptual confusion, most measures and their implementation mechanisms have a strong agriculture bias. Forestry receives a rather low priority, the main objective being the creation of plantations on marginal farmlands. Moreover, since the MS place their own priorities as regards EU aid, forestry lags behind in many countries with a weak forest sector. Administrative overload impedes the effective implementation of the EU rural development policy. Until recently, the measures were designed for private forests only, leaving outside a large area of public forests, mostly managed for social and environmental services.

Table 8.3   Evaluation of the contribution of current IFC in the EU Rural Development Policy towards SFM

Criteria & Indicators	Quantification	Description
Carbon cycles		The EU rural development policy promotes mainly afforestation measures that have resulted in a net increase of carbon stocks on agricultural land. Issues of non-permanence, leakage, additionality, etc. are not clarified, according to the IPCC rules.
General capacity	+	
Land Use & Forest Area	+	
Growing Stock	++	
Carbon Balance	0	
Health and vitality		Measures for improving the status of forests have been defined, but their application has been minimal.
Status of Forests	0	
Productive functions		Focusing more on agriculture than on rural development, the forestry measures are production-oriented mainly. The policy does not touch upon non-wood goods.
Wood Production	+	
Non-Wood Products	0	
Biological diversity		In some countries, the implementation of forestry measures has improved the mixture of forests. However, the implementation mechanism is vague and does not secure the creation of complex ecosystems. Some positive examples exist, but most measures are indifferent or even negative to biodiversity, since they focus on production.
General conditions	0	
Representative, Rare and Vulnerable Forest Ecosystems	0	
Threatened Species	0	
Biological Diversity in Production Forests	+	
Protective functions		As mentioned above, no such character is visible in the implementation of the forestry measures
General Protection	0	
Soil Erosion	0	
Water Conservation in Forests	0	
Socio-economic		
Significance of the Forest Sector	+	The forestry measures have definitely created more jobs and income in the forest sector. No other social service has improved significantly due to these measures.
Recreational Services	0	
Provision of Employment	++	
Research and Professional Education	0	
Public Awareness	0	
Public Participation	0	
Cultural Values	0	

(Contribution level: ++ very positive, + positive, 0 indifferent, - negative, very negative)

Table 8.3 offers a broad evaluation of the contribution of the current IFC in the EU rural development policy towards SFM using the Pan-European C&I. It contains an assessment of the level of reference to specific C&I in the rural development policy and a description of its effect towards SFM. The only positive effects of the forestry measures are on carbon stocks, employment and production. The social and environmental parameters of SFM are neglected. IFC in this policy does not contribute much to the achievement of SFM.

The Integration of Forest-Related Concerns in the EU Biodiversity Policy

The concept of biodiversity in forest ecosystems The terms 'biological diversity' or 'biodiversity' are broadly used and are encountered in international conventions, world summit reports, global and regional environmental action plans, etc. Biodiversity has become an important concept in conservation biology and other theoretical and applied sciences, such as forestry, agriculture and wildlife biology (Haila and Kouki, 1994). Despite its wide use and numerous definitions, the concept is extremely vague and difficult to use in the design of actual conservation measures. For this reason, the main characteristics of biodiversity in relation to forest ecosystems and SFM are described first before discussing the EU Biodiversity policy.

The first definitions of biodiversity appeared in the U.S. Office of Technological Assessment (1987), the common publication by IUCN, UNEP and WWF (1991) and the Convention on Biological Diversity (CBD) (UNEP, 1995). Their main characteristics are (Perlman and Adelson, 1997)

- biodiversity is viewed at three different levels: genes, species and ecosystems;
- biodiversity is used to describe the number, variety and variability of living organisms, embracing many different parameters, becoming essentially a synonym to 'life on earth'.

At first sight, these definitions appear simple. Yet, they face several conceptual problems concerning the boundaries between taxonomical entities, the parallel reference of different hierarchical levels and the over-simplified bias of considering species richness as the main measure of biodiversity.

The need to quantify biodiversity has led scientists and policy makers to oversimplify its content and ignore its dynamic character. Biodiversity is much more than just a measure of species abundance. Biodiversity conservation strategies should focus more on the dynamic character of nature rather than on the endless count and maintenance of biotic entities (Papageorgiou and Kasimiadis, 2004).

Forest ecosystems, like all terrestrial ecosystems, have a great diversity at all levels, species, genes, and ecosystems, which determines their adaptive, dynamic character. Because adaptation of forest ecosystems to environmental change is crucial, it is very important to ensure the conservation of forest biodiversity to maintain healthy and well adapted forests (Papageorgiou et al., 2003).

The global discussion on biodiversity conservation has influenced greatly the development of forest management concepts, especially after the launching of the CBD in 1992. Forest policy makers and managers have seen this 'invasion' of principles, ideas and rules as a threat to their understanding of 'sustainability' in forest management. Most foresters believe that multifunctional forest management can address the need for biodiversity conservation, since it can become one of the SFM targets, together with recreation, water and soil protection, etc. However, since forests are only managed when they provide profit, timber production is mostly seen as the main target and other benefits, including biodiversity, are considered as 'positive externalities' (Fig. 8.1a). On the other hand, many environmentalists have seen the forest sector as a threat for biodiversity. In several cases, they have developed complex forest management plans for the protection of a specific species (mostly easily recognizable, the so-called 'charismatic megafauna') or specific areas, where forest management is seen as a tool to conserve biodiversity while the other products and services are less important than the primary management goal, i.e. biodiversity conservation (Fig. 8.1b).

These two approaches have been in conflict at all levels of policy discourse. Conceptually, both suffer from the same problem; they consider biodiversity elements and not its dynamics. Thus, the object of primary or secondary management is usually a threatened species or an ecosystem of great 'ecological importance'. The dynamic character of biodiversity and its role in adaptive evolution are neglected. Furthermore, the application of both approaches on the ground is problematic. The forestry approach depends on the marketed value of forest products, placing low importance on non-marketed goods and services, and it cannot be utilized where ecosystems do not provide timber in adequate quantity and quality, such as the Mediterranean maquis. On the other hand, the environmentalists' approach applies only to protected areas while organizational and social constraints limit its generalized application.

A more dynamic approach towards SFM and biodiversity conservation considers the existence of biodiversity as a condition for the good functioning of forest ecosystems and their long-term ability to supply society with goods and services. Irrespective of target, forest management activities must maintain the dynamic cycles of nature to achieve sustainability. SFM concerns the integrity of specific biological systems that secure the adaptability of forest ecosystems to future changes (Papageorgiou et al., 2004). This section adopts the approach presented in Fig. 8.1c in order to assess the integration of forest elements in the EU biodiversity policy and to evaluate its effect on SFM.

The integration of forest-related concerns in the EU biodiversity policy The main documents describing the biodiversity policy of the EU are the EU Biodiversity Strategy and the Directive 92/43 (Habitats Directive). The overarching goals of the EU Biodiversity Strategy are: (a) to contribute to reversing present trends in biodiversity losses and (b) to place species and ecosystems in a satisfactory conversation status both within and beyond the EU territory.

a) The forestry approach; biodiversity as a positive externality of SFM

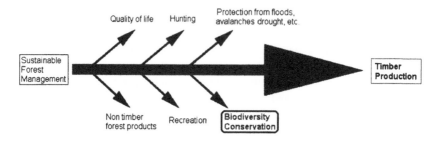

b) The environmentalist approach; SFM as a tool to conserve biodiversity

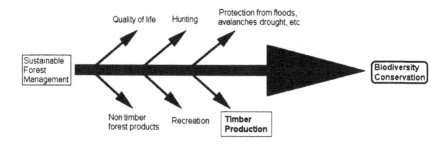

c) The 'dynamic approach'; biodiversity as a prerequisite for the production of goods and services

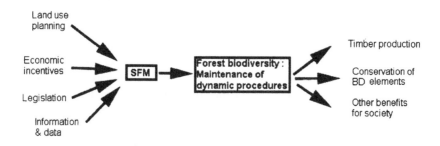

Figure 8.1   Different approaches concerning SFM and biodiversity conservation

Source: Papageorgiou et al., 2003

The Strategy, recognizing that globally, forests contain the greatest quantity of biological diversity' (European Commission, 1998b), includes a section that lists objectives for the conservation of biodiversity in European forests.

An important principle of the Biodiversity Strategy is the need for changes in the human patterns of consumption in the long term to achieve sustainable development. This is compatible with forest-related targets and strategies, due to a) the long time reference of forestry investments, b) the social character of forest management, which aims at long term benefits to society and c) the principle of sustainability. The rationale of the Biodiversity Strategy for protecting biodiversity (economic and environmental values, ethical principle of preventing avoidable extinctions) agrees with the main economic, environmental and social principles of SFM.

To implement this strategy, the Commission has issued Communications for Biodiversity Action Plans (BAPs) (European Commission, 2001b) for agriculture, fisheries, natural resources, and development co-operation. Forest biodiversity aspects were integrated in the natural resources and agriculture BAPs.

In 1992, the Council adopted the Habitats Directive, on the Conservation of Natural Habitats and Wild Fauna and Flora (European Council, 1992a), whose aim is to create a coherent European ecological network, the 'Natura 2000', consisting of a series of Special Areas of Conservation that will protect habitats and species of Community interest. Other important legislation relevant to forest biodiversity conservation is the Birds Directive (European Council, 1979) and the EU Regulation on the Convention on International Trade in Endangered Species of Wild Fauna and Flora (CITES) (European Council, 1997).

The European Commission is responsible for the establishment and management of the Natura 2000 network and for the enforcement of the Habitats Directive in the MS. It provides also financial support for protected forest areas through its programmes LIFE-Nature and LIFE-Environment.

The establishment of the Natura 2000 network has increased the attention given to forest biodiversity, both at the Community and the national level. The elaboration of lists of proposed Sites of Community Importance has generated considerable discussion, and in some cases conflicts, between the forest sector and conservation circles. However, the Habitats Directive can become a key tool to ensure protection of European forests. Six out of nine Natural Habitat Types of Community Interest, whose conservation requires the designation of Special Areas of Conservation, include forests. It is estimated that 30 per cent of all designated sites concern forest habitats while another 30 per cent partly contain woodland elements and related species. For this reason, DG-Environment has produced a guidance document, Forests and Natura 2000 (European Commission, 2003d), with extensive stakeholder consultation, designed to offer a better understanding of nature conservation in forests. It makes very clear that the establishment of the Natura 2000 network does not oppose forest economic activities and recommends identifying the measures required to maintain biodiversity through discussion with stakeholders and turning the outcome of this process into formal management objectives.

The implementation of the Birds and the Habitats Directives as well as the conservation and sustainable use of areas designated under this legislation require action within and outside designated areas, where forest management plays a major role. There is a lack of appropriate sanctions when measures are not properly implemented in all relevant sectors. The system of indicators and mechanisms for monitoring environmental conditions is of major importance for EU biodiversity policies and measures and can be linked with the C&I process in SFM.

Despite recent decisions and measures, the integration of biodiversity conservation in sectoral policies (including the forest sector) is still lacking. Provisions for integrating biodiversity issues in land-use planning exist only for protected areas and are in early stages of development. No reference is made to the use of existing land use regulations and forest legislation already developed through the forest policies of several countries.

Overall, several EU biodiversity goals, objectives and targets can be achieved through forestry measures. However, forests are still seen as a broad biodiversity element only. There is no clear reference to the potential use of SFM in the texts and forestry is considered part of agriculture: a related, but external, sector and often a threat. This attitude has started changing recently, mainly through the consideration of the SFM concepts and tools that are much more advanced than the biodiversity-related ones. The possibility of attaining biodiversity objectives through forest-related (and other sectoral) measures is gaining importance within the EU but results are still awaited.

Table 8.4 summarizes an assessment of the degree of IFC in the EU biodiversity policy based on the preceding analysis. The main findings are as follows:

- EU biodiversity policy documents mention forest ecosystems, mainly as biodiversity elements, or places where biodiversity is high;
- forestry is classified under agriculture and is often considered a threat;
- recent documents (European Commission, 2003d) have changed their view and see SFM as a tool to promote biodiversity objectives;
- the implementation effectiveness of this policy on the ground depends on MS action, as most policy documents do not suggest implementation mechanisms;
- the LIFE funding instrument is not adequate to promote the protection of biodiversity in most managed forests.

Table 8.5 presents an evaluation of the contribution of the IFC achieved in the EU biodiversity policy towards SFM. In general, the degree of IFC achieved has not contributed significantly to SFM, except for the protection of specific forest types and tree species. Some positive effects may also be recorded in the development of recreation services in forests.

Table 8.4   Degree of IFC in the EU Biodiversity Policy

Criteria	Description	Quantification
Policy object	Early approaches consider mainly biodiversity elements, ignoring the dynamic character of forests as ecosystems. Recent approaches have improved and come closer to forestry concepts.	**
Goals and objectives	The main objectives consider forests as "object" of biodiversity conservation only, neglecting the dynamic character of ecosystems. Recent texts show improvements.	*
Actors	DG Environment; few linkages to other DGs.	*
Structures, procedures and implementation mechanisms	The MS are responsible for policy implementation. Different priorities, administrative problems. Biodiversity elements are prioritized.	*
Instruments	LIFE-Nature and LIFE-Environment address only specific problems dealing mostly with species and specific areas. Linkages with other Funds, such as the EAGGF, are considered but not elaborated yet.	*

(Integration level: * = weak, ** = moderate, *** = strong)

On the other hand, most biodiversity policy documents see the productive side of forestry only and consider the management of forests as a threat against nature. In several cases, this has reduced the importance of the forest sector for the Union, especially since forestry is not supported by a common policy or a global UN Convention.

The Integration of Forest-Related Concerns into EU Water Resources Policy

In the last decade, a major revision of the EU water resources policy took place resulting in the Water Framework Directive 2000/60/EC (WFD) that attempts to reconcile the conflicting approaches of former directives focused on water quality into a more integrated approach, requiring MS to co-ordinate the protection of their waters under its scope (European Parliament and European Council, 2000). The WFD, premised on that 'water is not a commercial product like any other but, rather, a heritage which must be protected, defended and treated as such', adopts the ecosystem approach like the EUFS and SFM. It has several principles, goals and objectives in common with them including the sustainability, precautionary and, as regards policy implementation, subsidiarity principles.

Table 8.5   Evaluation of the contribution of current IFC in the EU Biodiversity Policy towards SFM

Criteria & Indicators	Quantification	Description
Carbon cycles		The EU biodiversity policy considers mainly negative afforestations that can contribute to carbon sequestration. A positive effect can be recorded in some MS as far as certain restrictions on the use of tree species and areas are concerned. The adaptability of forests towards climate change is not addressed.
General capacity	+	
Land Use & Forest Area	0	
Growing Stock	0	
Carbon Balance	0	
Health and vitality		No connection between the diversity of forest ecosystems and their health status is mentioned.
Status of Forests	0	
Productive functions		Forest production (wood) was considered, until recently, as a threat. SFM is not used as a tool to achieve biodiversity targets. Non-wood products have in some rare cases been promoted as alternative income sources, still without any major overall effect.
Wood Production	0	
Non-Wood Products	0	
Biological diversity		
General conditions	+	Integration was recorded only in cases of protected elements (areas, species) without considering the dynamic character of biodiversity in forests.
Representative, Rare and Vulnerable Forest Ecosystems	++	
Threatened Species	++	
Biological Diversity in Production Forests	0	
Protective functions		
General Protection	0	Mentioned, but not implemented.
Soil Erosion	0	
Water Conservation in Forests	0	
Socio-economic		The importance of the forest sector has definitely diminished after CBD-driven policies were introduced in the EU and the MS. Recreation services in protected areas have increased. Employment declines when the forest sector slows down, although some new professions may have been created in protected areas. Research, professional education and awareness have been stimulated probably by the numerous obligations towards the EU, although this impact has been rather small. In some cases, cultural values have been promoted through the establishment of protected areas.
Significance of the Forest Sector	-	
Recreational Services	+	
Provision of Employment	-	
Research and Professional Education	+	
Public Awareness	+	
Public Participation	0	
Cultural Values	+	

(Contribution level: ++ very positive, + positive, 0 indifferent, - negative, very negative)

WFD goals relating to forests and forestry are: a) protect and improve the aquatic environment, b) contribute to the provision of sufficient supply of good quality surface water and ground water as needed for sustainable, balanced and equitable water use, c) improve the status of aquatic ecosystems and d) mitigate the effects of floods and droughts.

Two WFD objectives are most related to the EUFS and SFM: (a) integrated water resources management based on river basins and (b) achieving 'good status' for all waters by a certain deadline. The first objective considers the river basin as the reference spatial unit for managing water resources holistically. Water quality and quantity is no more a problem of agriculture or consumption, but rather a complex, spatial management issue. River Basin Districts (RBDs) are the main administrative units for which a management plan and a programme of measures are to be prepared and used as a basis for water resources management in the river basin. This approach gives the floor to forest management to plan and apply significant and crucial measures, since the maintenance of a constant and 'healthy' vegetation cover is the most important factor influencing the improvement of water quantity and quality in a river basin. It focuses more on the role of protective forests, especially on steep slopes that characterize usually the upper parts of river basins.

The achievement of the second objective will then result not only from successful water resources management, but from a well functioning integrated management effort, where SFM plays an important role. Furthermore, this objective introduces a significant biodiversity component, which is again influenced by the 'naturalness' of ecosystem management, according to the principles of the EUFS and several international agreements for forests, especially the MCPFE procedure.

In general, both the WFD and the EUFS adopt holistic, long-term oriented approaches, considering bio-physical and socio-economic factors that affect the status of water and forest resources, and decentralized decision making, promoting broad public participation to account for local environmental and socio-economic particularities and needs.

Because the WFD has not been completely transposed to the national legal order yet, a prospective account of implementation issues is offered here. The description of the river basins, the design of the management plans and many WFD measures will inevitably concern forest issues. Several means for implementing the WFD are related to forest management. Forest authorities are, therefore, expected to participate in river basin management. Problems of overlapping responsibilities and conflicting interests, similar to those experienced in the case of protected areas, may arise (Papageorgiou and Catsadorakis, 2001). The identification of RBDs influences and is influenced by the existing planning of forest management activities in space. A change of forest management units may be necessary in some cases, especially where protective forests are concerned.

A crucial prerequisite for the success of the WFD is its integration with other sectoral policies, especially the forest-related ones. The WFD is not binding enough to guarantee this integration and MS have enough flexibility in implementing the directive. This can cause problems in countries whose forest sector is politically and economically important and there are strong pressures to leave the production and trade of forest products unregulated. In countries with less important forest sector, forest management could be ignored and the implementation of the WFD could fail due to the problems caused by overlapping legislation and administrative compartmentalization.

In sum, it is noted that the WFD and the EUFS share similar resource management approaches and spatial and temporal level of reference. The implementation of the WFD depends significantly on forest management activities in the MS. Although SFM can promote the goals of the WFD, the latter makes no reference to forest management or forests in general. The RBDs and the implementation mechanisms suggested by the WFD can act as integrative instruments for holistic resource management, but problems may arise due to administrative complexities and conflicting interests with existing forest authorities.

Considering the WFD as the central, most recent policy document of the EU water policy, whose implementation cannot be evaluated yet though, Table 8.6 presents a preliminary assessment of the potential degree of IFC that can be achieved. Table 8.7 presents a preliminary evaluation of the contribution of the IFC achieved in the WFD towards SFM.

Table 8.6   Degree of integration of forest-related concerns in the WFD

Criteria	Description	Quantification
Policy object	The WFD and the EUFS have compatible concepts and principles and congruent spatial and temporal levels of reference.	***
Goals and objectives	Several WFD and forest-related, EUFS objectives are complementary.	***
Actors	DG Environment. No forest-related actors are involved at the EU level.	*
Structures and procedures	The WFD implementation guidelines include participatory and integrated procedures, where forest-related actors could be involved. However, this is not obligatory for the MS.	**
Instruments	No instruments available yet.	*

(Integration level: * = weak, ** = moderate, *** = strong)

The Integration of Forest-Related Concerns in EU Natural Disasters Policy

Fighting against natural disasters is important for the EU in order to provide an acceptable level of security and well-being to its citizens. Several major natural disasters, such as floods, avalanches, soil erosion and forest fires are related to forests and forestry. Especially forest fires receive high priority by the EU and the MS that are mostly affected as their mitigation is a complex policy issue.

Table 8.7   Evaluation of the contribution of potential IFC in the WFD towards SFM

Criteria & Indicators	Quantification	Description
Carbon cycles		Slight benefits for the carbon balance are expected to occur indirectly, through regulations favouring forest cover protection in watershed provided in the RBPs, as the WFD does not mention afforestations and other measures for the maintenance and enhancement of carbon stocks.
General capacity	0	
Land Use & Forest Area	+	
Growing Stock	0	
Carbon Balance	+	
Health and vitality		No connection between the health status of forests and the WFD.
Status of Forests	0	
Productive functions		Forest production, although irrelevant to the priorities of the WFD, may indirectly benefit, if forest cover increases due to the potential use of forest management as a tool to improve watershed functionality.
Wood Production	0	
Non-Wood Products	0	
Biological diversity		
General conditions	+	The potential IFC may influence forest biodiversity positively through the possible reduction of management intensity in watersheds. Yet, these forests have been always managed with caution, thus, no significant added value is expected.
Representative, Rare and Vulnerable Forest Ecosystems	0	
Threatened Species	0	
Biological Diversity in Production Forests	0	
Protective functions		
General Protection	++	The implementation of the WFD will contribute to the protection of soil and water dynamic cycles in forests.
Soil Erosion	++	
Water Conservation in Forests	++	
Socio-economic		The implementation of the WFD is expected to increase the importance of the forest sector, since forestry activities will assume a new role, possibly increasing the employment possibilities for forest workers and influencing positively education, research and awareness in general. The WFD local participation mechanisms may encourage the participation of forest-related actors as well, such as forest owners and forest administrations, in watershed management. Conflicts may arise, especially if the experience of the relevant actors is ignored.
Significance of the Forest Sector	+	
Recreational Services	0	
Provision of Employment	+	
Research and Professional Education	+	
Public Awareness	+	
Public Participation	+	
Cultural Values	0	

(Contribution level: ++ very positive, + positive, 0 indifferent, - negative, -- very negative)

Most countries and international groups emphasize preparedness and prevention to limit damages as much as possible. This has led to a significant increase in spending for technologically advanced tools to predict, assess and suppress fires. However, the problem worsens; about 40.000 fires destroy 500.000 hectares of EU forests annually, mostly in the southern MS (Dimitriou et al., 2001). Despite the good predictability of forest fires and improvements in firefighting means, the area burnt in the EU-Mediterranean countries during the last decade is almost three times higher than that of the 1970s. This is an indication that the 'operational' approach adopted by most European countries so far, which is based on the improvement of the operational capacity of fire suppression mechanisms, is not as effective as expected.

Forest fires are a complex ecological and socio-economic phenomenon. The causes underlying their increase during the last decades are related to land abandonment, the decline of forest management and the ineffective institutional framework (Dimitriou et al., 2001). A more integrated approach is needed to manage fires as an ecosystem parameter, through the regulation of fuel, human activities and policies. This 'managerial' approach is gaining prominence in global discussions on forest fire mitigation (Alexandrian and Esnault, 1999).

The European Council adopted the Forestry Action Program in 1989 which was reviewed and strengthened in 1992 and included actions for protection against forest fires (European Council, 1992b). The main target was to identify and eliminate the causes of forest fires and to improve forest-monitoring systems. This program was oriented primarily to what the EU defines as 'fire prevention', promoting measures such as the provision of forest paths, firebreaks, water supply points, clearing equipment and monitoring facilities. Yet, these actions should be better described as 'pre-suppression', rather than 'prevention' (Dimitriou et al., 2001).

A milestone in the EU legislation against forest fires was the 1992 Inter-ministerial Seminar on Forest Fires in Lisbon that recommended (a) improved forest fire prevention policies within a framework of enhanced public awareness, environmental education and identification of the social and individual motivations and behavior patterns responsible for the majority of forest fires, (b) the compilation of a database of fire statistics and (c) the need to define and reverse the underlying political and social causes of forest fires in Europe. However, Regulation No. 2158/92, which followed this seminar, did not focus on these latter causes but included measures of a rather technical nature. Aiming to reduce the number of fire outbreaks and the extent of areas burnt, it provided for co-financing the collection of information on forest fires and the implementation of measures to protect forests against fire. Financial aid was distributed to the MS according to Regulations 1170/93, 1460/98 and 1727/99 that followed.

The Community has supported fire prevention actions in high or medium forest fire risk areas of six Member States with 10 million Euro annually, subject to existence of fire protection plans. These plans were also a pre-condition for funding forest fire prevention actions within the Structural Funds (Objective 1 and 5b regions), the Cohesion Fund and the Rural Development Regulation 1257/99 starting in 2000.

On 17 September 1998, alarmed by the wave of fires throughout the Mediterranean region, the European Parliament stressed the heavy human and economic costs and the environmental damage to forests. It called the Commission to make available, especially from the European Regional Development Fund

(ERDF), the resources required to alleviate the damage suffered by affected regions, to implement a Community forestry policy and to establish closer coordination between the MS on preventing and fighting forest fires.

The EU legislative and policy provisions, although recognizing the need for a different approach, usually end up with arrangements for suppression or pre-suppression financial aid (Dimitriou et al., 2001). EU legislation fails to connect forest fires with agricultural and other sectoral policies that are associated with the socioeconomic changes happening in the southern part of the Union. Similarly, the EUFS, based on a Central/Northern European forestry dogma, focuses mainly on production and promotes fire fighting at an operational level.

Forest fires are also included in the EU policy on civil protection whose approach is purely operational, aimed at the quickest and most efficient mobilization of MS to help after a disaster occurs, inside or outside the Union, through flexible legal and administrative provisions. The most significant achievements so far have been the establishment of several operational instruments, pilot projects and self-training workshops, the establishment of the Vade-Mecum of Civil Protection in the European Union and an extensive R&D effort. All EU civil protection initiatives are implemented on the basis of the subsidiarity principle.

The Community Action Programme on Civil Protection has resulted in the development of the European Forest Fire Information System[2] (EFFIS), aiming to give MS information of forest fire risk forecasts and assessments of fire damages. The scheme, set up by Council Regulation No 2158/92, provided significant co-financing for preventive measures implemented by MS, contributed to identifying forest fire-related causes and problems and helped reduce the average size of fires and the duration of single fires.

In 2002, the Commission submitted a proposal for a Regulation concerning monitoring of forests and environmental interactions in the Community, suggesting to combine the existing monitoring and information gathering under Regulations 3528/86 (atmospheric pollution) and 2158/92 (forest fires) into one scheme and a series of new monitoring activities to 'accompany the implementation in the forest sector of Community policies regarding biodiversity conservation, climate change mitigation and soil protection' (European Commission, 2001c). The Commission acknowledged, thus, the need to measure the effectiveness of EU policies, including measures to prevent and combat forest fires. The proposal further suggested the inclusion of isolated forest-related regulations and policies under a common 'umbrella' framework to better address the complexities of forests and forestry. Forest fires are no longer viewed as a natural disaster only. In 2003, Regulation No 2152/2003 (Forest Focus, FFR) was adopted which has four main pillars (European Parliament and European Council, 2003):

- monitoring of air pollution effects on forests;

---

[2] http://www.europa.eu.int/comm/agriculture/fore/fires/scif/index_en.htm.

- forest fire monitoring;
- forest fire prevention (complementary to measures financed by the RDR);
- studies to develop the scheme in relation to other environmental parameters such as biodiversity, soil condition, carbon sequestration, and climate change.

The scheme will run for the period 2003-2006. During 2004, the Commission will adopt the necessary implementing regulations with the assistance of the SFC.

Summarizing, the EU has so far followed an 'operational' approach towards forest fires, emphasizing suppression measures and activities that it calls, however, 'prevention'. Real prevention measures require a higher level of integration with existing forest-related policies and management techniques. A first step in this direction is the FFR, although much depends on the measures MS will take based on the information it will provide.

Table 8.8 presents a broad assessment of the degree of IFC in the EU natural disasters policy (forest fires). The level of the 'managerial' instead of the 'operational' character of this policy is evaluated mainly, exploring whether the EU considers forest fires as a natural disaster only, or if it adopts a more integrated approach.

Table 8.8   Degree of IFC in the EU Natural Disasters Policy

Criteria	Description	Quantification
Policy Object	The EU follows an 'operational' approach, emphasizing mainly suppression. The FFR changes this approach, acknowledging the complexity of the issue.	**
Goals and objectives	The goals and objectives of the Forest Focus aim at supplying information to contribute to better co-ordination between different policies.	**
Actors	DG Agriculture, DG Civil Protection, SFC. Co-operation with other actors, such as the UN Conventions and the DG Environment is considered.	**
Structures and procedures	Most MS follow the operational approach, which has proven inadequate. In several cases, they have used the funds provided for other purposes.	*
Instruments	The mechanism used for funding Rural Development through the SFs is used. More integration and control is needed.	*

(Integration level: * = weak, ** = moderate, *** = strong)

Finally, the contribution of the IFC achieved towards SFM is evaluated considering the current regulations, since the FFR has not had any effects on the condition of forests in Europe yet. Table 8.9 presents a very negative general picture, although much progress has been made in the field of scientific research and networking, because the approach adopted by the EU and the MS is deemed to be ineffective considering the net increase in the forest areas burnt all over the EU, especially in the south. The adoption of the FFR is expected to improve the situation, since it follows a holistic approach, stressing forest management and integration of several EU policies.

Table 8.9   Evaluation of the contribution of current IFC in the EU Natural Disasters Policy towards SFM

Criteria & Indicators	Quantification	Description
Carbon cycles		The 'operational' approach of the EU policy against forest fires has not managed to reduce the number of fires and the areas of forests burnt. An improvement is expected with the adoption and implementation of the new 'managerial' approach. Forest fires release carbon back into the atmosphere and forest areas burnt are recorded as emissions in the national reports sent to the UNFCCC.
General capacity	0	
Land Use & Forest Area	-	
Growing Stock	-	
Carbon Balance	-	
Health and vitality		The health status of EU forests declines since forest fires are not integrated in forest management.
Status of Forests	-	
Productive functions		Forest fires do not usually affect wood producing forests but they may reduce the production of non-wood forests that occur in the Mediterranean to a great extent.
Wood Production	0	
Non-Wood Products	-	
Biological diversity		Biodiversity concerns are included in the FFR. The current status of the EU policy on forest fires has not succeeded in reducing the threat against valuable species and ecosystems from wildfires.
General conditions	-	
Representative, Rare and Vulnerable Forest Ecosystems	-	
Threatened Species	-	
Biological Diversity in Production Forests	0	
Protective functions		The ineffectiveness of forest fire management has definitely negative effects on the ability of forests to protect soil and contribute to water balance, especially in the southern MS.
General Protection	-	
Soil Erosion	--	
Water Conservation in Forests	-	
Socio-economic		
Significance of the Forest Sector	0	
Recreational Services	-	The IFC in the EU policy against forest fires has promoted the distribution of accurate information and public awareness. However, recreation services are negatively affected.
Provision of Employment	0	
Research and Professional Education	+	
Public Awareness	+	
Public Participation	0	
Cultural Values	0	

(Contribution level: ++ very positive, + positive, 0 indifferent, - negative, -- very negative)

Summary Assessment and Evaluation

The results of the preceding analysis are synthesized to offer an overall evaluation of the current degree of IFC in selected EU policies and its contribution to SFM (Table 8.10).

Although policies adopt differing approaches, the IFC in their policy object is satisfactory. The EU rural development and forest fires policies make explicit reference to forests and forestry. An important objective of the biodiversity policy is the conservation of forest biodiversity.

The WFD makes no explicit reference to forests, but its potential for integration is considerable, since its principles and provisions relate directly to and agree with those of SFM. However, in each policy, forests are defined differently; instead of a common horizontal conceptual understanding of forests, their values and their needs, the view of forests in each policy is fragmented and sometimes contradictory or irrelevant (e.g. CAP vs. biodiversity policy). This fact impedes their overall synthesis into an integrated policy system that addresses forest issues robustly and effectively.

Table 8.10   Overall degree of IFC in EU policies

Criteria	Policies			
	Rural development	**Biodiversity Conservation**	**Water resources**	**Natural disasters**
Policy object	***	**	***	**
Goals and objectives	**	*	***	**
Actors	**	*	*	**
Structures and procedures	*	*	**	*
Instruments	*	*	*	*

(Integration level: * = weak, ** = moderate, *** = strong)

The IFC in policy goals and objectives is not complete, since these depend on the needs and interests of the associated sectors. The CAP sees forestry predominantly as afforestation and production, the biodiversity policy considers forests as elements and not as dynamic systems and the policy against forest fires follows the operational approach. The WFD adopts an approach that agrees with SFM, although it does not consider forests at all.

Many different actors are associated with the EU policies examined and almost all of them are not related with the forest sector. Therefore, they usually do not consider forest-related issues as a priority. This situation has partially improved through the SFC and some other thematic groups that present their view to all EU bodies. Yet, more research on the various actors and their role in IFC is needed.

As policy implementation rests with the MS, countries determine policy goals, procedures and instruments according to their needs and the influence of the relevant sectors. The generally weak implementation of EU environmental policies carries over to forest protection and management, because of the weak integration of forest-related concerns into extant EU policies. Again, lack of compatible definitions of 'forests' in EU policies, MS or even sectors within the same country, is among the main reasons for the overall ineffectiveness of the measures taken. Integration at the implementation stage is, therefore, weak, except for the prospective implementation of the WFD.

The main financial instruments used for policy implementation are the SFs. Forests and forestry receive direct support in the case of the CAP and the policy against forest fires, but the activities funded for forestry differ among countries because of different national priorities. As a result, funds dedicated to multifunctional forest management may finally support other activities such as the construction of rural roads (Papageorgiou and Mantakas, 2004b).

Based on the Pan-European C&I, Table 8.11 presents an assessment of the contribution of the current IFC in EU policies towards SFM which has been influenced, however, by differences in the degree of 'maturity' among policies and the availability of adequate data for all policies. Thus, since the WFD has not been implemented yet, its 'potential' integration is evaluated mainly introducing a positive bias for the WFD, since policy integration may be 'lost' during implementation. The FFR is more promising and compatible to SFM than the WFD. However, because forest fires continue unabated following actually a global trend, the EU policy against forest fires receives a low score.

The evaluation results suggest the following:

- Overall the current level of IFC does not contribute to SFM. Complementarity is achieved in some cases, but effects from different policies are contradictory mostly. Each policy valuates and prioritizes forests differently, following its particular conceptual approach. Where the approach is closer to SFM (e.g. WFD, rural development) the influence is greater;
- The WFD seems to have the best contribution to SFM, but its evaluation is prospective;
- With respect to the carbon cycle, positive influences are expected from policies focusing on the maintenance of forest cover and production. The failure of the policy against forest fires to adopt a more managerial approach reduces its contribution to SFM, since it causes a severe reduction in the EU forest cover, thus decreasing the 'permanence' of its carbon stocks;
- The productive functions of forests are influenced positively by the CAP, as regards wood only. This is far from the SFM approach;

- The biodiversity policy does not influence the overall maintenance of the dynamic character of forest ecosystems and acts in a fragmented and inadequate way as it considers biodiversity elements only;

Table 8.11   Evaluation of the contribution of the IFC in EU policies towards SFM

C&I	Policies			
	Rural development	**Biodiversity Conservation**	**Water resources**	**Natural disasters**
Carbon cycles				
General capacity	+	+	0	0
Land Use & Forest Area	+	0	+	-
Growing Stock	++	0	0	-
Carbon Balance	0	0	+	-
Health and vitality				
Status of Forests	0	0	0	-
Productive functions				
Wood Production	+	0	0	0
Non-Wood Products	0	0	0	-
Biological diversity				
General conditions	0	+	+	-
Representative, Rare and Vulnerable Forest Ecosystems	0	++	0	-
Threatened Species	0	++	0	-
Biological Diversity in Production Forests	+	0	0	0
Protective functions				
General Protection	0	0	++	-
Soil Erosion	0	0	++	--
Water Conservation in Forests	0	0	++	-
Socio-economic				
Significance of the Forest Sector	+	-	+	0
Recreational Services	0	+	0	-
Provision of Employment	++	-	+	0
Research and Professional Education	0	+	+	+
Public Awareness	0	+	+	+
Public Participation	0	0	+	0
Cultural Values	0	+	0	0

(Contribution level: ++ very positive, + positive, 0 indifferent, - negative, -- very negative)

The WFD will influence positively protective functions, when it is implemented. So far, the ineffectiveness of the EU policy against forest fires acts very negatively on protective functions and contributes also to desertification.

One of the main reasons explaining these poor results is the lack of a common conceptual approach that is commonly agreed by all EU and national level actors.

Although the objects of most policies are multidimensional, their goals and implementation mechanisms are still influenced by narrow sectoral interests. Furthermore, the hesitation of the private forest sector, mainly in the countries with significant wood production, to co-ordinate with other sectors makes IFC towards SFM even more problematic. In general, the forest sector has adopted a rather defensive attitude towards other policies. At the international level, a Forest Convention was denied in 1992; in the EU a common forest policy has been turned down several times by the MS, and public and private forest actors try to prevent the interference of other actors in their business to keep the forest sector intact and independent from other sectors. Forest actors (policy makers, scientists, owners, etc.) have always considered their multifunctional approach as sufficient to address all issues relevant to the broader forest environment. However, this attitude has brought the opposite results. International conventions and regional strategies have strengthened significantly other sectors that have 'invaded' the forest policy area; hence, the current dispersion of forest-related issues in different policies detracting from the achievement of SFM. This situation raises the question whether the achievement of SFM could be promoted by improving the integration of EU policies, which incorporate forest-related elements.

SFM is a complex undertaking whose achievement requires the effective coordination among its various components. Even if IFC in EU policies is successful, each policy, say the CAP, will inevitably address one or a few forest-related parameters only. Moreover, there will always be conflicts among the different SFM components, as each policy will focus on a different component. There are three ways to deal with this problem: (a) to impregnate all EU policies with the totality of SFM components and resolve conflicts within each policy separately, (b) to promote the integration of EU policies (incorporating forest-related concerns), or (c) to design an integrated EU common forest policy. The first option is not realistic, as it implies a complete IFC and the domination of the forestry approach in all policies. The other two options are further elaborated in the concluding section below.

Integration of EU Policies or an Integrated EU Common Forest Policy?

So far, the EU-15 has failed to establish a Common Forest Policy. The only common political document is the EUFS, that contains important principles but no implementation mechanisms. Forest management remains a competence of the MS (European Commission, 2003a). Although some countries (e.g. the Netherlands) have supported the idea of an EU forest policy in order to support forest expansion, carbon sequestration and forest health, most MS resist the idea for different reasons as discussed in the beginning.

However, the recent accession of the 10 new MS may change this situation. Most of them have a long tradition in forestry and are important producers of forest products (Hanzl and Urban, 2000). Biodiversity conservation is a major pillar of forest policy in many countries and emphasis is given to securing societal benefits from forests probably because of their predominantly public ownership status, (Hjortsø and Stræde, 2001; Kallas, 2002). Therefore, while in EU-15 public forests and public forest administration were definitely neglected, their role now may gain importance. The political agenda of EU-25 is expected to better prioritize non-marketed benefits from forests. For countries with a significant public forest sector, a common forest policy may not be a bad idea. For example, even before its official accession to the EU, Cyprus, had appealed for the creation of an EU common forest policy (Hellenic Ministry of Agriculture, 2003).

The timing seems right for re-opening the dialogue on the formulation of an EU common forest policy. Besides the accession of the new MS and the change in the nature of forestry in Europe, the EUFS is currently under evaluation. Furthermore, the international forest policy dialogue has started reconsidering the possibility of creating a Forest Convention after all. The 2003 UNFF in Geneva has decided the formation of an *expert ad hoc* group for the 'Consideration with a view to Recommending the Parameters of a Mandate for Developing a Legal Framework on all Types of Forests' (UN/ECOSOC, 2003). In the perspective of these new trends, the two options suggested previously should be explored; either to promote the (horizontal) integration of EU policies (containing forest-related components) or to design an integrated EU common forest policy to achieve SFM. Preliminary suggestions are offered below.

The horizontal integration of EU policies (FPI) is a demanding task for several reasons. The first requirement is that each policy should sufficiently cover one or more components of SFM, as described by the C&I, so that they act in a complementary way towards SFM, based on a horizontal understanding and agreement on basic forest-related concepts. However, the EU policies examined address the priorities and needs of the (strong) sectors they serve, adopt their particular approaches and prioritize non-forest issues. The current situation depicted in Tables 8.10 and 8.11 reveals that a complete integration of all components of SFM may not be feasible. Some aspects of SFM are not addressed at all and some others receive conflicting influences from different policies. On the other hand, despite the progressive development of SFM concepts in the EUFS and other policy documents, the forest sector has been the force shaping the Union's attitude towards forests that focuses mainly on market issues and private forestry. Moreover, the application of the subsidiarity principle produces non-coordination of national policies and problematic implementation of EU policies.

Recently, the EU has made significant, procedural, efforts towards the co-ordination of forest-related EU policies, aiming actually at a FPI. These include the EUFS, the establishment of the Standing Forestry Committee and of various Advisory Committees as well as of the Inter-Service Group on Forestry. However, the results so far show that, while co-ordination, communication and co-operation between the Commission, the MS, interest groups and stakeholders occurs within the above committees, the actual co-ordination structures in forestry still reflect the existing organizational frameworks and formal EU competences on forestry matters. Thus, different policies cannot be linked procedurally only, through committees and contact groups, as long as there are substantive differences among them.

Instead of trying to promote FPI, the other option is to design an integrated EUCFP. The dialogue between policies is indispensable because of their inevitable linkages. However, it is argued that a spatially integrated, long term, multi-purpose EUCFP is necessary to define the proper environment for SFM and to guide FPI towards SFM, since forest concerns will always fall back in EU policies that support non-forest sectors. The concepts and mechanisms that will be developed through a EUCFP will greatly assist the development of other policies as well, as it has happened with the principle of sustainability at the international level.

The question that needs to be addressed is how to promote a EUCFP when there is so much resistance from the forest sector. A way forward may be to disseminate a common understanding of SFM to all forest-related sectors as, besides its ecological and social relevance, SFM is the only economic and political option for the forest sector. However, SFM requires a broad spatial and temporal frame of reference and coordination among policies. Most benefits deriving from forests are not marketed and market mechanisms do not secure their maintenance and supply to society. Thus, something more than a production-oriented system is needed to achieve a comprehensive SFM. The forest sector should continue producing and supplying forest products in the free market subject, however, to existing restrictions associated with the modern approach to SFM. Such restrictions may be strengthened further if they are imposed by other non-forest policies (e.g. the biodiversity policy). If all parties in the forest policy dialogue accept SFM, then SFM should be promoted through a broad and strong policy scheme such as the proposed EUCFP. In this context, and given the current evaluation of the EUFS, the EU may have to reconsider the, until recently neglected, role of public forests. But SFM cannot apply to public forests only. Public forests can be the initial ground for a strong and targeted policy for the non-market benefits of forests. This would improve the attitude of the MS towards a EUCFP and reduce the resistance of the private forest sector.

Recapitulating, to operationalize and achieve SFM, to contribute, among others to combating desertification, ideally all three conditions should be satisfied: (a) integration of forest-related concerns into non-forest policies (IFC), (b) integration among policies (FPI) and (c) formulation of a EUCFP. Future research should concentrate on these broad issues borrowing from the ideas suggested in this chapter.

References

Alexandrian, D. and Esnault, F. (1999), 'Public policies affecting forest fires in the Mediterranean basin', *FAO Forestry Paper*, Vol. 138, pp. 39-46.

Arévalo, J.P.D. (2003), *The EU Enlargement in 2004: Analysis of the Forestry Situation and Perspectives in Relation to the Present EU and Sweden*, Rapport 10, National Board of Forestry, Sweden.

Canadian Forest Service (1997), *The Montreal Process Progress Report*, Ottawa.

Dimitriou, A., Mantakas, G. and Kouvelis, S. (2001), 'An analysis of key issues that underlie forest fires and shape subsequent fire management strategies in 12 countries in the Mediterranean basin', WWF/Mediterranean Programme Office and IUCN, Gland, Switzerland.

Elands, B.A. and Freerk Wiersum, K. (2001), 'Forestry and rural development in Europe: an exploration of socio-political discourses', *Forest Policy and Economics*, Vol.3, pp. 5-16.

European Commission (1997), *Situation and Outlook: Rural Developments; A CAP 2000 Working Document*, DG Agriculture, Official Publications of the European Communities, Luxembourg.

European Commission (1998a), *Communication from the Commission to the Council and the European Parliament on a Forestry Strategy for the European Union*, COM 649, 03/11/1998, Office for Official Publications of the European Communities, Luxembourg.

European Commission (1998b), *Communication to the Council and Parliament on a European Community Biodiversity Strategy*, COM (98) 42, Office for Official Publications of the European Communities, Luxembourg.

European Commission (1999), *The Common Agricultural Policy: 1999 Review, European Commission*, DG Agriculture, Office for Official Publications of the European Communities, Luxembourg.

European Commission (2001a), *Evaluation of the Community aid scheme for forestry measures in agriculture of Regulation N° 2080/92*, AGRI/2001/33002-00-00-EN, DG Agriculture, Office for Official Publications of the European Communities, Luxembourg.

European Commission (2001b), *Communication from the Commission to the Council and the European Parliament - Biodiversity Action Plans in the Areas of Conservation of Natural Resources, Agriculture, Fisheries, and Development and Economic Co-operation*, COM/2001/0162, Office for Official Publications of the European Communities, Luxembourg.

European Commission (2001c), *Community Information System on Forest Fires*, 2001 Report, DG Agriculture, Brussels.

European Commission (2003a), *Sustainable Forestry and the European Union: Initiatives of the European Commission*, Office for Official Publications of the European Communities, Luxembourg.

European Commission (2003b), *Rural Development in the European Union, Office for Official Publications of the European Communities*, Luxembourg.

European Commission (2003c), *Conclusions of Second European Conference on Rural Development in Salzburg*, DG Agriculture, Brussels.

European Commission (2003d), *Natura 2000 and Forests, Challenges and Opportunities*, DG Environment, Office for Official Publications of the European Communities, Luxembourg.

European Council (1979), *Council Directive 79/409/EEC of 2 April 1979, on the conservation of wild birds*, OJ. L 103, 25.4.79, Office for Official Publications of the European Communities, Luxembourg.

European Council (1989), *89/367/EEC: Council Decision of 29 May 1989 setting up a Standing Forestry Committee*, OJ L 165, 15.6.1989, Office for Official Publications of the European Communities, Luxembourg.

European Council (1992a), *Directive 92/43/EC on Conservation of Natural Habitats and of wild Fauna and Flora*, OJ L 206/7, 22.7.92, Office for Official Publications of the European Communities, Luxembourg.

European Council (1992b), *Regulation (EEC) No 2158/92 on Protection of the Community's Forests Against Fire*, LJ L 217, 31/07/1992, Office for Official Publications of the European Communities, Luxembourg.

European Council (1997), *Council Regulation (EC) No 338/97 on the protection of species of wild fauna and flora by regulating trade therein*, OJ L 61/40, 9.12.1996, Office for Official Publications of the European Communities, Luxembourg.

European Council (1998), *Council Resolution of 15 December 1998 on a Forestry Strategy for the European Union*, Office for Official Publications of the European Communities, Luxembourg.

European Council (1999), *Regulation (EC) No 1257/1999 on support for rural development from the European Agricultural Guidance and Guarantee Fund (EAGGF) and amending and repealing certain Regulations*, Office for Official Publications of the European Communities, Luxembourg.

European Parliament and European Council (2000), *Directive 2000/60/EC of 23 October 2000 Establishing a Framework for Community Action in the Field of Water Policy*, OJ L327, 22.12.2000, Office for Official Publications of the European Communities, Luxembourg.

European Parliament and European Council (2003), *Regulation (EC) No 2152/2003 of 17 November 2003 concerning monitoring of forests and environmental interactions in the Community (Forest Focus)*, OJ L324/1 11.12.2003, Office for Official Publications of the European Communities, Luxembourg.

Glück, P. (1999), 'National Forest Programs - Significance of Forest Policy Framework', in P. Glück, G. Oesten, H. Schanz, and K-R. Volz, (eds), Formulation and implementation of National Forest programs, *EFI Proceedings*, Vol. 30, pp. 39-52.

Haila, Y. and Kouki, J. (1994), 'The Phenomenon of Biodiversity in Conservation Biology', *Ann. Zool. Fennici*, Vol.31, pp. 5-18.

Hanzl, D. and Urban, W. (2000), *Competitiveness of Industry in Candidate Countries: Forest-Based Industries*, The Vienna Institute for International Economic Studies, Vienna.

Hellenic Ministry of Agriculture (2003), *Hellas 2003: Events and Activities in the Forest Sector*, CD-ROM, Hellenic Ministry of Agriculture, Athens, Greece.

Hjortsø, C.N. and Stræde, S. (2001), 'Strategic Multiple-Use Forest Planning in Lithuania – Applying Multi-Criteria Decision-Making and Scenario Analysis for Decision Support in an Economy in Transition', *Forest Policy and Economics*, Vol.3, pp. 175-88.

Hogl, K. (2000), 'The Austrian Domestic Forest Policy Community in Change? Impacts of the Globalization and Europeanization of Forest Policies', *Forest Policy and Economics*, Vol. 1, pp. 3-13.

Hogl, K. (2002), 'Patterns of Multi-Level Co-ordination for NFP-Processes: Learning from Problems and Success Stories of European Policy-Making', *Forest Policy and Economics*, Vol. 4, pp. 301-12.

Humphreys, D. (2001), 'Forest Negotiations at the United Nations: Explaining Cooperation and Discord', *Forest Policy and Economics*, Vol. 3, pp. 125-35.

International Tropical Timber Organization (2002), *Manual for the Application of Criteria and Indicators for the Sustainable Management of Natural Tropical Timber Forests*, ITTO, Yokahama.

International Union for the Conservation of Nature (IUCN), United Nations Environmental Programme (UNEP) and World Wide Fund for Nature (WWF) (1991), *Caring for the Earth: A Strategy for Sustainable Living*, IUCN, Gland, Switzerland.

Kallas, A. (2002), 'Public Forest Policy Making in Post-Communist Estonia', *Forest Policy and Economics*, Vol. 4, pp. 323-32.

Kennedy, J.J., Dombeck, M.P. and Koch, N.E. (1998), 'Values, Beliefs and Management of Public Forests in the Western World at the Close of the Twentieth Century', *Unasylva*, Vol. 49, pp. 16-26.

Kneesaw, D.D., Leduc, A., Drapeau, P., Gauthier, S., Paré, D., Carignan, R., Doucet, R., Bouthillier, L., and Messier, C. (2000), 'Development of Integrated Ecological Standards of Sustainable Forest Management at an Operational Scale', *The Forestry Chronicle*, Vol. 76, pp. 481-500.

Koch, N.E. and Rasmussen, J.N. (1998), *Forestry in the Context of Rural Development*, Final Report of COST Action E3, Danish Forest and Landscape Research Institute, Horslom.

Mayer, P. (2000), 'Hot Spot: Forest Policy in Europe: Achievements of the MCPFE and Challenges Ahead', *Forest Policy and Economics*, Vol. 1, pp. 177-85.

McDonald, G.T. and Lane, M.B. (2004), 'Converging Global Indicators for Sustainable Forest Management', *Forest Policy and Economics*, Vol. 6, pp. 63-70.

MEDACTION (2004a), 'Regional and Industrial Policy of the European Union', Module 4: Design of a Desertification Policy Support Framework, Deliverables 33&34, Part B, Ch. 2. European Commission, DG-XII, Contract No. ENVK2-CT-2000-00085 (www.icis.nl/medaction).

MEDACTION (2004b), 'Rural Development Policies', Module 4: Design of a Desertification Policy Support Framework, Deliverable 35, Part B, Ch. 2. European Commission, DG-XII, Contract No. ENVK2-CT-2000-00085 (www.icis.nl/medaction).

Merlo, M. and Paveri, M. (1997), 'Formation and Implementation of Forest Policies: a Focus on the Policy Tools Mix', in Proceedings of the Eleventh World Forestry Congress, 13-22 October, Antalya, Turkey, pp. 233-54.

Ministerial Conference for the Protection of Forests in Europe (1993), *Resolution H1, General Guidelines for the Sustainable Management of Forests in Europe*, MCPFE Liaison Unit, Helsinki.

Ministerial Conference for the Protection of Forests in Europe (1998a), *Resolution L1: People, Forests and Forestry, Enhancement of the Socio-Economic Aspects of Sustainable Forest Management*, MCPFE Liaison Unit, Lisbon.

Ministerial Conference for the Protection of Forests in Europe (1998b), *Resolution L2: Pan-European Criteria, Indicators and Operational Level Guidelines for Sustainable Forest Management*, MCPFE Liaison Unit, Lisbon.

Ministerial Conference for the Protection of Forests in Europe (2003a), *The Vienna Declaration. Living Forest Summit: Common Benefits – Shared Responsibilities*, MCPFE Liaison Unit, Vienna.

Ministerial Conference for the Protection of Forests in Europe (2003b), *Resolution V1: Cross-Sectoral Co-operation and National Forest Programmes*, MCPFE Liaison Unit, Vienna.

Ministerial Conference for the Protection of Forests in Europe (2003c), *State of Europe's Forests 2003 - The MCPFE Report on Sustainable Forest Management in Europe*, MCPFE Liaison Unit, Vienna.

Papageorgiou, A.C. and Catsadorakis, G. (2001), 'Political and social aspects of protected areas management in Greece', *Law and Nature–Journal of Environmental Legislation*, Vol. 7, pp. 161-88.

Papageorgiou, A.C. and Kasimiadis, D. (2004), 'The genetic component of biodiversity in forest ecosystems', *Göttingen Research Notes in Forest Genetics* (in press).

Papageorgiou, A.C., Arabatzis, G. and Tampakis, S. (2003), 'Forest Development and Biodiversity in the Enlarged European Union', in 'Hellas 2003: Events and Activities in the Forest Sector', CD-ROM, Hellenic Ministry of Agriculture, Athens, Greece.

Papageorgiou, A.C., Kasimiadis, D., Arabatzis, G. and Tampakis, S. (2004), 'Genetic Diversity, Biodiversity and Management of Forest Ecosystems', in Proceedings of the 1st Hellenic Environmental Congress, 7-9 May 2004, Orestiada, Greece, Greek Geotechnical Chamber, Thessaloniki (in press).

Papageorgiou, A.C. and Mantakas, G. (2004a), 'Forest policy of the European Union', in MEDACTION, Module 4: Design of a Desertification Policy Support Framework, Part B (EU), Chapter 8, Final Report, MEDACTION (Policies for Land Use to Combat Desertification), EC, DG-XII (ENVK2-CT-2000-00085).

Papageorgiou, A.C. and Mantakas, G. (2004b), 'Forest Policy in Greece', in MEDACTION, Module 4: Design of a Desertification Policy Support Framework, Part C (Greece), Chapter 9, Final Report, MEDACTION (Policies for Land Use to Combat Desertification), EC, DG-XII (ENVK2-CT-2000-00085).

Perlman, D.L. and Adelson, G. (1997), *Biodiversity: Exploring Values and Priorities in Conservation*, Blackwell Science, Malden, Massachusetts, U.S.A.

Pons, A.O. and Quézel, P. (1985), 'The History of the Flora and Vegetation and Past and Present Human Disturbance in the Mediterranean Region', *Geobotany*, Vol. 7, pp. 25-43.

Pülzl, H. (2002), 'IPF/IFF Proposals for Action and their Implementation by National Forest Programmes by National States and the European Community', in I. Tikkanen, P. Glück and H. Pajuoja (eds), *Cross-Sectoral Policy Impacts on Forests*, EFI Proceedings Vol. 46, pp. 59-74.

Repetto, R. (1988), 'Overview', in R. Repetto and M. Gillis (eds), *Public Policies and the Misuse of Forest Resources*, Cambridge University Press, Cambridge, pp. 1-42.

Rojas Briales, E. (2002), 'Forest Management Planning as an Instrument for Monitoring, Optimizing and Incentivating Sustainable Forest Management', Proceedings of the Research Course 'The Formulation of Integrated Management Plans (IMPs) for Mountain Forests', Bardonecchia, Italy, 30 June-5 July 2002, European Observatory of Mountain Forests, Italy, pp. 55-62.

United Nations (1992), *Report of The United Nations Conference on Environment and Development*, A/CONF.151/26, Rio de Janeiro.

United Nations/Department of Economic and Social Affairs (1992), *Statement of Principles to Guide the Management, Conservation and Sustainable Development of All Types of Forests*, Report of The United Nations Conference on Environment and Development, A/CONF.151/26, Vol. III, Rio de Janeiro.

United Nations–Economic Commission for Europe/Food and Agriculture Organization (2000), *Temperate and Boreal Forest Resources Assessment*, UN/ECE Trade Division, Timber Section, Geneva.

United Nations / Economic and Social Council and European Commission (2000), *Forest Condition in Europe*, 2000 Executive Report, UN/ECE and EC, Geneva and Brussels.

United Nations / Economic and Social Council (2004), *United Nations Forum on Forests*, Report on the Fourth Session, UN Official Records 2004, Suplement No. 22, New York.

United Nations/Economic and Social Council (2003), *United Nations Forum on Forests*, Report on the Third Session, UN Official Records 2003, Suplement No. 22, New York.

United Nations Environmental Programme (UNEP) (1995), *Global Biodiversity Assessment*, Cambridge University Press, Cambridge, U.K.

United States Office of Technological Assessment (1987), *Technologies to Maintain Biological Diversity*, United States Government Printing Office, Washington D.C.

Van de Veelde, J. (2003), 'Forest Issues and the Environment at the EC', in E. Hedrick (ed.), *Forest regulation – a threat to production forestry?*, Proceedings of the ITGA/COFORD Seminar, 14 November 2002, COFORD, Dublin.

World Commission on Environment and Development (1987), *Our Common Future*, Oxford University Press, Oxford.

Worrel, A.C. (1970), Principles of Forest Policy, McGraw-Hill C., New York.

Wulf, M. (2003), 'Forest Policy in the EU and its Influence on the Plant Diversity of Woodlands', *Journal of Environmental Management*, Vol. 67, pp. 15-25.

Chapter 9

Policy Integration:
Bringing Space Back In

Helen Briassoulis

...Because the world does not exist on the head of a pin

Introduction

It is widely acknowledged and sufficiently documented by now that the achievement of sustainable development, the simultaneous satisfaction of social, economic and environmental goals, requires integration. Policy integration, specifically, figures prominently as policies impinge on interlinked and interdependent human activities, environmental resources, and human-environment relationships. Ordinary people, engaging in activities to satisfy their needs, and public and private sector administrators and managers, carrying out tasks to meet their organizational mandates and achieve collectively agreed goals, combine various means which originate in, or are influenced in some way by, numerous policies. In this process, laypersons and managers alike, intentionally or unwittingly, are actively practicing, or attempt to practice, policy integration within concrete geographic units, as their goals cannot be achieved, to whatsoever extent, if the policy-derived means are at conflict with one another within these units.[1] Frequently also, the use of policy-related means (e.g. subsidies) to achieve local goals causes impacts (pollution, competition, migration, etc.) elsewhere, hindering or facilitating the achievement of local goals in other places. The unending list of examples suggests that the test bed of the lack of policy integration is implementation on the ground. Only then all stakeholders and the resources affected are revealed, conflicts arise and solutions are sought (Pressman and Wildavsky, 1992). The quest naturally arises then that, when policies are conceived and formulated at higher levels, they should be integrated on a spatial basis to make a genuine contribution to local or regional sustainable development.

---

[1] Persson (2002, p.23) notes "integration does not always take place at the political level, where important normative judgments should be made, but is sometimes dealt with by civil servants (see SEPA, 1999 and Peters, 1998)".

The crosscutting nature of contemporary, complex socio-environmental problems,[2] the incidence of spatial impacts of Community policies and the problems arising from their non-coordination have revived the time-old theme of integration over and across administrative and organizational levels (the EU, national, regional and the local). The achievement of social and economic cohesion, the supreme goal of the European integration project, is inconceivable in a spatial void. The quest for sustainable spatial development, reflected, among others, in the three 'spatial development guidelines' of the European Spatial Development Perspective (ESDP, 1999),[3] encapsulates the idea that only spatially fit, balanced, interconnected development patterns can assure socio-economic welfare, eliminate socio-spatial inequalities, and protect and make wise and efficient use of the resource base of development. The transition to sustainable spatial development requires the development of functional spatial synergies, cross-sectoral linkages (horizontal policy coordination) and close co-operation amongst the authorities responsible for territorial development at all levels (vertical coordination) (SUD, 2003).

The need for the spatial integration of Community policies floats in the air but the question has not been rigorously conceptualized, articulated and satisfactorily addressed yet. With the exception of calls for (vertical) coordination among spatial/organizational levels and for spatial planning, on the part of spatial experts who have spurred the discussion mostly, its study and actual practice are at relatively early stages of development. Its horizontal aspects, in particular, are less researched compared to its vertical aspects. This chapter aims to provide an overview of the current state of affairs as regards this subject and of theoretical frameworks that can meaningfully inform its further elaboration to help address complex policy problems. The second section presents the interest, the evidence and the justification for the spatial integration of Community policies, while the third reviews current proposals on this subject. The fourth section explores theoretical frameworks directly or indirectly related to spatial development. The last section summarizes the discussion, suggests broad approaches to the spatial integration of Community policies and highlights open questions and issues that the political process and future research are called to resolve.

Spatial Integration of Community Policies: Interest, Evidence, and Justification

The lack of spatial integration of Community policies and its negative implications for the achievement of sustainable development in the European territory,

---

[2] See Chapter 1, this volume.
[3] Polycentric spatial development and a new urban-rural partnership, parity of access to infrastructure and knowledge, and wise management of the natural and cultural heritage (ESDP, 1999).

especially after the new enlargement,[4] acceded to the political agenda in the last decade following a series of efforts aiming at raising awareness on the issue. This section documents the evolution of the informal and of the official interest in the problem, reviews pertinent evidence on its effects that justify the need to address it, and probes into its root causes that the literature suggests. The aim is to produce a first conceptualization of the spatial policy integration (SPI) issue and to derive the principal policy-relevant questions to be addressed.

The first signs of an interest in the spatial repercussions of Community policies appeared in the late 1980s and early 1990s when the planning ministers of the Member States (MS) first endeavored informally to urge the European Commission to assume a planning role and improve its policies by placing them in a spatial perspective (Faludi, 2001). The European Commission contributed to the initial debate with two reports, "Europe 2000" in 1991 and "Europe 2000+" in 1994 (EC, 1998). In 1997, the first official draft of the European Spatial Development Perspective (ESDP) was approved at the informal meeting of Ministers responsible for spatial planning of the MS of the European Union held at Noordwijk, Holland. To produce essential technical elements needed for the finalization of the first ESDP and address a limited number of priority themes, specialized research institutes initially carried out a Study Program on European Spatial Planning (SPESP). At around that time Liberatore (1997), examining the integration of sustainable development objectives into EU policy-making, made reference to the 'space and time' dimension that requires the study of:

> ...the interactions between different spaces: the *geographical space* of the affected environment, the *economic space of the activities* that have an impact on the environment, the *institutional space of the relevant authorities and policy instruments*, and the *cultural space of values* (p.117).

and to the 'organization' dimension that concerns the mismatch between the territorial competences of environmental authorities with the affected environment.

In parallel, the European Commission took the initiative to organize, in co-operation with the MS, the European Parliament, the Economic and Social Committee and the Committee of the Regions, a series of transnational seminars centred on the main themes indicated by the ESDP. In 1998, representatives from 19 Directorates-General (DG) of the European Commission, the EUROSTAT and the European Environment Agency prepared a working document entitled "Report on Community Policies and Spatial Planning" to enhance the awareness of the territorial dimension in formulating new policy guidelines and in implementing current Community policies as well as to strengthen co-ordination and co-operation (EC, 1998).

In 1999, the final form of the ESDP was approved at the informal meeting of the Ministers responsible for spatial planning at Potsdam, Germany. The ESDP

---

[4] That came into effect on May 1, 2004.

identified a number of Community policies with spatial impact on the EU territory including the CAP, the Structural Funds (SFs), Environmental Policy, the Trans-European Networks (TENs), Community Competition Policy, Research, Technology and Development (RTD) Policy, and the loan activities of the European Investment Bank (ESDP, 1999). Section 2.3 of the ESDP, "For an improved spatial coherence of Community policies", recommends searching for 'functional synergies' and an integrated and multisectoral spatial development approach in various Community programs.

The launching of the ESPON Programme (European Spatial Planning Observation Network) followed, under the INTERREG Community Initiative, to fill the considerable gaps in knowledge about spatial development trends and the spatial effects of sectoral policies on the EU and the European level, anticipating also the impacts on the spatial structure of an enlarged Union (ESPON, 2002). Since 1999, the ESPON programme, following an integrated approach and with a clear territorial orientation, has commissioned several projects that include studies of the territorial impacts of selected sectoral and structural policies.[5] It is expected that the results of these projects will contribute, among others, to the improvement of the spatial co-ordination of EU sectoral policies.

In 2000, the European Conference of Ministers responsible for Regional Planning of the Council of Europe (CEMAT), meeting in Hannover, adopted the Hannover Document entitled "Guiding principles for Sustainable Spatial Development of the European Continent" that concerns the entire continent of Europe. Like the ESDP, this document stresses the need for close horizontal cooperation between spatial planning and sectoral policies with significant geographical impacts as well as vertical cooperation among the various administrative levels to promote the balanced and sustainable development of Europe (CEMAT, 2000).

In 2001, several official documents demonstrated and heightened the interest in the territorial dimension of Community interventions. The Second Cohesion Report (EC, 2001) introduced the concept of 'territorial cohesion',[6] presenting it as a complementary part of economic and social cohesion policy and proposing the strengthening of the territorial coherence of other policies. It highlighted the particular challenges presented by regions with specific geographical characteristics (islands, remote and mountainous regions) and suggested key areas that should include a territorial dimension such as: the operation of the SF, the adaptation of cohesion policy and its relations to other policies, the review of the TENs, and the ongoing review of the CAP (SUD, 2003).

Similarly, the White Paper on European Governance emphasized the need to address the territorial impact of EU policies in areas such as transport, energy and

---

[5] For information visit: http://www.espon.lu/.
[6] Article 16 of the EC-Treaty and article 36 of the Charter of fundamental rights of the EU had already mentioned the promotion of territorial cohesion as a common task of the Community and the MS, without defining it, however (SUD, 2003).

the environment, to provide that policies form part of a coherent whole and avoid a sector-specific logic, and to encourage more coherent territorial governance (CEC, 2001a). In both a Consultation paper for the preparation of the EU Strategy for Sustainable Development (EUSSD) (CEC, 2001b) and the European Commission's Communication on the EUSSD (CEC, 2001c), the poor cross-sectoral policy coordination and its costly implications for territorial development were underscored as key factors hindering progress towards sustainable development.

In terms of empirical evidence, the findings of several studies document the lack of and the need for spatial integration of Community policies. The following summary examines why Community policies cause what impacts, what the root causes of this problem are, and why it is necessary to address it, suggesting the principal policy-relevant questions concerning this issue. Policies that have been studied so far, in decreasing order, include: the SFs, the CAP, transport policy, competition policy, environmental policy, telecommunication policy, energy policy, Single Market, Economic and Monetary Union, economic and social cohesion, Structural fisheries policy, RTD policy, industrial competitiveness and enterprise policy. Other policy domains with an emerging European spatial dimension are culture, tourism, distributive trade, education and training as well as the broader issue of urban-rural relationships (CURS, 2002).

To promote economic and social cohesion, the Community has increasing responsibility or competence over various components of the territorial system (environmental resources and sinks, economic sectors, social groups, culture, etc.) and territorial development, which is shared, in practice, with the MS. Its interventions vary from distributing financial resources from the Community budget (to support incomes, fund research, etc.), to promoting specific initiatives (such as regionalized and horizontal structural measures), and introducing legislation that MS are obliged to enforce (ranging from the provisions of the Treaty to sectoral policies and the development of guidelines) (EC, 1998; SUD, 2003). Table 9.1 gives examples of relevant Community interventions.

Nadin (2000, p.6 cited in CURS, 2002) identified three types of EU policies that can be considered 'spatial': (a) EU sectoral policies that are defined in spatially-specific ways such as the TENs, (b) EU sectoral policies that are not expressed in spatial terms but have an influence on spatial structure and development of the EU territory such as financial instruments under the CAP and (c) EU policies that are intended to influence or be implemented through spatial planning policy in the MS such as horizontal environmental policy (e.g. the EIA Directive). As a result, Community policies, like all policies, cause spatial impacts because they influence, directly or indirectly, the conduct of human activities and the use and state of environmental resources which occupy certain territories, occur in particular places, and relate among them in socio-spatially-specific ways.

Table 9.1   Examples of Community interventions in socio-spatial development

Community policies	Types	Examples
Policy Statements	White Papers, Council statements and EU Parliament debates Communications and recommendations	Cohesion reports on Regional Policy, Transport White paper, 6t^h Environmental Action Programme, Integrated Coastal Zone Management Recommendation, Council statement on sustainable development
Legislation	Directives	Strategic Environmental Assessment, EU Water and Waste Directives
Research Programmes	Reports and recommendations	5th Framework research programmes, European Spatial Observatory Network (ESPON)
Funding Programmes	Studies and projects on the ground	Structural Funds Objective 1 and 2, INTERREG, Urban Community Initiatives, Community Framework for Co-operation to promote Sustainable Urban Development
Expert Groups	Reports and recommendations	CDCR and the Subcommittee SUD, the Urban Environment Expert Group, Expert Group on the Environment and Land Use Communication

Source: SUD (2003)

The lack of spatial integration of Community policies as well as between them and national or sub-national policies inevitably manifests itself during implementation. A region[7] usually experiences the impacts of more than one policy simultaneously. Community policies influence individual components of the regional system and/or their interrelationships, inevitably modifying its structure and functioning, and its relationships to other regional systems. At the same time, national or sub-national policies may be affecting the same components of the regional system, directly or indirectly. Moreover, a region may experience the spillover effects of Community and national/subnational policies targeting other regions with which it is economically, socially, and environmentally connected.

---

[7] The term 'region' is loosely used here to refer to a locale, an administrative region, a nation, or a trans-regional entity.

Several of the unwanted impacts of Community policies have been attributed to the fact that they do not account sufficiently for intra-regional and interregional linkages and for the influence of other Community as well as national and subnational policies in the same or in other regions. The end result is either inefficient, wasteful use of financial and human resources and negative environmental and socio-economic impacts at more than one spatial/organizational levels,[8] or policy ineffectiveness. At the EU level, these unwanted impacts detract from the official goal of achieving social and economic cohesion in the European territory and of promoting sustainable development (ESDP, 1999; CEC, 2001b; Leygues, 2001; Robert et al., 2001).

Beginning with the "Report on Community Policies and Spatial Planning" (EC, 1998), the spatial impacts of various Community policies have been, and are currently being investigated, in greater depth and detail in the context of the ESDP process and of ESPON-commissioned studies. Although this undertaking faces considerable methodological difficulties that may put in question the results obtained, several broad findings are plausible and are summarized in the following.[9] The focus is mostly on impacts resulting from lack of spatial integration among policies either in a horizontal (Community with Community policies) or in a vertical (Community with national/subnational policies) sense.

Several Community policies, such as the SFs, the CAP, transport and communication policy, and competition policy, induce changes in accessibility, the regional concentration of economic activities, trade patterns and the mode of production in both urban and rural areas.[10] These produce 'asymmetrical' shocks in the wider European territory and may be strong enough to destabilize the economic base of entire regions and harm their competitiveness, reflecting the failure to anticipate territorially differentiated effects of European integration (ESDP, 1999; Leygues, 2001; SUD, 2003). In addition, they generate social problems, environmental damage and crime, among others, as well as congestion of major roads and motorways that cannot be counteracted without the implementation of other policies (ESDP, 1999; SUD, 2003).[11] The impacts depend, to a significant

---

[8] Usually when policies work at cross-purposes and implementation of one violates rules pertaining to the other.

[9] Some studies are on-going at the time of this writing.

[10] For example, market liberalization can increase spatial competition, which often favours areas with locational advantages. In recent years, the Commission has attempted to concentrate aid on the least favoured regions and to maintain a differential in aid intensity between regions to allow the weaker regions to compensate for their structural handicaps (EC, 1998).

[11] For example, the CAP has produced intensification, concentration and specialization of production in agriculture with negative effects on spatial development such as monotonous landscapes, abandonment of traditional land management practices, use of large areas of wetland, moorland and natural rough pasture, ground water pollution by increased use of pesticides and fertilizers, and reduction in biological diversity.

extent, on the strength of interregional linkages and on whether policies are coordinated on a spatial (or, territorial) basis.

The allocation of Community resources through fragmented and territorially incoherent measures may prove sub-optimum, counteracting the objectives of economic and social cohesion, in case of locally conflicting policy impacts or lack of synergies between regional/national and Community policies (Leygues, 2001; SUD, 2003). These impacts are more readily expressed in pecuniary, cost inefficiency terms as Robert et al. (2001) attempted to do. Uneven or targeted allocation of funding may create greater regional imbalances and augment the interregional prosperity gap, as it was the case, for example, of the CAP set-aside measures (ESDP, 1999). Moreover, sectoral policy objectives and instruments may contradict sustainability objectives for spatial planning at national and sub-national levels as, for example, the provision of subsidies for the location of new development in less sustainable locations and tax incentives that encourage long distance commuting (Nadin, 2003).

Major European, mainly sectoral, crises has been attributed also to the fact that sectoral policies were not coordinated on a territorial basis. These include the BSE crisis and its link with increased productivity obscuring the territorial origin of components, and the ecological disasters in the EU's coastal regions that followed the deregulation of maritime transport (Leygues, 2001; SUD, 2003).

The other side of the coin is that the objectives of territorial cohesion and polycentric development require that Community and other policies be integrated on a spatial basis both across and over space. The SFs are broadly considered to have strengthened and empowered the regional and local levels of governance and to have stimulated partnerships and bottom up policy-design. This is evidence of their potential to contribute to European-wide cohesion and competitiveness by encouraging spatially-sensitive allocation of Funds through the tailoring of policies to the needs and preferences of those living and operating in affected territories (Nordregio, 2003).

The CAP is considered to have favoured the development of already rich rather than of poor, underdeveloped rural areas (ACRDR, 2003). In the face of increasing world trade competition, rural areas with weak economies will continue to be exposed to economic pressures to intensify agricultural production, while preserving a certain level of environmental quality, thus increasing the need for spatial policy integration (ESDP, 1999). More broadly, endogenous development requires a thorough understanding of regional potential and a long-lasting commitment towards integrated approaches based on local participation and the development of regional institutional capacity. It is encouraging that the urban-rural relationships are receiving particular attention as explanatory factors and determinants of the spatial impacts of Community policies (CURS, 2002). Separate, isolated, sectorally and spatially unrelated policies, on the one hand, produce negative spillover effects from rural to urban areas and vice versa and, on the other, they cannot tackle effectively the problems of both the urban and the rural areas. To support comprehensive strategies for integrated regional development, Community policies should look at both areas as parts of a wider territory (CURS, 2002).

All studies recognize the considerable methodological problems that prevent an unambiguous assessment of the spatial impacts of Community policies. These include the complexity of spatial relationships, the diverse modes of Community interventions, quantitative and qualitative differences among the various policy measures, differing national priorities, geographic variations due to the location of assisted regions (as measured by an EU-wide peripherality index), specific environmental, cultural, and socio-economic conditions, and the effects of market forces that are, in their turn, reinforced by the policies themselves (EC, 1998; ESDP, 1999; ACRDR, 2003).

The root causes of the SPI problem have to be sought, however, in the mismatch between the supply of and the demand for spatial coherence of Community interventions. The dynamic and evolutionary nature of contemporary socio-spatial phenomena to which policies refer together with the attendant transformations in governance structures and approaches, create a demand for policy integration on a spatial basis to serve efficiently the development problems of particular territories and to ensure the social and economic cohesion of the EU as a whole. However, the extant EU policy apparatus, characterized by sectoralization and a relative neglect of the spatiality of policy problems, and the place and role of spatial policy and planning in the MS, respond slowly to this demand. The following discussion negotiates these interdependent explanatory factors.

Policy problems more often than not concern facets of broader socio-spatial phenomena from which they cannot be cut off without producing inadequate and incomplete analyses and less efficient and sustainable solutions. A coarse analysis even of a simple, however uni-dimensionally and narrowly defined, local problem, say local unemployment, destruction of a wetland, low tourist demand, inevitably reveals its spatially interlinked and multi-level social, economic, environmental, cultural, political and many other determinants. This is so because human activities and environmental resources have variable spatial reaches and influence, and spatial relationships of varying strength and quality develop within and between them. These are frequently expressed as accessibility and proximity of activities to resources, spatial dependencies among economic activities, spatial conflicts and competition among activities for location, resources, markets, etc.

The coming to prominence of the spatial aspects of contemporary policy problems owes to broader socio-economic developments, frequently cited in the literature by now. Socio-economic globalization, the liberalization of world trade, the post-1970s economic crises, macroeconomic convergence, technological change, international migration, and the blurring of national borders, among others, create functional interdependencies in space, among economies, resources, places, and regions, but also opportunities for networking (EC, 1998; ESDP, 1999). Contemporary socio-spatial phenomena are crosscutting, horizontal, complex and dynamically evolving. The real origins of problems may be at long distances from the places where impacts are felt as the impacts of a local action may reverberate throughout the globe. It is frequently difficult to distinguish the causes of problems from their effects, as their roles are interchanging and mediated by complex, multi-level spatial relationships.[12]

With certain exceptions perhaps, most policies have an inherent 'customer orientation' (Peters, 1998; Avery, 2001; Shannon, 2002; Hertin and Berhout, 2003), as policy problems usually refer to client groups, which inevitably imparts them a conspicuous or inconspicuous spatiality and territoriality. In certain problems, such as those of urban, rural, and less-favoured areas, regions experiencing industrial decline, etc., client groups are overtly territorial. In others, such as those of particular economic sectors, environmental problems, etc., the spatial dimension is covert although they do involve client groups attached to and operating in places that are associated with particular sectors, linked through spatially referenced networks, and using (and degrading) local and supra-local resources.[13]

The systemic, spatial nature of contemporary phenomena and of the associated policy problems inevitably imparts a systemic character to policy interventions. Policy decisions and their impacts are not restricted to actors from only one spatial level, or place or sector, but encompass diverse actor networks that develop, change, evolve and interact on and across multiple spatial levels both in the short and in the longer term. Thus, policy impacts are difficult to predict and anticipate. Environmental complexity is intricately interwoven with socio-economic and institutional complexity and interplay (Dryzek, 1987; Brown, 2000; Young, 2002). In other words, policies have now a much broader and complex spatial scope than it was the case in the less environmentally and socio-economically interconnected world of the past.

The MS of the EU, faced with the challenges of managing multisectoral, crosscutting policy problems, which transgress the established functional specialization of extant political/administrative systems, realized the limitations of their regulatory function, especially within a general climate of fiscal austerity that required efficient use of scarce resources. The national/central level was often too large-scale and monolithic to ensure appropriately differentiated and customized regulatory responses to societal problems and, at the same time, it was too small-

---

[12] The relationships, for example, between transport and territorial development are nonlinear and complex. At the local level, a certain density threshold is necessary to render the provision of public transport infrastructure profitable. On the other hand, high levels of infrastructures and transport services produce land value appreciation, segregation of urban functions and, in the long term, transport flows increases that generate demand for more infrastructure. Provision of transport infrastructure to boost weak economies may prove ineffective or produce undesirable impacts (population and labour outmigration and further local economic decline).

[13] This becomes clearer if one analyses a particular mode of policy intervention, such as agricultural subsidies. The actors as well as the resources involved are linked to particular places and relate through particular spatial relationships.

scale and increasingly unable to regulate these problems (EC, 1998; Leygues, 2001). Recent, however slow, changes in perception and orientation in various EU policy arenas as regards the conception of rural areas, rural development and of urban-rural relationships heighten the awareness of the need for a territorial approach to policy making for sustainable development (CURS, 2002).

Gradually, a simultaneous 'double differentiation' of the political systems of the EU MS started to occur. A trend towards increased territorialization of public interventions, delegation of operational responsibility to the local and the regional levels, and clearer division of governmental functions between the different spatial/organizational levels increasingly gains momentum, bringing official actors and social partners together in development and underscoring the suitability of decentralized local and regional management (Leygues, (2001). This trend runs parallel to a process of increased Europeanization of national policies (Bache, 2003). As a result, public action is now conceived and implemented in the context of polycentric, highly complex networks of actors, characterized by considerable interorganizational dependency and, consequently, create a growing need for coordination between the players involved (ESTIA, 2000, Leygues 2001).

While these broader socio-economic developments and transformations in governance mentality and practice are taking place for some time now, the response of the EU policy apparatus appears to be lagging behind in providing effective mechanisms of spatial policy coordination to promote optimal territorial functionality and allocation of resources as well as efficient public services (EC, 1998). This owes to the inherently sectoralized and departmentalized structure and operation of EU policy- making (Avery, 2001) and of the associated Community administrative apparatus that is still predominantly space-blind. This is not accidental, however, as the following discussion demonstrates.

The formal objectives of Community policies derive from and have to comply with the provisions of the EU Treaties which mostly lack a territorial orientation, apart from the economic and social cohesion provisions (Art. 158-162) (EC, 1998). Historically, the Community institutions have developed along sectoral lines and their policymaking and administrative culture is sector-based (Avery, 2001; Leygues, 2001; CURS, 2002). Community policies impose the same, often very detailed, rules on all MS irrespective of their different environmental, geographic and socio-economic diversity and development needs that have increased with the recent enlargement (SUD, 2003). Leygues (2001) observes:

> Strangely enough, the deepening of European integration and the resulting extension of the common policies has translated into an over-specialization of functions and competences within the Community bodies, particularly the Commission. This specialization can be observed not only at Directorate-General level but also at the level of directorates, units, and even groups of officials or individual officials (p.41).

As a result, the Community approach on matters of horizontal coordination is generally weak, reflected in meagre inter-administrative structures and procedures, and with no practical effect (Leygues, 2001; SUD, 2003). The substantive, financial and administrative inability of the Union to possess its own implementation apparatus together with the considerable record of implementation problems of EU policies at national and sub-national levels, have led to the devolution of implementation powers and responsibility to lower levels, in accord with the subsidiarity principle, and a shift in policy style towards 'framework' directives[14] and procedural measures.[15] The point is, however, that issues of horizontal coordination cannot be resolved at the lower level of management alone but require policy coordination at higher levels in the first place (Peters, 1998; CEC, 2001b; Lehtonen, 2004).

It is not surprising then that, although the awareness of the territorial dimension of policy problems is gaining ground in EU policy circles, the changes in governance that have occurred in the MS have hardly made any inroads into the Community institutions and the territorial coordination and integration of Community policies is lagging behind demand (SUD, 2003). This is evident in several occasions as in the management of urban-rural interactions though integrated policy approaches (CURS, 2002).

Only a number of selected Community practices manifest the gradual introduction of territorial concepts in various sectors of Community policy. These include horizontal initiatives such as integrated rural development (the LEADER initiative), INTERREG programmes, actions financed by Art. 10 of the ERDF Regulation, such as the TERRA programme (EC, 2000), integrated development of coastal zones,[16] the Structural Fund partnership/programming principle, etc. However, for the time being, their effects are still fairly limited and the attempts at cultural change towards diversified, space-based approaches have a long way to go before they affect the dominant sectoral mentality and trends (SUD 2003). A more detailed account of current proposals to address the SPI issue is offered in the next section.

Although the Community, through its policies, intervenes in spatial matters, the competences over territorial development issues are essentially in the hands of the MS, as these are matters of national sovereignty and the subsidiarity principle applies. Therefore, the role of national spatial policy and planning systems to contribute to SPI as well as their relationships with the Community policies should not be underestimated. Leygues (2001) notes that the success of spatial coordination of public policies at the national and lower levels is influenced by (a) basic political agreement on the major objectives, (b) the role and place of territorial policy within the national political/administrative system and the quality of conflict resolution and consensus procedures, and (c) the availability of political and financial resources to organize communication and put in place consensus and

---

[14] Such as the Water Framework Directive.
[15] Such as the Strategic Environmental Assessment Directive, the Extended Impact Assessment (CEC, 2002), etc.
[16] See, for example: http://www.globaloceans.org/globalinfo/eu/eu.html.

compromise processes. Community policies with a strong territorial dimension encounter implementation problems as they usually fall within the remit of national and sub-national authorities with different, frequently overlapping and conflicting, competences, mandates and spatial jurisdictions in the same sectors. Moreover, in several MS, the political position of planning vis-à-vis sectoral policies is weak diminishing its potential role as a lower level spatial coordinating mechanism. Finally, the vertical relationships between the EU, national and sub-national administrative levels are not always friction-free and coordinated, blocking the smooth application of the reciprocity principle.[17] Box 1 offers a glimpse of this situation in the case of rural development policy.

Box 1

The three current forms of EU intervention on rural development represent quite different approaches. The rural development programmes of the 2nd pillar of the CAP have substantially a sectoral function. In contrast, mainstream Structural Funds programmes and the Community Initiative Leader Plus are territorial, with a multi sectoral and integrated approach. .... Because interventions are fragmented in different types of programmes, the coordination among them is weak and mostly left to Member State preferences. The result is that EU rural development policies are not managed or implemented as a package of integrated measures and there is little coordination between the different administrations involved (SUD, 2003).

Summing up, the problem is that, despite their 'spatial content' and isolated efforts at integrated approaches,[18] Community policies still appear to be spatially blind and frequently produce other problems, ignoring the territorial dimensions of the policy problems they are supposed to solve, most of which are problems of particular regions or of interregional relationships.[19] As a result, regional and local authorities and their representative associations, but also the general public, find it increasingly difficult to see any coherence in and positive impact of Community policies in their territories and are increasingly expressing their concern and confusion with the European institutions (Leygues, 2001). A considerable demand for action on the territorial coordination of Community policies is on the rise that the Union cannot continue to ignore[20] (Leygues, 2001). The sustainable spatial

---

[17] The process allowing policy priorities and their territorial impact to be gradually and continually harmonized.

[18] Which are mostly incidental and non-systematic.

[19] For example, subsidized agriculture in a region may set in motion rural-to-rural migration; extra revenues, in the absence of other development options, may be invested in urban areas, raising land values, intensifying land speculation, causing land use change, etc.

[20] The Parliament opined on May 28th, 1998 that: "… the Commission […] has immediate responsibility for and the opportunity of improving the complementarity and consistency of Community policies, in particular by establishing the internal mechanisms for co-ordination between its various departments…".

development of the EU requires an integrated approach to its territory through better management of the spatial impacts of Community policies (SUD, 2003). This necessitates the elaboration of a territorial development reference framework with interlinked horizontal and vertical components. The former should provide for horizontal, inter-sectoral policy linkages. The latter should provide for vertical, cross-scale linkages among policy makers and stakeholders, as no single spatial/organizational level is exclusively relevant in tackling territorial development issues (Leygues, 2001).

From the policy point of view, the EU faces a critical choice between two different approaches to respond more efficiently, effectively and realistically to this demand. One approach sees the development of an integrated Common spatial policy (as implied by the ESDP) while the other, dispensing with the need of a common spatial policy, focuses on the elaboration of spatial integration schemes for current and planned Community policies that might be customized to fit the particular sustainable development needs of European regions at various spatial levels while ensuring the socio-economic and territorial cohesion of the EU. This issue is detailed in the last section. The discussion now turns to current proposals for the spatial integration of Community policies.

Spatial Integration of Community Policies: Current Proposals

The need for the territorial coordination of Community policies to support the sustainable and balanced development of the enlarged European territory is being discussed for more than a decade in various fora. This section offers a selective account of the main proposals that have been advanced so far, aiming to identify the broad types, features and directions of proposed actions. Although the discussion concerns *Community policies*, it necessarily implicates national policies that have an important role to play in facilitating the overall task. The proposals considered place significant emphasis, not unduly of course, on the central role of spatial (physical and land use) planning as an instrument of policy integration at national and sub-national levels mostly but also for cross-scale policy integration, as the majority of their authors are spatial experts (ESTIA, 2000; Nadin, 2003). The proposals considered here do not concern the issue of spatial integration, i.e. the integration of different geographical regions (ESTIA, 2000); such proposals are relevant to spatial planning for the sustainable development of the EU territory and its regions mainly (see, e.g. ESTIA, 2000).[21]

All proposals start from three premises: (a) strategic territorial development is indispensable to achieve the goals of social and economic cohesion of the EU; (b)

---

[21] The following statement suggests the division of labour between spatial planning and policies: "Spatial integration is directly connected to spatial planning and spatial development, processes that directly depend on the relevant institutions which designate the goals, policies and the strategies for their implementation." (ESTIA, 2000).

the processes of intersectoral cooperation are currently limited to satisfactorily treat territorial development issues;[22] and (c) new unifying conceptual frameworks are needed to secure convergence and co-ordination among various sectoral policies (EC, 1998; ESDP, 1999; CEC, 2001b; CEC, 2001c; SUD, 2003). The recognition of the critical influence of urban-rural dynamics on the effectiveness of policy interventions in both urban and rural areas (CURS, 2002) and the emphasis on integrated approaches to rural and urban development is particularly important in policymaking for complex socio-environmental problems, such as desertification. Such problems cannot be properly analyzed and addressed outside the broader context of this dynamics.

The following discussion distinguishes between general and sector-specific proposals. The thrust of the literature is on the former as systematic sector-specific studies to support sensible proposals are in very early stages of development. Most proposals can be considered as 'framing proposals', indicating the goals of SPI (what it should strive for) and procedures to establish it as an official process in the Union, paying less attention, for the time being, to concrete ways to realize these goals because of the complexity of socio-spatial relationships and dynamics.

Given the incomplete state of systematic knowledge with respect to spatial structure and evolution in the EU and the economic, social and environmental impacts of Community policies and investments in the various parts of the EU, emphasis is placed on *launching new or continuing on-going relevant studies and spatial research programmes*, as necessary preconditions for the elaboration of concrete and meaningful SPI proposals (SPESP, 2000; ESPON, 2002). The themes of the proposed studies include the analysis of impacts of major Community policies, urban-rural relationships, cross-border and international cooperation, the development of new territorial indicators at NUTS III level,[23] analysis and evaluation of the outputs of the INTERREG IIIA, IIIB and IIIC programmes (the Interact programme), etc. (ESPON, 2002; CURS, 2002; SUD, 2003).

Elaborating the goals of SPI is another necessary precondition for action, which is predicated, however, on the necessity to change the culture of sectoral compartmentalization of Community policies[24] (Leygues, 2001). Towards this purpose it is proposed to refine and update the ESDP objectives (ESDP, 1999),[25] to incorporate these objectives in Community policies and to elaborate the territorial dimension of the Lisbon strategy based on the outcomes of ESPON-commissioned

---

[22] Such as the better co-ordination of the various public activities that influence the organization and use of the European territory.

[23] These should reflect issues such as: demography, degree of polycentric development, accessibility, research and innovation potential, education and training, ICT presence and critical mass of services, as well as basic economic, social and environmental information (SUD, 2003).

[24] Since the Commission has a monopoly on the right of initiative, its policy proposals necessarily bear the mark of this sectoral specialization culture (Leygues, 2001).

[25] The ESDP groups policy options under three spatial development guidelines (see, footnote 5).

studies. Polycentric development, coupled with improved accessibility, is also considered an instrument for improving the co-ordination between territorial and sectoral policies, promoting the objectives of competitiveness, cohesion and sustainable development (SUD, 2003).

The realization of the goals of SPI requires the establishment of an enabling *territorial governance* framework that could be managed on the basis of the Open Method of Coordination (SUD, 2003). Relevant actions include: (a) institutionalizing spatial planning, spatial development and territorial cohesion through inclusion of the terms in the provisions of the draft Constitutional Treaty of the European Union,[26] (b) setting up a Territorial Development Council with a clear mandate, (c) strengthening the role and position of the ESPON programme, and (d) more effective and efficient allocation of resources (Leygues, 2001; SUD, 2003). Particularly important is the development and strengthening of the horizontal and vertical links among administrative levels such as those between (a) the DGs in Brussels, (b) administrative bodies in the MS and (c) the three levels of territorial actors: infra-national/national/Community (Leygues, 2001; SUD, 2003). The means to achieve this include, among others: (a) launching a local and regional dialogue upstream of decision-making, (b) more effective involvement of the Committee of the Regions and (c) communication and network building through social and political networks (Leygues, 2001; CURS, 2002).

More detailed *procedural improvements in the EU policy process* have been proposed such as Leygues' (2001) four-phase process for coordinating Community policies and their impact on sustainable development and cohesion within the EU. It comprises (a) an indicative, periodic strategic orientation document from the Commission aiming to promote the coordination of Community policies and their impact, presented jointly with the financial perspective proposals, (b) upstream of the Commission's framework and orientation documents, a Strategic Impact Assessment (SIA) with consultation of the regional and local authorities, (c) presentation of proposals from the Commission to other institutions and (d) requests from the Commission to the Council and the Parliament to promote the necessary organizational measures, such as arbitration procedures within each of the various sectoral decision-making bodies for cross-checking the decisions of the different sectoral Councils to arrive at coherent decisions while taking into account the impact of sectoral decisions on one another.

To carry out the policy coordination process successfully, *administrative changes in the Commission as well as within the Parliament and the Council* are deemed necessary. These include reform in the Commission for greater integration and coherence in its decisions, assigning a greater coordination and arbitration role

---

[26] The "Council of Experts for European Spatial Development, drawn from seven different countries, has recommended the inclusion of spatial planning in the Constitutional Treaty of the European Union. Based on Part I of the draft Constitutional Treaty (CONV 528/03 of 6 February 2003), the Council of Experts proposes to add the term 'territorial' cohesion in Art.3 (The Union's Objectives) and the term 'spatial development' in Art.15 (on Supporting Action)." (SUD, 2003)

to the Secretariat-General and the Commission President, a dedicated office to administer the proposed SIA, the organization of interdepartmental teams that are genuinely coordinated, and consultation procedures with regional, local and transnational authorities (Leygues, 2001).

Proposals concerning *instruments for SPI* include the creation of a Community legal instrument for cross-border, transnational and interregional cooperation[27] and the gradual abandonment of detailed general EU policies in favour of flexible 'framework policies' that can be implemented in a more territorially-specific way (Leygues, 2001; SUD, 2003). In addition to the SIA proposal, other proposals emphasize the need to incorporate the requirement for assessing the territorial impacts of Community policies in extant or proposed assessment procedures,[28] on improving territorial impact assessment methods, and integrating them with risk and other assessment methods (ESDP, 1999; SUD, 2003). However, the most firmly and extensively supported instrument is *strategic integrated spatial planning*, which is relevant for the national and sub-national level, especially in promoting balanced and mutually reinforcing urban-rural development and contributing to spatial integration (ESDP, 1999; CEMAT, 2000; CURS, 2002). It is noted, however, that spatial integration *is not* identical to policy integration on a spatial basis although it is true that urban, regional, and land use planning has the potential to integrate and coordinate various policy resources to promote spatial development goals at various spatial/organizational levels.

For *the integration of specific policy sectors* with profoundly spatial impacts and significance, proposals concerning the integration of transport with structural and agricultural policy dominate, aiming to exploit synergies among the three policy areas to promote spatial development goals for the EU and its regions. The formulation and implementation of policies based on common references, drawn from the ESDP, is believed to ensure cohesion in the territorial development strategies of the MS (EC, 1998; ESDP, 1999). Vertical integration among policy levels is considered important and relevant measures are suggested. For example, DG-Environment encourages pilot tripartite schemes focused on sustainable transport and land use (SUD, 2003).

The integration of structural policy with other policies is considered of fundamental importance for the social and economic cohesion of the EU and the strengthening of its competitiveness. Although not explicitly addressing specific other policies, except for vague references to the CAP and the transport policy, a more territorially targeted decision making is proposed mainly so that cohesion investments,[29] which may be associated with several policies, enhance elements of key importance in the European territory, reinforce the territorial potential (human,

---

[27] The Third Cohesion Report refers to it as a "form of a European cooperation structure ('cross-border regional authority')" (EC, 2004, p.xxxi).

[28] Such as the Integrated Impact Assessment (IIA) that will be obligatory for all EU initiatives after 2004.

[29] These are various kinds of resources allocated to promote cohesion in the EU; structural policy belongs to these investments (SUD, 2003).

physical, environmental, institutional), promote polycentric development and support Europe's urban centers (SUD, 2003). Because the mode of operation of structural policy relies heavily on the subsidiarity principle, with national and sub-national levels making important decisions, emphasis is placed on combining sectoral and spatial approaches in drafting regional development plans as well as in the vertical integration among policy levels.[30] Examples of such project-oriented efforts are the INTERREG IIIB and IIIC programmes that demonstrate the added value of joint management of natural, cultural and other resources in cross-border and trans-national contexts (SUD, 2003). The Third Cohesion Report (EC, 2004, p.xxx) mentions that:

> ...building on the experience of present INTERREG Initiative, the Commission proposes to create a new objective dedicated to furthering the harmonious and balanced integration of the territory of the Union by supporting cooperation between the different components on issues of Community importance at cross-border, trans-national and interregional level.

Evidently, the emphasis is on integration at lower levels of government rather than the EU level.

Proposals for integrated approaches to managing urban-rural relationships, through combination and coordination of sectoral and territorial policies, focus on national and sub-national levels mostly and offer rather vague prescriptions for the goals of spatial integration (ESDP, 1999; CURS, 2002; SUD, 2003). An important precondition for the success of these efforts is a sense of common purpose among urban and rural players that is not as strong as desired (SPESP, 2000).

Recapitulating, the proposals that have been advanced so far focus more on desiderata and goals for the task of integration rather than on concrete approaches and procedures. They highlight, however, the importance of the coordination of Community policy goals and priorities as the necessary condition to guide the elaboration of policy coordination schemes on a spatial and geographical basis. The Third Cohesion Report (EC, 2004) is characteristically vague on the coordination and complementarity of cohesion with other policies. It advances proposals for individual policies that concern principally the delivery of these policies ignoring their multiple relationships with other policies.

There is frequently a confusion of spatial policy integration with spatial integration. The majority of proposals focus on vertical integration among spatial/organizational levels and on the related procedural aspects of the task (communication, participation, partnership, networking, linking players, etc.).[31]

---

[30] In addition, DG-Regio is promoting 'policy proofing' of Community policies to take account of the characteristics of different regions.

[31] Curiously enough, however, the Third Cohesion Report, although it emphasizes the development of partnerships among organizational levels, it suggests the separation of decisions in the case of large projects adopted by the Commission but managed by separate programmes at lower levels (EC, 2004, p.xxxv).

However, as Leygues (2001, p.42) observes "It would ... be wrong to think that new rules on the distribution of powers between different levels of government would suffice for Community policies to stop having a territorial impact across the EU", alluding to the need for horizontal coordination among Community policies.

Several procedural improvements are proposed but most of them refer to the national and sub-national levels and they have a strong project or programme, and not policy, orientation. Moreover, the procedural focus of the proposals leaves open the question of the substantive content of SPI, which is not easy to specify for two main reasons. First, integration in general requires a strong and sustained political will and the formation of a perception of common development goals among all those involved in development matters. Second, the complexity, diversity and fluidity of socio-spatial relationships, which grows as one moves down the spatial hierarchy, does not allow the concretization of integration proposals at higher policy levels. Human activities, resources, and administrative bodies have variable and intersecting spatial reaches, which may change over time and cannot be known with certainty in the present or anticipated in the future. It is no accident then that several proposals prioritize spatial planning at national and sub-national levels as the proper vehicle to draw integrated plans for particular territories whose effectiveness, however, hinges critically on SPI at the Community level.

The proposals reviewed do not address directly the choice between an integrated spatial policy and the development of SPI schemes as proper vehicles to the task. However, it seems that the supporters of the ESDP opt for the first alternative whereas the second option may be more viable, feasible and realistic (SUD, 2003). Whatever the preferences, the question is who will take the initiative to promote actively the spatial integration of Community policies. The obvious answer is the Commission, given its mandate and role but political realities cannot be ignored, an issue discussed in the concluding section.

Finally, with the exception of the use of certain territorial concepts indirectly referring to spatial development theories, the proposals lack a rigorous and solid theoretical basis to justify the SPI thesis and to provide coherent and well-founded guidance for the attendant choices of approach and design. The next section turns to this essential topic to provide a basis for its future discussion and elaboration.

In Search of a Theoretical Framework for Spatial Policy Integration

"Policies imply theories. Whether stated explicitly or not policies point to a chain of causation between initial conditions and future consequences" (Pressman and Wildavsky, 1992, p.xxiii). Drawing an analogy to this oft-cited statement, all policies, including the Community policies, embody a spatial theory; this is overt in those which are explicitly spatial or it is revealed after the fact from the incidence of their spatial impacts. This section addresses the question of a theoretical framework which will provide justification and guidance and will frame the practical aspects of the spatial integration of Community policies; how this might be achieved in practice. The term 'theoretical framework' is used instead of 'theory' because multiple disciplinary theoretical perspectives[32] that inform the study of the spatiality[33] of socio-environmental phenomena are inevitably engaged.[34]

An appropriate theoretical framework should encompass the wide and diverse spectrum of components of the socio-spatial system and the relationships among them on and across various hierarchical levels that policies should respect and observe both in a substantive and in a procedural sense. It should suggest how policies influence these components and their relationships to produce various spatial impacts. Moreover, it should be 'translatable'; i.e. it should be possible to translate the theoretical guidance offered into SPI design principles, procedures, and instruments. Finally, it should offer flexibility in formulating tailor-made theories for particular territorial/geographic entities[35] to support and inform spatially-sensitive integration of Community policies, both at the level of the EU as well as across spatial/organizational levels.

A detailed review of explicit or implicit theories embodied in Community policies is beyond the present purposes. This section first considers two basic spatial notions that are explicit or implicit in most official policy texts, 'sustainable development' and 'region'. Then it offers a selective, brief account of theories from the broader socio-spatial as well as from the narrower nature-society literature, all of which underscore the integrative function of space, thus, offering the obvious and indubitable foundation for the formulation of the requisite theoretical framing of SPI. An appropriate theoretical framework is then suggested for further elaboration.

Sustainable development, an inherently spatial concept, is the overarching rhetoric that straddles the theoretical framing of most contemporary Community policies. Its interpretation and operationalization, however, varies with and is colored by the fundamental theoretical origins and orientations of individual policies. Economic, transport, agricultural and regional development policies draw heavily on economic theories and the market paradigm, endorsing a 'weak' or 'very weak' sustainability perspective and mostly discounting the spatial dimension of policy problems (O'Riordan & Voisey, 1998). Environmental policies, informed generally by natural science theories,[36] adhere to the 'strong' or 'very strong' sustainability view and account somehow for the spatiality of policy problems. However, the transition to sustainable development through balancing

---

[32] Such as geography, sociology, economics, psychology, etc.

[33] Explained later in this section.

[34] On the differences between 'theoretical framework' and 'theory', see Ostrom (1999, pp.39-40).

[35] Several Community policies have been criticized that they reflect socio-spatial relations and problems of the European North, this partly explaining their problematic adaptation and implementation to other geographical settings (Hatzimichalis, 1995).

[36] See, e.g., the ecosystem approach of the WFD.

economic, environmental and social goals is inconceivable in a vacuum of space[37] and it is better achieved within particular territories. As EC (1998) puts it:

> Territories, ... play everywhere the same roles as (1) the physical base for productive activities, (2) the life support system for people and natural resources, and (3) the place where the impacts of most policies can be seen or felt. The territory, therefore, provides a unique medium for developing a crosscutting, multi-sectoral perspective, for reconciling sometimes conflicting objectives, setting mutually compatible targets and ensuring that interventions affecting its organization, structure and use are coherent (p.7).

Regional sustainable development is the main thrust of several Community policies and initiatives.[38] The 'region' is a central, theoretical and operational notion, not equivalent in general, however, to the broader notion of 'territory' that is also used in policy texts. Although the precise definition and delineation of the region, either homogeneous or functional, remains a contested issue,[39] it is considered the proper and meaningful, spatial entity and spatial level for the management of environmental issues and for policy interventions, as it occupies an intermediate position between the local (neighbourhood, commune) and the national and supra-national level, on which environmental and socio-economic processes and associated decisions are satisfactorily integrated, and spatial relationships can be adequately addressed and managed (Campbell, 1996; Selman, 2000; Berger, 2003; Lehtonen, 2004). It represents a minimum level of governance where practical implementation of (development) policies takes place, ensuring local decision making autonomy that does not detract from supra-local harmonious, coordinated development (Hardy and Lloyd, 1994; Roberts, 1994; Kruseman et al., 1996; Hakkinen, 1999; Scleicher-Tapesser, 1999; Vonkeman, 2000; Theys, 2002).

Several policies are already couched in terms of, or contain direct references, to sustainable land use, regional development, rural development, and spatial development more generally, resounding variously the ecological equilibrium theoretical perspective where the spatial or land use structure is the key factor mediating the dynamic equilibrium between population, resources, technology and institutions in a region (Coccossis, 1991).[40] This perspective is reflected in several

---

[37] EC (1998, p.13) notes characteristically "integrated policies cannot exist without a territorial reference".

[38] For more information, see the REGIONET site: http://www.iccr-international.org/regionet

[39] Depending on the issue and the level of analysis; moreover, for a number of reasons, political/administrative regions are used that may not be the best choice for all types of problems (Berger 2003).

[40] "Availability and use of natural resources of a durable quality are closely related to land use patterns, which in turn are responsive to economic development policies and requirements. In other words, sustainable environment and sustainable spatial development condition each other, most of the negative environmental impacts having their origin in

EU policy choices aiming to assist regions overcome barriers to achieving sustainable development. These include: (a) delimitation of areas eligible for financial support and modulation of assistance rates, (b) improvement of basic infrastructures, (c) differentiation of policies and measures on the basis of specific territorial criteria, (d) development of functional synergies and (e) design of integrated approaches[41] (ESDP, 1999).

A cursory look at the literature, however, reveals a wealth of theoretical approaches, frameworks, paradigms, particular theories and empirical studies of the nature-society system, with either an explicit or an implicit spatial content and orientation, which have not been exploited yet in policy making for complex socio-spatial problems (see, for example, Briassoulis, 2000). Sociological and economic theories that negotiate the driving forces of socio-spatial phenomena mainly possess implicit spatialities and spatial references. More directly relevant are socio-spatial development theories from the fields of geography, regional science and planning. These negotiate the spatiality of socio-environmental phenomena[42] and the integrating function of space in explaining them, the socio-spatial relations linking economic activities to particular places (and relations among places), resources and socio-cultural practices, the complexity that draws on spatial interdependencies, and the spatial relationships and linkages between their socio-economic and environmental dimensions. Hence, their prescriptions can be translated relatively more readily into cross-policy linkages and address more meaningfully the indispensable linkages between policy design (SPI here) and policy implementation.[43]

In geography, space has been conceptualized as absolute, relative and relational (Gregory and Smith, 1994).[44] The relative and relational conceptions have more

---

territorial imbalances induced by competing and conflicting demands on given spatial potentials" (EC, 1998, p.19).

[41] Such as the LEADER and ICZM initiatives.

[42] That is, theories that do not employ abstract or point representation of activities, resources, etc., that eliminates the spatial and social attributes of the associated phenomena and emphasizes excessively the physical properties of space.

[43] "The study of implementation requires understanding that apparently simple sequences of events depend on complex chains of reciprocal interaction. Hence, each part of the chain must be built with the others in view. The separation of policy design from implementation is fatal" (Pressman and Wildavsky 1992, p.xxv).

[44] "*Absolute space* 'distinct physical and eminently real or empirical entity in itself (empty, Sayer 1985); *relative space* 'a relation between events or an aspect of events, and thus bound to time and process' (space only exists where it is constituted by matter, Sayer, 1985), it is constituted by objects but it is not reducible to these objects; *relational space* (Harvey) 'space is contained in objects in the sense that an object can be said to exist only insofar as it contains and represents within itself relationships to other objects. To analyze the social production of space and the spatiality of phenomena, a distinction is necessary between objective and social space; the former refers to 'objects in space' while the latter concerns relations of class, race and gender, which are inscribed in (and in part constituted through) its places, regions and landscapes" (Gregory and Smith 1994, pp. 573-5).

essential and practical meaning as they concern economic, social, political and cultural relations among activities and resources, in their characteristic spatial arrangements, and point to the indivisibility of the social, economic and environmental aspects of phenomena. Sayer (1985) correctly asserts that it is not possible to abstract content from form and "space makes a difference but only in terms of the particular causal powers and liabilities constituting it" (Sayer, 1985, p.52).

The concepts of 'spatiality', 'territory' and 'territoriality' are also relevant here. Spatiality denotes the social implications of space and, although its definitions vary with the epistemological tradition (Gregory and Smith, 1994), all oppose the separation of space and society.[45] As human experiences are not cognitive abstractions of separate objects but "constellations of relations and meaning" encountered in everyday activities (Pickles, 1985 cited in Gregory and Smith, 1994, p.582), human spatiality cannot be understood independently of the beings that organize it; hence the connections and correspondence between social and spatial structures, social and spatial forms[46] (Lipietz, 1977; Massey, 1984; Gregory and Urry, 1985).[47] The term 'territory', interpreted as an area with characteristic socio-economic organization, relationships and physiognomy (Kox, 2002; Johnston et al., 1994) captures these relationships that apply not only to the social but also to the environmental realm.[48] Moreover, 'territoriality', as "intimately related to how people use the land, how they organize themselves in space, and how they give meaning to place" (Sack 1986, p.2) is, in several respects, akin to the notion of 'spatiality'.

Theoretical frameworks and paradigms that provide explanations of socio-spatial phenomena and related problems range from global level theories, such as the core-periphery and the unequal exchange and dependency theories (Cooke, 1983; Emmanuel, 1972; Amin, 1976) to national-regional level theories such as the uneven development and capital logic theories[49] (Lipietz, 1977; Cooke, 1983; Smith, 1990). Uneven development, a systematic process of economic and social development that is uneven in space and time, is "a basic geographical hallmark of the capitalist mode of production .... combining the opposed but connected processes of development and underdevelopment" (Johnston et al., 1994, pp.648-9). Lipietz's theory of unequal regional exchange (Lipietz, 1977) examines how different modes of production connect across inter-regional space. The emergent

---

[45] "Social space ... physical extent fused through with social intent" (Smith (1990) cited in Gregory and Smith 1994, p.584). An equivalent definition may apply to environmental space.

[46] As Massey (1985, p.12) puts it "Space is a social construct – yes. But social relations are also constructed over space, and that makes a difference".

[47] An analogous definition may apply to the spatiality of environmental phenomena, as perceived by human agents, at least.

[48] As the term 'territory' is also used in the biological and environmental sciences.

[49] The label 'capital logic' used by Cooke (1983) owes to the common explanatory framework of these theories that relates to the logic of *capital accumulation*.

spatial division of labour, based on unequal exchange between underdeveloped regions, characterized by pre-capitalist mode of production, and developed regions, dominated by the capitalist mode of production, leads to differences in wage levels. Underpayment and the subsequent squeeze upon living standards stimulate rural depopulation and migration to urban-industrial agglomerations.

The broader theoretical shell *of uneven development* addresses the uneven nature of socio-spatial development arguing that space and social process are integrated at various levels through the medium of capital. In the words of Smith (1990, p.xiv):

> In its constant drive to accumulate larger and larger quantities of social wealth under its control, capital transforms the shape of the entire world. No God-given stone is left unturned, no original relation with nature unaltered, no living thing unaffected. To this extent, the problems of nature, of space, and of uneven development are tied together by capital itself. Uneven development is the concrete process and pattern of the production of nature under capitalism.[50]

This theory may contribute to the theoretical framing of SPI as it suggests how the articulation of economic, social and environmental systems gives rise to spatial patterns of development and underdevelopment which underlie most contemporary socio-environmental phenomena.

The economic relationships linking the processes of production, consumption (reproduction) and resource use are essentially socio-spatial relationships. Places where these processes occur are connected through dynamically changing flows and transfers of value, commodities, resources, information, etc. several of which are embodied in various forms of landesque capital (road and communication networks, irrigation networks, terraces, etc.). Because of the spatial forward and backward linkages among economic sectors, sectoral crises (e.g. de-industrialization) become territorial crises. Similarly, because "the social and the natural are impossible to conceptualize separately and then 'put together'" (Williams, 1981), environmental crises are also territorial crises linked to particular modes of production and socio-spatial relationships.

Certain traditional schools of thought in geography offer holistic, spatially-explicit approaches to society-nature relationships. The so-called *Berkeley School*, marked by Carl Sauer's "interests in landscape creation as a representation of culture and ... his emphasis on studies of the evolution of the cultural landscape", "involved appreciation of the 'natural environment', reconstruction of past landscapes, and processes of change through the spread (diffusion) of human agency"[51] (Johnston et al., 1994, p.33).

---

[50] See also Briassoulis (2000), Ch. 3.
[51] See, Sauer's classic essay 'The morphology of landscape' (Sauer, 1996; first appeared in 1925).

Cultural ecology, an interdisciplinary field of study of the nature-society relationships, drawing upon ecology and systems theories with contributions from sociology, anthropology and geography, provides comprehensive descriptions of the complex interactions between people and their bio-physical environment (Sack, 1990; Butzer, 1990). As Sack (1990) observes:

...the primary, though by no means only, device that ecologists employ to connect human and natural systems is, to put it positively, to focus on characteristics that both systems possess, or to put it slightly negatively, to reduce human actions to physical ones" (Sack, 1990, p.665).

Central in most studies is the concept of *adaptation*, "an on-going process of adjustment as people cope with internal and external impulses, in the short or long term, whose basic function is to maintain a balance between population, resources and productivity" (Butzer, 1990, p.696). The 'ecological equilibrium' framework mentioned earlier is a related theoretical perspective whose general premises imply the possibility and existence of multiple equilibria of socio-spatial systems.[52]

These relatively older theoretical approaches have their counterparts in contemporary interdisciplinary thinking on spatially heterogeneous, linked social-ecological systems that is embedded within the sustainability discourse and feeds into theorizing on sustainable land use, regional development and spatial development more generally. Some approaches are explicitly informed by Complexity theory thinking while others, although not overtly mentioning it, resound its basic premises. The diffusion of Complexity thinking in geography, economics and the natural and the social sciences, among others, in the 1980s has helped revive the interest in the importance of space and spatial processes in explaining socio-economic and socio-environmental phenomena and in designing policy interventions to resolve development problems (see, for example, Krugman, 1997; Berkes and Folke, 1998; Portugali, 2000; Wilson, 2000; Gunderson and Holling, 2002; Reitsma, 2002).[53] Thrift (1999, p.32) notes, "here ... is a body of theory that is preternaturally spatial: it is possible to argue that Complexity Theory is about, precisely, the spatial ordering that arises from injections of energy".

The main difference of complexity-informed from conventional approaches to resource policy and management problems is that they drop the reductionist, linear-change world view of scientific inquiry and adopt a holistic, non-linear, evolutionary perspective. The linked, socio-ecological and institutional (decision) systems are conceptualized as complex adaptive systems (CAS) and are treated by means of integrated approaches. The policy implications of this perspective are

---

[52] Similarly, Rappaport (1968) drawing upon structural-functionalist and ecological concepts proposed "an ecosystem approach that related culture to abiotic and biotic components of the environment as a spatially bounded unit" (cited in Merchant 1990, p.674).

[53] To avoid repetition, the following discussion of complexity-informed approaches draws heavily on Chapter 1 of this volume.

what make it most relevant to the present purposes. The following discussion summarizes the main features of CAS that have been introduced in Chapter 1, presents a recent integrated, complexity-informed, theoretical framework and suggests the form of a SPI theoretical framework.

CAS are characterized by *non-linear relationships* among their components and by *multiple equilibria*. They *adapt* to internal or external changes through *self-organization* and *learning*. *Positive feedback mechanisms* account for the transition between multiple states and for the sensitivity of CAS to initial conditions. Consequently, in CAS, history matters; their path dependence may lead to lock-in situations, where change is irreversible, or to surprises. CAS are organized in nested hierarchies on various spatio-temporal scales; at each scale, their properties are *emergent*, a synthesis of the properties of systems at lower scales but not reducible to them. They co-evolve with their environment as their changes affect its structure which in its turn triggers changes in the individual systems. Their evolution is governed by slow processes of gradual change and fast processes of abrupt, discontinuous change.

The literature amply demonstrates that urban, regional, rural, and human-environment systems in general possess the properties of CAS, thus making the use of complexity-informed approaches indispensable to their study (see, e.g., Wilson, 2000; Batten, 2001; Gunderson and Holling, 2002). As a result of their non-linear, self-organizing and emergent character, the properties of these systems cannot be explained by knowing their individual components; their causality structure and future evolution are difficult to predict, but they possess an inherent diversity and flexibility to adapt to changing conditions. Thus, their management faces considerable uncertainty that dictates the employment of adaptive, learning and precautionary approaches.

The synthesis of these ideas into an integrated theoretical framework is best exemplified by the Panarchy model of Gunderson and Holling (2002). This is briefly summarized below, based on Holling (2001), because it offers valuable guidance for policy interventions that foster both ecological and social resilience and sustainability within a system, drawing on analyses of the institutions that link social and natural systems.

The term 'panarchy', used to describe the evolving nature of CAS, denotes the hierarchical structure in which systems of nature[54] and humans[55] as well as combined human-nature systems[56] and social-ecological systems[57] are interlinked in never-ending adaptive cycles of growth, accumulation, restructuring, and renewal taking place in nested space/time hierarchies ranging from the micro- to the macro-level. Understanding adaptive cycles and their scales facilitates the evaluation of their contribution to sustainability and the identification of the points

---

[54] E.g., forests, grasslands, lakes, rivers, and seas.
[55] E.g., structures of governance, markets, settlements, and cultures.
[56] E.g., agencies that control the use of natural resources.
[57] E.g., co-evolved agricultural management systems.

at which a system is capable of accepting positive change and the points where it is vulnerable.

A CAS hierarchy is not a top-down sequence of authoritative control but a sequence of semi-autonomous levels formed from the interactions among a set of variables with similar speeds and geometric/spatial attributes. Each level of a dynamic hierarchy communicates a small set of information or quantity of material to the next higher (slower and coarser) level and serves two functions (a) to conserve and stabilize conditions for the faster and smaller levels and (b) to generate and test innovations by experiments occurring within a level. Holling (1986) has called this latter function 'an adaptive cycle' which is shaped by three general[58] properties (a) the *inherent potential* of a system that is available for change, loosely thought of as the 'wealth' of a system; (b) the *internal controllability* of a system, a function of the connectedness among its components, reflecting its degree of flexibility or rigidity and, thus, sensitivity to perturbations; and (c) the *adaptive capacity*, i.e. the resilience of the system, a measure of its vulnerability to unexpected or unpredictable shocks.

An adaptive cycle comprises four stages: growth (exploitation of resources), r, accumulation (conservation of resources), K, restructuring (release of accumulated potential), Ω, and renewal (reorganization), α. The transition between these stages – from r to K, from Ω to α, and so on – is schematically depicted as an eight-shaped figure (Figure 9.1) that represents the relationship between the potential and the connectedness of the system for a given level of adaptive capacity. It involves both slow processes of change, such as a long period of (predictable) slow accumulation and transformation of resources or, equivalently, from exploitation r to conservation K, and fast processes, such as a short period of restructuring and reorganization, from Ω to α. The latter, that are inherently unpredictable and highly uncertain, create opportunities for innovation, what Schumpeter (1950) has called "creative destruction".

Sudden changes, such as from K to Ω, may occur when the system develops rigidities due to over-connectedness that render it vulnerable to disturbances caused by natural or socio-economic agents (e.g. diseases, droughts, price changes, political or other decisions, new technologies, etc.). During reorganization, potential may leak out of the system which then flips to a less organized state, denoting an exit from the present adaptive cycle. If opportunities are offered to capture the potential available, however, the cycle is repeated, i.e. a fast movement from α to r occurs. In a sense, the adaptive cycle embraces growth and stability, on the one hand, and change and variety, on the other. Adaptive cycles are contextual and contingent depending on the system studied.

The resilience of the system, the third property of the adaptive cycle, expands and contracts as the phases of the cycle proceed. Under conditions of low system connectedness and high system potential, i.e. in the α phase, resilience is high

---

[58] Because they are found at all scales; from the cell and the individual to the biosphere and culture.

offering opportunities for innovations and experimentation. Successful innovations bring the system again to r and the adaptive cycle is repeated.

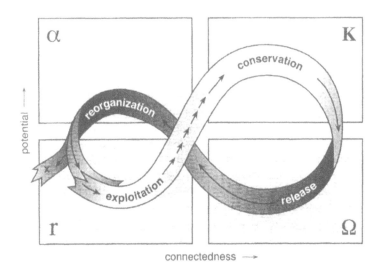

Figure 9.1 Stylized representation of an adaptive cycle

Source: Reprinted from Gunderson and Holling (2002, p.34)

The 'panarchy' is a nested set of adaptive cycles. It represents the ways in which a healthy social-ecological system can invent and experiment, benefiting from inventions that create opportunity while being kept safe from those that destabilize it because of their nature or excessive exuberance. The whole panarchy is therefore both creative and conserving. While each level is allowed to operate at its own pace, simple interactions across levels maintain system integrity. The interactions between cycles in a panarchy combine learning with continuity. This process can serve to clarify the meaning of 'sustainable development'. Sustainability is the capacity to create, test, and maintain adaptive capability. Development is the process of creating, testing, and maintaining opportunity. 'Sustainable development', that combines the two, therefore, refers to the goal of fostering adaptive capabilities while simultaneously creating opportunities.

Several contemporary approaches to sustainable spatial and regional development echo complexity-theoretic ideas, although they are not explicitly cast in such terms. Regional sustainable development is viewed as a dynamic process of adaptation and change, and not as a precise objective that may be achieved with the proper allocation of resources. Multiple equilibria of regional development exist, satisfying the 'sustainability' criterion, because of the evolutionary nature of human goals, the inevitability of short-term unsustainable situations, and the systemic nature of contemporary socio-environmental problems that place limits on the ability to achieve sustainable development at the regional level when the reasons for unsustainability are global (Vonkeman, 2000).

Complexity-informed approaches in general, and the Panarchy model in particular, imply a different, space-sensitive approach to policy making for complex socio-environmental problems, thus providing a fertile ground to develop a theoretical framework for SPI on various scales, starting from that of the EU as a whole. The task of fully developing and implementing such a framework is undeniably onerous but essential. At the EU level, it should indicate principles for spatially-sensitive EU policy design[59] and criteria to judge whether SPI has already been achieved or how it might be achieved as well as to judge its contribution to sustainable development.[60]

Guided by the conceptualization of multi-level socio-ecological systems as CAS, this framework should synthesize notions and approaches from sociological, economic and nature-society theories of spatial development to give substance to their components, their relationships, and the processes governing their change and evolution over space and time. Focus should be placed on policies affecting strategic resources and their influence on both the fast and the slow variables and processes of the system to identify policy impacts and guide policy diversification and modulation to facilitate adaptation to local conditions and societal learning.

The co-existence of multiple spatial equilibria and the self-organizing properties of spatial systems necessitate spatially fine-tuned and coordinated policies that manipulate locations to achieve advantageous effects and foster social and ecological resilience. More specifically, they allow enough room for local initiatives, exploit locally available potential and avoid detrimental spatial shifts of socio-economic activity that might increase spatial inequalities[61] and produce unwanted surprises.

The complexity-informed spatial integration framework should also guide the procedural aspects of the task. More specifically, an adaptive management approach to the design of multi-level territorial governance systems should strive for both their horizontal and vertical spatial integration, providing for multiple points of intervention to address uncertainty and cope with surprises. Territorial representation at all levels is a necessary prerequisite to achieve this objective. The principal task should be to transform the current governance systems from complicated, or at times chaotic, to complex by properly elaborating hierarchically nested formal (and informal) structures as suggested by the Panarchy model (Gunderson and Holling, 2002) and related approaches. The nested hierarchy of the policy system will secure the requisite flexibility for adaptation to changing internal and external conditions and will secure also accountability that may be lost when too much power and responsibility is been devolved to lower levels.

---

[59] Some of which are already in place such as the precautionary principle.

[60] As defined by Holling (2001) above.

[61] I.e. the development-underdevelopment gap.

Synopsis–Bringing Space Back In

Most of the impetus and the proposals for the spatial integration of Community policies come, not unjustifiably, from spatial experts and the DGs to which they relate, while a relative neglect on the part of the Policy Sciences cannot go unnoticed.[62] This chapter examined the interest, the evidence, the justification and current proposals related to SPI and explored germane theoretical frameworks. This account revealed a relative consistency between the analysis of the problem and the proposals offered but it uncovered also the poor theoretical framing of both. This closing section discusses the political support offered to the issue, offers proposals as regards the choice of approaches to realizing the spatial integration of Community policies, suggests future research directions, and comments on the difficulties of the whole undertaking.

Signs of political support for the spatial integration of Community policies are weak at best. This pessimistic prognosis draws from a deeper, critical examination of the available evidence. The mostly abstract discussion of the issue, with certain exceptions that are mentioned later, make plain the highly politicized nature and the complexity of the spatial dimension of socio-environmental problems, a point supported by several theories of social and spatial development. The suggestion of SUD (2003) for "…. (the need to) define territorial cohesion more precisely and develop it as a more policy relevant concept" reveals the crux of the problem. Who will define the concept and will be, at the same time, willing (and powerful) to act on it? Otherwise, it is meaningless to discuss alternative definitions on the political level. The critical question, therefore, is who has an interest in the spatial aspects of socio-economic development and who is persuaded that it is to the common good of the EU as a whole to strive for more spatially integrated Community policies? Without shared purpose, the whole discussion reduces to wishful thinking.

The sectoralized character and culture of Community policy making together with the considerable differences in policy making styles among the various Community bodies prove to be strong barriers to the horizontal coordination of EU policies on a spatial basis, as the recent theoretical developments would suggest (CEC, 2001b). Because the origin of interest and the concern for spatial issues and SPI lay in DG-Regional Policy mostly, there is weak interest and support from other DGs especially those that are traditionally indifferent to spatial issues.[63] Stronger incentives, backed by strong political will, might be needed to induce the process at the level of the EU. Efforts to promote spatially integrated policy initiatives (e.g. the INTERREG programme and the LEADER initiative) are rather

---

[62] The Policy Sciences do not appear particularly overtly interested in the issue perhaps because perhaps their traditional orientation is not adequately sensitive to spatial issues. However, indirectly they do address the spatial integration question because simply the spatiality of social phenomena penetrates policy analysis.

[63] Lenschow (2002) makes a similar argument for the case of environmental policy integration.

isolated, fragmented and limited to particular issues and geographic areas,[64] leaving essentially untouched the issue of providing a solid, common basis for the spatial integration of all Community policies.

More generally, the Community seems to adopt a 'hand-off' approach to the issue, trying to by-pass it while demonstrating a more general turn towards less regulatory and interventionist modes of policy making. The promotion of the Open Coordination Method to delivering several Common policies (social, environmental, etc.), the preference for framework directives[65] and market-based instruments, the devolution of significant powers to lower administrative levels, the strengthening of the subsidiarity principle and the encouragement of third sector involvement indicate a weakening political will to decisively intervene in matters of Community interest such as those of socio-spatial development. Evidence of evading the issue *at the level of the EU* is the rather excessive insistence on vertical communication, cooperation and coordination among spatial/organizational levels and the incentives offered to encourage vertical partnerships as well as cooperation at lower spatial levels.[66] Although commendable, these efforts introduce further institutional complexity[67] that overshadows the actual complexity of spatial development problems. Less direct evidence, such as the emphasis on procedural internal housekeeping,[68] assessment methods[69] and indicators, suggests also the relative neglect of the more essential harmonization of Community policies on a spatial basis.

These tendencies reveal a contradiction, however; the political choices of the Community go counter to claims for promoting social, economic and territorial cohesion and for making the EU a competitive, knowledge-based society (Lisbon Strategy). The evidence presented in this chapter demonstrates that without a spatially coherent policy system the Community cannot guarantee achievement of this vision for Europe.

The encouragement of integrated approaches to address territorial development issues, promote territorial cohesion and reinforce the spatial content and orientation of several EU policies are signs of a positive change in the right direction.[70] The

---

[64] Given the complexity of socio-spatial relationships, the effectiveness of these integrated programmes can be questioned also; such as the fact that the success of a programme in a locality hinges on what happens in other, supra-local or infra-local, geographic areas.

[65] And directives instead of regulations more generally.

[66] Characteristically, the Third Cohesion Report (EC, 2004) contains a section on territorial cohesion that refers to lower level efforts and does not touch the EU level (DGs and other administrative bodies).

[67] As more actors get involved, their relationships increase exponentially, functions are specialized further, all of which hinder integration (Zahariadis, 2003).

[68] Notably, the Better Regulation Action Plan (BRAP) (CEC, 2002a).

[69] Notably the Extended Impact Assessment of policy proposals (CEC, 2002b) which is an action of the BRAP.

[70] "Some initial aspects of territorial cohesion can already be identified. It can be seen as an umbrella concept and an integrated part of the cohesion process. It covers the territorial dimension of social and economic cohesion and is closely linked to the fundamental EU

ESDP is the most publicized attempt to address the issue but it is an indicative document that suggests the goals of spatial integration and offers advice and ideas but no visible links or persuasive arguments how these will materialize through horizontal coordination of Community policies at the EU level.[71] Moreover, its suggestions are heavily skewed towards spatial planning at national and sub-national levels which may not be the only avenue to spatial policy integration of Community policies in general, especially in countries where spatial planning is not a powerful government sector. Overall, the question has not been comprehensively addressed yet. Some preliminary proposals are sketched below.

The spatial integration of Community policies can take place either at the level of the EU or at national and sub-national levels or through on-going, institutionalized dialogue, collaboration and cooperation between the EU and the lower levels (Figure 9.2). A main prerequisite in all these cases is the existence of a coherent theoretical framework to support analysis, institutional design and practical applications, an issue discussed later as a topic of future research.

At the level of the EU, the choice basically involves either the development of an integrated spatial policy or of spatial policy integration schemes (SPIS). The ESDP (1999) seems to favour the first choice, at least in the long run. The ESDP has been developed, however, on the basis of a loose theoretical framing and a rather limited analysis of complex spatial issues.[72] As a consequence, it proposes a rather vague, rigid policy structure, relying heavily on spatial planning, that does not possess the requisite flexibility to operate in the present European multi-level governance system. Overall, the prospects of a Common integrated spatial policy are not bright for a number of reasons.

Socio-spatial development problems are complex, both in an environmental and in a socio-cultural and institutional sense, cutting across several thematic areas, administrative boundaries and spatial levels (see, Chapter 1, this volume). The spatial planning systems of the MS differ widely as do the respective national planning cultures and traditions, and administrative and political systems. The growing decentralization trends diversify further the formal and informal actors involved in societal governance. The European Union already intervenes in societal affairs with a variety of policies and policy initiatives all of which have significant spatial impacts in the short or in the longer term. The systemic character

---

objective of 'balanced and sustainable development' (Art. 2 EU-treaty). It demands a more integrated approach, from a territorial perspective, to both EU investments directly relevant to the cohesion of the European territory (structural funds/cohesion fund) and other EU policies also relevant to territorial cohesion (e.g. TENs, environmental policy and the CAP)." (SUD, 2003).

[71] Note also: "The ESDP lacks the indicators, data and analyses that can be used to link policies and investments more effectively to specific territorial conditions and development potential" (SUD, 2003).

[72] Not to mention that it reflects a North European-centered view of spatial development issues and practices that has to be broadened to represent other socio-spatial and geographic contexts of the Union.

of these impacts makes difficult their precise definition, assessment and timely management especially within democratic, multilevel governance structures where communication and coordination is not easy given the large number of spatio-temporally dispersed actors and of their relationships.

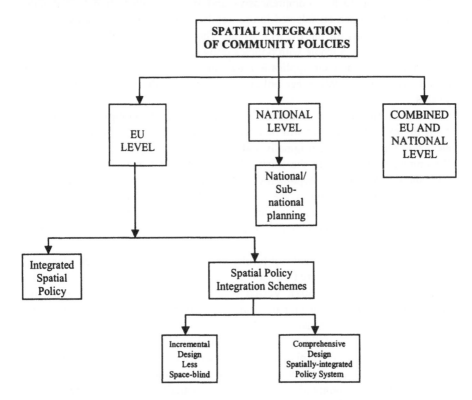

Figure 9.2   Alternative approaches to the spatial integration of Community policies

Under these conditions it is obvious that a Common integrated spatial policy for the EU-25 is not a feasible or viable response to the management of socio-spatial affairs. It might be preferable to elaborate SPIS that exploit and add value to the interdependencies and complementarities, while removing the conflicts, among extant Community policies. If this avenue is chosen, two main approaches can be pursued. The first is an incremental approach which focuses on making extant policies less space-blind, introducing certain spatial concepts, measures and criteria in policy design and delivery. OSFA[73] policies are abandoned,[74] irrespective of potential short-term benefits, in favour of highly heterogeneous in space and fluid

---

[73] OSFA: One-Size-Fits-All.
[74] Because they erode ecological and social resilience (Carpenter and Brock, 2004).

in time policies (Carpenter and Brock, 2004) that transcend administrative boundaries and sectoral divisions and focus on concrete spatial development issues. This approach is currently being practiced, in various guises, in several policy areas such as regional and agricultural policy.

The other approach is comprehensive and strategic, attempting to elaborate alternative SPIS based on selected future trends scenarios as well as (sustainable) development visions for the EU as a whole. The emphasis of policy design is on the policy system as a whole with policies designed so as (a) to be sensitive to the 'when and where' of their application and, more importantly perhaps, (b) to be spatially coordinated with other policies impacting on the same or other spatial systems.

The goal, therefore, is to design a spatially-coherent, multiple management, policy regime, whose output will be alternative SPIS. These can be developed for selected groups of policies originally and then be extended gradually to encompass all Community policies. In the spirit of the adaptive management approach, the SPIS should be seen as hypotheses to be tested and their implementation as experiments whose outcomes are used to update the SPIS to adapt them to changing socio-economic and environmental circumstances. This is no easy task especially if certain preconditions are not satisfied as discussed in closing this section.

At the national and sub-national levels, SPI can be promoted through planning, an approach strongly favoured and advocated recently (see, for example, ESDP, 1999; Nadin, 2003; EC, 2004). Vested with adequate power and resources, the lower levels of government prepare spatial development plans whose realization requires the proper combination of Community and national policy measures. However, the success of this mostly procedural approach is not guaranteed for three reasons, at least. First, planning traditions differ widely among MS, some of which may be rarely using planning as a guide for the rational development of their territory. Second, if Community policies are not spatially integrated by design and the sectoral rigidities barring horizontal cooperation remain, these will reflect on lower levels, reinforcing the sectoralization of the national policy systems. Third, national and subnational plans cannot contribute to the social, economic and territorial cohesion of the Union, especially where cross-cutting, EU-wide issues are at stake (such climate change, desertification, industrial decline, etc.), if they are not coordinated among them, a function that belongs to the competences of the Community and is served by its policies.

A more viable approach might be to attempt the spatial integration of Community policies through a mixed system of ground level coordination and higher level transformations in Community policy making. The encouragement of dialogue and more active involvement of all levels of government and the civil society in decision making initiated by the Community may facilitate this solution and eventually may assist in removing the compartmentalization and reinforce the horizontal restructuring of EU policy making (see, for example, CEC, 2001a).

In closing this discussion, issues that the political process and future research have to resolve, irrespective of approach pursued, are briefly mentioned. A common 'vision', based on a common understanding and a common frame of mind in addressing development issues, is the *sine qua non* of policy integration in general and the spatial case is no exception. The Community should take the initiative to formulate, promote and keep updating a spatial vision[75] on the basis of scientific evidence[76] and by strengthening its current efforts not only towards multi-level, vertical, dialogue (EC, 2004), but also towards horizontal communication and cooperation among its Services. All policy making bodies of the Union should gradually adopt and adapt to a commonly agreed vision so that, even without communicating on specifics, their actions are harmonized, compatible and congruent.

The common vision should draw on a common, shared, explicit theoretical framework of socio-spatial development to guide action and practice, not just serving decorative and symbolic purposes. As suggested before, complexity theory-informed approaches can be exploited to carry out this highly demanding task; i.e. to provide a platform for the synthesis of theories from the natural, the social and the policy sciences to guide the conceptual as well as the operational integration of policies on a spatial basis.

Parallel developments in analytical tools are necessary to support the shift towards spatially-sensitive, integrated policy making. Community, but also national, policies should adopt similar or comparable, compatible and consistent spatial and temporal systems of reference, classification systems, data and information collection systems, and assessment methods. The complexity-informed theoretical framework suggested should guide their selection and required synthesis. Methodological pluralism is strongly recommended because only a combination of quantitative and qualitative methods can comprehensively account for the features and functioning of socio-spatial systems. The requirement for extended impact assessment of Community policy proposals (CEC, 2002b) could include a SPI criterion in its provisions.

Adopting a complexity-informed policy outlook necessarily raises the question of suitable institutional structures at both the Community and the national levels. The multi-level governance model that materializes gradually in the Union provides an appropriate basis that should be, however, refined into a nested organizational hierarchy, as already suggested previously, through future research and political action.

Finally, spatially-relevant policy instruments should be promoted. Spatially-sensitive tools, such as SEA or Sustainability Appraisals (George, 2001), should be improved and gradually replace the EIA. New instruments developed in one policy area should attempt to achieve spatial complementarities and avoid conflicts with

---

[75] This is not a one-off exercise as socio-spatial dynamics will necessitate changing and adapting this vision as well as modifying policy choices accordingly.
[76] Such as the findings of ESPON-commissioned studies.

those developed in other areas. However, their use outside, or in the absence, of strategic planning at various spatial levels should not be encouraged. Although the choice of policy instrument mixes may be a lower level task, the Commission may experiment with such mixes specifically designed and adapted to the particularities and needs of selected geographic areas to serve as examples for further developing spatially integrated policy instruments. The desertification-prone regions of the Southern EU are opportune cases given the complexity and urgency of their problems.

The spatial integration of Community policies is not an easy task with guaranteed results, broadly the achievement of sustainable socio-spatial development.[77] The complexity of spatial relationships characterizing contemporary socio-environmental problems and the dynamic character of the policies themselves that keep changing and evolving as their objects and actors change preclude optimism on the existence of policy fixes[78] and one-off solutions. There will be always limits to the effectiveness of the policy system to deliver the desired results, however well it may be spatially integrated. The spatial integration of Community (and national) policies is not a panacea but a requirement that should be satisfied, among many others, to steer the policy system towards facilitating the transition to socio-spatial sustainable development paths. In this perspective, the implications of complexity-informed approaches, such as the Panarchy model, for the development of SPIS for the EU but also for the MS, should be explored in greater depth and detail.

References

ACRDR (2003), *The Territorial Impact of CAP and Rural Development Policy*, Third Interim Report, ESPON Project 2.1.3, Arkleton Centre for Rural Development Research University of Aberdeen, Aberdeen, Scotland.

Atkinson, G. and Oleson, T. (1996), 'Urban sprawl as a path dependent process', *Journal of Economic Issues*, 30, pp. 609-15.

Avery, G. (2001), 'Policies for an Enlarged Union', Report of Governance Group 6, *White Paper on European Governance* Area no. 6 Defining the framework for the policies needed by the Union in a longer-term perspective of 10-15 years taking account of enlargement, European Commission, Brussels.

Bache, I. (2003), 'Europeanization: A Governance Approach', Paper presented at the EUSA 8th International Biennial Conference, Nashville, March 27-29, 2003.

Batten, D.F. (2001), 'Complex landscapes of spatial interaction', *The Annals of Regional Science*, 35, pp. 81-111.

---

[77] See the spatial development guidelines of the ESDP in footnote 5.
[78] In the present case, ideal SPIS.

Berger, G. (2003), 'Reflections on governance: Power relations and policy making in regional sustainable development', *Journal of Environmental Policy and Planning*, 5(3), pp. 219-34.

Briassoulis, H. (2000), 'Analysis of land use change: Theoretical and modelling approaches', in Scott Loveridge (ed), *The Web Book of Regional Science*, Regional Research Institute, West Virginia University, USA.
(http://www.rri.wvu.edu/WebBook/Briassoulis/contents.htm).

Butzer, K.W. (1990), 'The Realm of Cultural-Human Ecology: Adaptation and Change in Historical Perspective', in B.L. Turner, II, W.C. Clark, R.W. Kates, J.E. Richards, J.T. Mathews, and W.B. Meyer (eds), *The Earth as Transformed by Human Action*, Cambridge University Press, Cambridge, pp. 685-701.

Campbell, S. (1996), 'Green Cities, Growing Cities, Just Cities? Urban Planning and the Contradictions of Sustainable Development', *Journal of the American Planning Association*, 62 (3), pp. 296-312.

Carpenter, S.R. and Brock, W.A. (2004), 'Spatial complexity, resilience and policy diversity: fishing on lake-rich landscapes', *Ecology and Society*, 9(1), p.8 [online] URL: http://www.ecologyandsociety.org/vol9/iss1/art8.

CEC (2001a), *European Governance; A White Paper*, COM(2001) 428 Final, Brussels.

CEC (2001b), 'Consultation paper for the preparation of a European Union strategy for sustainable development', Commission Staff Working Paper, SEC(2001) 517, Brussels.

CEC (2001c), *A Sustainable Europe for a Better World: A European Union Strategy for Sustainable Development*, COM(2001) 264, Brussels.

CEC (2002a), *Simplifying and improving the regulatory environment'*, Action plan Communication from the Commission, COM(2002) 278 final, Brussels.

CEC (2002b), *Communication from the Commission on Impact Assessment*, COM(2002) 276 final, Brussels.

CEMAT (2000), *Guiding principles for Sustainable Spatial Development of the European Continent*, 12th Session of the European Conference of Ministers responsible for Regional Planning (CEMAT), Hannover, 2000.

Cooke, P. (1983), *Theories of Planning and Spatial Development*, Hutchinson, London.

CURS (2002), *Urban-rural relations in Europe*, First interim report, October 2002, ESPON 2006 Programme (Coordinator: Centre for Urban and Regional Studies).

Dryzek, J.S. (1987), *Rational Ecology, Environment and Political Economy*, Blackwell Publishing, Oxford.

EC (1998), *Report on Community Policies and Spatial Planning*, Working document of the Commission Services, European Commission.

EC (2000), *TERRA, An experimental laboratory in spatial planning*, European Commission, DG-Regional Policy, Brussels (ISBN 92-828-3021-7).

EC (2001), *Second report on Economic and Social Cohesion*, European Commission, DG-Regio, Brussels.

EC (2004), *Third Report on Economic and Social Cohesion*, COM(2004) 107, Office for Official Publications of the European Communities, Luxembourg.

ESDP (1999), *European Spatial Development Perspective: Towards Balanced and Sustainable Development of the Territory of the EU.* European Commission, Office for Official Publications of the European Communities, Luxembourg.

ESPON (2002), *The ESPON 2006 Programme. Programme on the spatial development of an enlarged Union, Summary*, Luxembourg (For more information and regular updates see http://www.espon.lu).

ESTIA (2000), *Spatial Planning Priorities for Southeast Europe*, ESTIA INTERREG IIC (European Space and Territorial Integration Alternatives: Spatial development strategies and policy integration in SE Europe), Thessaloniki, September 2000.

Faludi, A. (2001), 'Introduction: The European Spatial Development Perspective', Paper presented at the 92nd National Planning Conference of the American Planning Association, March 10-14, 2001, New Orleans.

George, C. (2001) 'Sustainability appraisal for sustainable development: integrating everything from jobs to climate change', *Impact Assessment and Project Appraisal*, 19(1), pp. 95-106.

Gibbs, D., Jonas, A. and While A. (2003), 'Regional Sustainable Development as a Challenge for Sectoral Policy Integration', Paper presented at Workshop II of the EU Thematic Network Project REGIONET, Regional Sustainable Development - Strategies for Effective Multi-level Governance, Lillehammer, Norway, 29-31 January 2003.

Gregory, D. and Urry, J. (eds) (1985), *Social Relations and Spatial Structures*, St. Martin's Press, New York.

Gunderson, L.H. and Holling, C.S. (eds) (2002), *Panarchy; Understanding Transformations in Human and Natural Systems*, Island Press, Washington, DC.

Hadjimichalis, K. (1995), 'The southern margins of Europe and European integration', in *Proceedings* of the scientific conference "Regional Development, Spatial Planning and Environment in the context of a Unified Europe", Vol. I, pp. 74-92. Panteion University, Athens, December 15-16, 1995, (in Greek).

Hakkinen, L. (ed.) (1999), *Regions – Cornerstones of Sustainable Development*, Proceedings of the Second European Symposium. Joensuu, 13-14 September, 1999. Edita, Helsinki.

Hardy, S. and Lloyd, G. (1994), 'An impossible dream? Sustainable regional economic and environmental development', *Regional Studies*, 28(8), pp. 773-80.

Holland, J.H. (1995), *Hidden Order: How Adaptation Builds Complexity*, Helix, Reading, MA.

Holling, C.S. (1986), 'The resilience of terrestrial ecosystems: local surprise and global change', in W.C. Clark, and R.E. Munn, (eds), *Sustainable Development of the Biosphere*, Cambridge University Press, Cambridge, pp. 292-317.

Holling, C.S. (2001), 'Understanding the complexity of economic, ecological and social systems', *Ecosystems*, 4, pp. 390-405.

Johnston, R.J., Gregory, D. and Smith, D.M. (eds) (1994), *The Dictionary of Human Geography*, 3rd edition, Blackwell, Oxford.

Kox, K. (2002), *Political Geography: Territory, State and Society*, Blackwell Publishers, London.

Kruseman, G., Ruben, R., Kuyvenhoven, A., Hengsdijk, H. and van Keulen, H. (1996), 'Analytical framework for disentangling the concept of sustainable land use', *Agricultural Systems*, 50, pp. 191-207.

Krugman, P. (1997), *How the Economy Organizes Itself in Space: A Survey of the New Economic Geography*, The Economy as an Evolving Complex System II, Sante Fe Institute, Addison-Wesley.

Lenschow, A. (2002), *Environmental Policy Integration: Greening Sectoral Policies in Europe*, Earthscan, London.

Lehtonen, M. (2004), 'The environmental – social interface of sustainable development: capabilities, social capital, institutions', *Ecological Economics*, 49, pp. 199-214.

Leygues, J-C. (2001), *Multi-level governance: Linking and Networking the various regional and local levels*, Report of Working Group 4c (White Paper on European Governance). European Commission, Brussels.

Liberatore, A. (1997), 'The integration of sustainable development objectives into EU policy-making: Barriers and prospects', in S. Baker, M. Kousis, D. Richardson and S. Young (eds), *The politics of sustainable development: Theory, policy and practice within the European Union*, Routledge, London.

Lipietz, A. (1977), *Le Capital et son Espace*, Maspero, Paris.

Massey, D. (1984), *Spatial Divisions of Labour*, Macmillan, London.

Massey, D. (1985), 'New directions in Space', in D. Gregory, and J. Urry (eds), *Social Relations and Spatial Structures*, St. Martin's Press, New York, pp. 9-19.

Nadin, V. (2003), 'Spatial Planning and Sustainable Development', Paper presented at the Policy Seminar, European Regional Sustainable Development Network, Asturias, 3-4 April 2003.

Nordregio (2003), *Territorial Effects of Structural Funds*, Second Interim Report. ESPON 2.2.1. Nordregio, Stockholm, Sweden.

O'Riordan, T.R., and Voisey, H. (eds) (1998), *The Transition to Sustainability*, Earthscan, London.

Ostrom, E. (1999), 'Institutional rational choice: An assessment of the Institutional Analysis and Development framework', in P. Sabatier, (ed.), *Theories of the Policy Process*, Westview Press, Boulder, Co, pp. 35-72.

Portugali, J. (2000), *Self-organization and the City*, Springer, New York.

Pressman, J.L. and Wildavsky, A. (1973), *Implementation: How Great Expectations in Washington Are Dashed in Oakland*, University of California Press, Berkeley, CA.

Reitsma, F. (2002), 'Measuring Geographic Complexity', On-line paper. (http://www.glue.umd.edu/~femke/written/GeographicComplexityPaper.pdf)

Robert, J., Stumm, T.J., de Vet, M., Reincke, C.J., Hollanders, M. and Figueiredo M.A. (2001), *Spatial Impacts of Community Policies and the Costs of Non-Coordination*, EC, DG-Regional Policy, ERDF Contract 99.00.27.156, Brussels.

Roberts, P. (1994), 'Sustainable regional planning', *Regional Studies*, 28(8), pp. 781-7.

Sack, R.D. (1986), *Human Territoriality: Its Theory and History*, Cambridge University Press, Cambridge.

Sack, R.D. (1990), 'The Realm of Meaning: The Inadequacy of Human-Nature Theory and the View of Mass Consumption', in B.L. Turner, II, W.C. Clark, R.W. Kates, J.E.

Richards, J.T. Mathews, and W.B. Meyer (eds), *The Earth as Transformed by Human Action*, Cambridge University Press, Cambridge, pp. 659-71.

Sayer, A. (1985), 'The difference that space makes', in D. Gregory and J. Urry (eds), *Social Relations and Spatial* Structures, St. Martin's Press, New York, pp. 49-66.

Schleicher-Tappeser, R. (1999), *INSURED–Instruments for Sustainable Regional Development*, Final Report, Commission of the European Communities, DG XII. Environment and Climate Research Programme, Contract No. ENV4-CT96-0211.

Schumpeter, J.A. (1950), *Capitalism, Socialism and Democracy*, Harper & Row, New York.

Smith, N. (1990), *Uneven Development: Nature, Capital and the Production of Space*, 2nd edition, Blackwell, Oxford.

SPESP (2000), *Study Programme on European Spatial Planning*, Final Report, Nordregio, Brussels/Stockholm, (http://www.nordregio.se; http://www.bbr.bund.de).

SUD (2003), *Managing the Territorial Dimension of EU Policies after Enlargement*, Expert Document, Subcommittee on Spatial and Urban Development (SUD) of the Management Committee for EU Regional Policy (CDCR), European Commission, Brussels, (europa.eu.int/comm/regional_policy/debate/document/futur/member/esdp.pd).

Theys, J. (2002), 'L' amenagement du territoire face au developpement durable: Sens et limites d' une integration', in Proceedings of the Third European Symposium "Regions-Cornerstones for Sustainable Development Research and Sustainable Regional Development", 18-19 December, 2000, Tours, France, pp. 27-45.

Vonkeman, G.H. (ed.) (2000), *Sustainable Development of European Cities and Regions*, Kluwer Academic Publishers, Dordrecht.

Williams, S. (1981), 'Realism, Marxism and Human Geography', *Antipode*, 13(2), pp. 31-8.

Wilson, A.G. (2000), *Complex Spatial Systems: the modelling foundations of urban and regional analysis*, Prentice Hall, London.

Young, O. (2002), *The Institutional Dimensions of Environmental Change. Fit, Interplay, and Scale*, MIT Press, Cambridge, MA.

Chapter 10

Policy Integration:
Realistic Expectation or Elusive Goal?

Helen Briassoulis

Introduction

Integration is a constitutive element of sustainable development. Its central function is to maintain the coherence and integrity of dynamic human-environment systems to perpetuity. In this spirit, policy integration (PI) seeks to bind together currently departmentalized, disparate and uncoordinated policies that fail to tackle contemporary, cross-cutting, complex socio-environmental problems, sometimes being among the forces producing these problems. Removing conflicts and overlaps among policies and providing for their proper coordination improves the coherence of the policy system, which is, thus, expected to facilitate the transition to more sustainable modes of functioning of human-environment systems. In European Union policy making, the requirement of Article 6 of the 1999 Amsterdam Treaty, to integrate environmental protection requirements into the definition and implementation of Community policies and activities, has contributed to a both vague and narrow conceptualization of PI, commonly referred to as environmental policy integration (EPI). The majority of the relevant literature adopts more or less this interpretation but it has addressed also its limitations, such as the lack of clarity and specificity, and has questioned its ultimate effectiveness. This book posited that a broader conceptualization of PI, as integration of policies, or inter-policy integration, is more appropriate to capture the full meaning of the term, to provide a basis for its operational expression and to elaborate proper means to realize it.

A full-fledged, comprehensive analysis based on this broader conceptualization of PI was a rather utopian goal for the present endeavor because both primary and secondary essential information is still missing to give substance to its several and multiple aspects. The contributions included in this volume had thus the more modest aim to examine certain aspects of PI from the perspective of selected EU policies. Although their conclusions are not final and uncontested, they do make useful suggestions and show the way ahead for a more complete analysis that could serve as the basis to make more comprehensive and informed decisions on the issue. This concluding chapter summarizes first their preliminary findings that assess the present state and future prospects of PI in terms of its principal dimensions. Then, it negotiates important questions upon which the feasibility and effectiveness of PI hinge critically, comments on the implications of the present state of PI for combating desertification and suggests future research directions.

The Present State of Policy Integration in the European Union

The analysis of PI undertaken in the individual book chapters concerned the level of the EU but often reference was made to national and sub-national levels because of the particular mode of EU policy implementation that relies heavily on subsidiarity. The selection of the EU policies was based on two principal considerations; first, their importance for socio-spatial development and, second, their role in combating desertification in Mediterranean Europe, a particularly complex, although low profile, policy problem. The policies considered were: regional, rural, transport, social, economic, environmental, water, and biodiversity policy. Forest management was treated as a special case where PI could potentially help compensate for the lack of a common EU forest policy to date. Finally, spatial PI was treated also as a crosscutting issue of fundamental importance for achieving meaningful and effective integration of policies over and across spatial/organizational levels.

The following discussion summarizes the preliminary findings of the individual contributions as regards the present state and future prospects of PI organized according to the principal dimensions of PI; substantive, analytical, procedural and practical (see Chapter 2, this volume). The findings should be read within the limitations of the present analysis that relied on limited secondary data and a few interviews with individuals (Greece-based) from related policy contexts. Moreover, only a few pairs of policies have been analyzed and selected dimensions were covered depending on the available information. Many other EU policies, their relationships and their implementation in a variety of EU member states remain to be explored. However, because the afore-mentioned policies constitute important EU policies, it is conjectured that the trends identified may apply reasonably well to several other policy contexts at the EU level.

The Substantive Dimension

The substantive dimension of PI has to do essentially with whether policies view a policy issue – such as regional, rural, or spatial development, the environment-development relationship, environmental and resource protection, desertification, etc. – through the same or congruent and compatible theoretical and conceptual lenses, drawing on similar or compatible value systems. If this is the case, their objects, concerning selected features of the policy issue – environmental, social, economic, cultural, political, and other – will be congruent by implication, their goals will be most often than not compatible, and the relationships among the respective policy actors will be characterized by cooperation and coordination. Naturally, if policies share common actors by design, the compatibility among value systems follows directly.

The analysis of the relationships among the EU policies considered revealed a broad pattern of lack of congruence among their objects as well as among their theoretical and conceptual foundations and framing. The worldviews, assumptions, underlying value systems and orientations, and consequently the goals, of policies based on economics-informed paradigms, such as the regional, rural, transport, and macro-economic policy, differ widely from those based on social sciences and/or environmental sciences-informed paradigms. Even when integrated policies are allegedly formulated, in other words, when vertical PI is promoted, as it is the case with regional and rural policy (Chapters 3 and 4, this volume), there seems that little or no effort has been made to reconcile the different theoretical underpinnings of their constituent elements (e.g. economic with social and environmental goals and instruments).

The sustainability rhetoric frequently frames the contemporary policy discourse and thus appears to offer a loose framework to compensate for the substantive gap in PI and to unify technically disparate policy objects. However, the record until now is not encouraging. Most policies appear to be paying simply lip service to the notion of sustainable development while focusing on and prioritizing particular aspects of policy problems. Essential theoretical treatment of the linkages among the economic, environmental and social dimensions of particular issues, such as rural or spatial development, is missing except for loose references to notions such as the 'multi-functional use of space' and the promotion of 'multi-functional landscape' approaches. What is missing is the development and adoption of integrated, interdisciplinary theories, which overcome the fragmentation of reality, the compartmentalization of space and spatial development, and the separate treatment of interlinked activities that characterizes uni-disciplinary theories. Such theories should indicate which relationships among the characteristics of a problem are important and should be addressed by respective policies, constituting the substantive basis of their integration. Because it is rather improbable that a single, interdisciplinary theory of socio-environmental problems will be ever formulated and adopted, an alternative view holds that pluralism and dialogue among different theories might be a more effective means towards PI, given that it is guided by genuine efforts to reconcile and coordinate conflicting points of view.

Another trend that appears to provide fertile ground for substantive PI is the territorialization of several policies, such as the regional, rural and water policy, that is backed by related theoretical notions, such as regional or territorial cohesion, and theories of socio-spatial development, such as New Regionalism, ecological modernization, and the ecosystem approach. However, the sectoral and uni-disciplinary orientation of several policies continues to dominate their workings and relationships to other policies despite the territorial cohesion rhetoric.

Related to the lack of agreement in the theoretical framing of most policies are deeper conceptual disagreements and vague or loose definitions of terms that open up the way for a multiplicity of interpretations (and mis-interpretations) of nominally similar concepts in various policy contexts. Sustainable development is a case in point; economic and regional policies adopt a weak sustainability interpretation while social and environmental policies abide mostly by the strong sustainability variant. As a consequence, different policies adopt different operational expressions of the same concept that eventually reflect on the corresponding policy decisions (policy instruments, implementation structures and mechanisms, etc.) thus deepening the chasm among them. For example, the economic interpretation of environmental protection suggests the use of economic instruments (subsidies, levies, etc.) while a more comprehensive interpretation suggests a variety of other instruments (e.g. standards, procedural regulation, communication/education).

Although the analysis of policy actors and their networks was limited by the availability of adequate secondary information, the contributions make clear the fundamental role of the socio-culturally determined value systems of policy actors in the substantive divergence among the policies studied. Due to their predominantly sectoral origin, policy actors hold particular sectoral views of policy objects and of the associated concepts and favour particular approaches that may not be compatible or congruent with those of other policies. This becomes especially acute in the case of integrated policies. In rural development policy, for example, where long-established agricultural interests strongly support the sectoral orientation of rural development at the expense of its territorial dimension.

The sustainability rhetoric and the territorialization trends account also for the noticeable, although still hesitant, present convergence among policy goals. The analyses of the separate policies confirmed the findings of other studies that it is easy to achieve some agreement at the abstract level of policy goals (Lenschow, 2002; Lafferty and Hovden, 2002). However, this convergence is rather superficial concealing important ideological and practical differences in the way policies operationalize their goals and their relationships with those of other policies. More often than not, the goals of one policy are added to rather than combined and integrated with those of another. The different dates at which policies have appeared partly explain this situation. Earlier policies could not have anticipated those that were formulated later.

Sometimes goal integration is asymmetric and 'one-way' in the sense that one policy aims at integration with another but the reverse is not true if the priorities of one policy are unrelated or even contradict those of the other as the cases of the regional-transport policies and the Rural Development Regulation-EU biodiversity policy demonstrate (Chapters 3 and 7 respectively, this volume). Even worse, some policies may be internally inconsistent as they contain conflicting goals. The rural development policy is a notorious example; its competitiveness-related goals dominate its non-competitive goals, thus essentially undermining its aim of being an integrated policy. Finally policy integration – construed as integration among policies – does not appear to be a goal of most policies. The exception is the requirement for EPI that is, however, loosely and vaguely stated.

The Analytical Dimension

The analytical dimension of PI concerns the congruence among the spatial and temporal frames of reference and the methodological approaches associated with different policies. These are closely related to their theoretical and conceptual framing. With the exception of the spatial dimension, these concerns are little addressed in the literature. The spatial frames of reference[1] of the EU policies analyzed are not always congruent raising serious questions as regards the effectiveness of socio-environmental management and governance in general and of PI in particular. Social and economic policies, including the regional, rural, and transport policy, refer to administrative subdivisions as defined by the NUTS classification system while environmental policies refer to ecological units. The EU Water Framework Directive (EU WFD) specifically adopts the river basin as the spatial unit of reference, which follows hydrological boundaries. Administrative units are not necessarily suitable for addressing cross-cutting socio-environmental problems, among others, as "the territorial competences of authorities in charge of environmental protection do not always match with the affected environment" (Liberatore, 1997, p.117).

Two considerably different proposals are offered for the resolution of the spatial mismatch problem as a means towards facilitating PI. The first sees to the delineation of hybrid spatial systems of reference reconciling the administrative with the biophysical and other dimensions while the second sees to the constructive exploitation of the boundary problems that imperfect spatial matches create through coordination and integration mechanisms.

One of the reasons for the relative neglect of the spatial coordination issue among EU policies is that their implementation is delegated to the national and sub-national levels. Structural and procedural provisions referring to these levels are offered[2] with the expectation that the spatial mismatch problems will be resolved there. However, the present contributions make clear and confirm the conclusions of previous studies (Peters, 1998; Avery, 2001) that lower level spatial policy coordination and planning are infeasible and problematic in the absence of spatial coordination provisions formulated at higher (the EU) level.

Some attention is being paid to the issue of vertical spatial integration but in the confines of individual policies with initiatives being offered to improve the communication, cooperation and coordination among the administrative agencies involved at the EU, national and sub-national levels. However, the cross-level relationships among the diverse policy actors involved at each level are not usually being addressed.

---

[1] I.e. spatial classification schemes, spatial units of reference and delineation of spatial areas.
[2] See next section.

Little attention is being paid also to the lack of integration among the temporal frames of reference of the policies studied here. Rarely do the temporal units, time intervals, time horizons, timing of actions that policies adopt coincide by design. Regional policy has a fixed time frame, the length of the programming period (seven years), while all other policies leave largely unspecified the temporal details of their provisions. This may be one of the causes of ineffective and wasteful policy interventions as rarely do policies deliver results in isolation and in a timeless space!

The methodological aspects of PI have been minimally addressed or have not been addressed at all. This is why the present contributions could not cover the topic. The theoretical underpinnings of the policies are usually associated with particular methodological approaches and techniques of policy analysis. It is highly probable that methodologies used in different policy contexts will not be always congruent, producing incompatible analyses of the various aspects of the policy problems under consideration.

The Procedural Dimension

The policies analyzed score better in terms of procedural integration although this is not generally true for several of them and in all respects. Procedural integration is delegated to lower levels, however, where policies are implemented while relatively few provisions exist for the EU level (among the competent DGs and other EU policy making bodies).[3] This is readily explained by the emphasis placed on decentralized decision making and the subsidiarity principle that are widely promoted for various reasons. The EU level is remote from the level where problems arise, actors are concrete and solutions are sought and implemented; hence, PI is delegated to lower levels with the belief that better and more expedient results will be obtained. However, the evidence does not support fully this expectation as it is discussed below.

Several structural and procedural arrangements as well as instruments are provided for the coordination among policies at the national and subnational levels. Structural measures include, for example, provisions for the establishment of Management Authorities and of River Basin Authorities to deliver regional and water policy respectively, and of other administrative bodies with a more or less coordinating role. Funding requirements and regulations are frequently used to achieve coordination between regional and rural with environmental policies. Legal instruments, mainly the EIA and the SEA requirements, are widely applied

---

[3] One such EU level provision is the Commission Communication for Impact Assessment concerning all major initiatives, i.e. those which are presented in the Annual Policy Strategy or later in the Work Programme of the Commission, as a tool to improve the quality and coherence of the policy development process and to contribute to a more coherent implementation of the European Strategy for Sustainable Development (CEC, 2002a). Impact assessment is an action of the Better Regulation Action Plan (CEC, 2002b).

as means to secure compliance with environmental protection and sustainable development goals of regional and rural development as well as transport and other policy measures. In addition, the cross-compliance requirement adds extra weight on the need to harmonize the requirements of different policies on the ground. Good farming practice codes are tools that may facilitate greater coordination in achieving differing policy goals, such as productivity improvements coupled with environmental protection.

The recent emphasis on procedural regulation, i.e. the specification of required procedures rather than of specific results, underlines the growing recognition of the importance of planning as a coordinating mechanism as well as of public consultation, indicators, evaluation and monitoring in fostering PI during implementation. The EU WFD suggests integrated river basin planning and management (IRBPM), for example, as the main vehicle for integrating water-related goals into regional and rural development decisions. IRBPM thus provides a platform for integrating the WFD with SF programming and Rural Development Plans and for the coordinated use of the financial resources provided through EU regional and agricultural policies.

Finally, economic instruments, such as water pricing, whose inclusion is mandatory in River Basin Plans, are suggested as potentially integrative instruments between economic and environmental policies.

Procedural integration, like substantive integration, is frequently asymmetric with one policy providing provisions for PI which are not matched by similar movements in other policy areas (cf. Lenschow, 2002). One reason is that policies are formulated by different DGs where different actors participate and different procedures apply, at different dates and the timing of decisions differs. As usual, however, implementation provides the ultimate testing ground of the effectiveness of procedural PI in resolving complex socio-environmental problems. The enforcement of the subsidiarity principle and the decentralization trend in decision making, which the EU promotes rigorously, are the principal determinants of how EU advice and measures for PI materialize at lower levels. On the positive side, the encouragement of several structural and procedural arrangements, like partnerships and public participation in decision making, have had positive results but mostly in northern EU member states (MS). In southern MS, where frequently institutional capacity to implement PI measures is deficient or missing, the results are not encouraging and depend considerably, if not exclusively, on the will of national and regional authorities and of individual policy recipients, like farmers and their associations, to implement them properly.

The Europeanization of policies and the mode of policy delivery[4] account also for the success or failure of procedural PI provisions on the ground. More Europeanized policies, like the regional and the environmental, are more probable to be implemented properly. On the contrary, less Europeanized policies, like the social, where national governments have the upper hand because the OMC applies,

---

[4] Mainly, the use of the open and the close coordination methods (OMC, CMC).

are less probable to be well integrated with other policies during implementation. In fact, social policies are the least integrated, both substantively and procedurally, with other EU policies. In addition, the use of directives or of framework directives (e.g. the WFD) does not secure the implementation of PI provisions because wide discretion is left to lower levels and several clauses for derogations are provided (the WFD is a case in point). In any event, several problems still remain as regards the implementation of various arrangements, like public participation, which owe to the political culture of each MS and other contextual and idiosyncratic factors.

On the negative side, decentralization and the extensive application of the subsidiarity principle weaken an already weak EU capacity to control actual outcomes, hence threatening PI during implementation. This is especially serious for the southern MS, where economic development goals still dominate over environmental protection while lack of effective monitoring and control by the EU distorts the proper implementation of several policies. In general, it seems that a more centralized control of policy implementation could create a more solid ground for the integration of EU policies.

Despite the advances made in terms of procedural PI, several issues remain to be settled and improvements are necessary to exploit the potential created. The internal consistency and effectiveness of several procedural arrangements provided even in allegedly integrated policies (e.g. rural and regional development policy) is questionable, owing in large part to the sectoralization and departmentalization of EU and national policy making, the inertia of policy views that fail to recognize the importance of PI and the bureaucratic prerogatives that develop to defend the status quo. Procedural rationality in decision making which would minimize the costs of compliance with policy requirements and enable the smooth and unhindered participation and involvement of all actors concerned remains to be achieved in many MS. The lack of EU guidelines on issues such as public participation and the role of different types of actors, the design of rural development and river basin plans and the coordination of policy instruments produce an implementation deficit that does not serve the cause of PI. The preliminary assessment of the effectiveness of extant procedural PI arrangements points once more to the need to develop more coherent PI at the EU level and not leave it entirely to the discretion and will of lower level administrators and other types of implementers (Avery, 2001). Lastly, procedural integration will be always ineffective in the absence of more essential, substantive integration among policies as it is discussed in the next section.

The Practical Dimension

The practical aspects of PI are the least explored in general and the present contributions did not cover this issue. These concern the diverse information requirements that are necessary for a proper analysis of policies and of their integration. The well-known deficiencies in the availability, compatibility, consistency and congruence of spatially, temporally and conceptually integrated data and information systems do not facilitate the proper analysis of the object of

PI (Briassoulis, 2001). Hence, limited essential analytical support can be provided for the elaboration of PI options at all spatial/organizational levels.

Summarizing, in the case of the policies studied in this volume, procedural integration fares better among the four dimensions of PI but its long-term effectiveness, at least, is questionable in the absence of more essential substantive integration. The practical dimension of PI is the least neglected while the analytical dimension receives some treatment with respect to its spatial component.

Policy Integration: Realistic Expectation or Elusive Goal?

In concluding this chapter and book the most prominent questions with respect to PI are negotiated and the question of whether PI is a realistic expectation or an elusive goal is addressed in the sense of the eventual provision of an adequate supply to meet the demand for PI generated by contemporary complex socio-environmental. The implications of the present state of PI for combating desertification in Mediterranean Europe are sketched and future research directions are proposed.

EPI or PI for Complex Policy Problem Management and Sustainable Development?

The thrust of the current literature on PI revolves around the concept of EPI that still remains vague. The implicit assumption is that by introducing environmental considerations in some way in sectoral policies will contribute to the achievement of sustainable development. The direct implication of this assumption is that the environmental is the most important component of sustainable development which does comply with the broader conceptualization of sustainable development as involving a balance among all three dimensions – the economic, the environmental and the social. Moreover, this environment-biased conception of sustainable development ignores the indubitable linkages that exist between the environmental and the social as well as between the social and the economic facets of policy problems.

More importantly, the way of introducing environmental concerns in sectoral policies remains unspecified reflecting either a vague conception of the term 'policy' or an instrumental identification of policies with their goals and instruments. In both cases, however, the point is that environmental concerns relate to more than one environmental policy. Therefore, a meaningful introduction of, say water quality considerations, in sectoral policies would necessitate the coordination of a sectoral policy with water policy. Moreover, because even if isolated policy instruments are considered, these are not delivered by themselves; they are administered by specific administrative bodies, concern particular policy and other actors and relate to other instruments (e.g. environmental regulations cannot properly enforced in the absence of financial instruments). It follows that essentially the effort to specify and operationalize EPI inevitably leads to PI as it was construed in this book; i.e. as integration among policies in terms of their objects, goals, actors, procedures and instruments.

A related question concerns the choice between vertical PI and horizontal PI. The notion of EPI reflects essentially a vertical conception of PI. In the same spirit, it can be argued that social and economic considerations can be incorporated in policies to formulate integrated policies. Evidently, an integrated policy, the outcome of a PI process, may fare better compared to a uni-dimensional and narrow policy. However, the effectiveness of a vertically integrated policy remains dubious and contested because policies do not function in isolation; instead they exist within socio-political milieus inhabited by and contended with several other policies. For example, in the case of EPI, conflicts among non-environmental sectors may generate negative environmental impacts not accounted for totally by sector-based EPI; conflicts between rural and energy policy may result in environmental impacts; so do conflicts between rural and regional policy (e.g. water shortages). Although vertically (i.e. internally) consistent and coherent policies may be more likely to be well integrated with other policies, it turns out that only horizontal PI, or inter-policy integration as it might be called, can guarantee effective support for sustainable development.

Contemporary socio-environmental problems are characterized by bio-physical, socio-economic and institutional complexity that introduces important non-linearities in the operation of both the human-environment and the policy system. The most pronounced and serious implication of complexity is the uncertainty, unpredictability, contextuality and contingency of policy interventions to resolve these problems. Policy integration, particularly spatial PI, may contribute to the management of complex policy problems variously. Several policy systems are not complex but complicated[5] and chaotic or highly rigid and inflexible. PI can turn such policy systems into ones that possess organized, and thus manageable, complexity by exploiting the growth of multi-level governance systems and developments to transform them into nested organizational hierarchies that hold the human-environment system together while allowing it the flexibility to adapt and innovate.

By eliminating redundancies (among actors, procedures and instruments), constraining conflicts and promoting policy coupling, PI reduces the number of system elements and their interactions, thus reducing the degree of (unwanted) system complexity, especially of institutional complexity. Strategic simplification can be utilized towards this purpose that, at the same time, should secure beneficial redundancies, open alternatives and options. This can be achieved through an adaptive management approach, based on communication and cooperation,

---

[5] Reitsma (2002) notes that complication is a quantitative escalation of the theoretically reducible; thus, it is different from complexity which implies that the whole cannot be fully understood by analyzing its parts. The dictionary definition of 'complicated' usually identifies it with the 'complex'.

encouraging learning and adaptation to new situations as well as to unforeseen contingencies. Procedural integration may provide for those positive feedback mechanisms that reinforce beneficial and attenuate undesirable policy impacts and foster novelty and creativity in coping with complex socio-environmental problems.

Substantive vs. Procedural PI

The thrust of policy activity and research on PI focuses on procedural integration. This is not surprising as it is much easier to obtain consensus on and compliance with procedures rather than goals, values, preferences, and priorities. Hence, the pursuit of procedural rationality dominates policy making; i.e. actors in a decision context, instead of aiming at rational goals, aim at fair and rational procedures. Such procedures should favour the discovery of alternative ways for different groups to reach acceptable solutions (Simon, 1979). In the case of environmental policy measures, for example, such as the EIA, the SEA and the WFD provisions, this might mean that procedures for integrating environmental requirements in sectoral policies should be fair, rational and compatible with other sectoral policy-specific procedures irrespective of whether environmental and sectoral policy goals agree.[6]

Naturally, the question arises whether procedural PI is adequate to secure a well-functioning policy and human-environment system. The fact that structures and procedures exist or are provided through procedural arrangements does not automatically imply their suitability for all cases of PI, their adoption and implementation, and their effectiveness with respect to the goals of PI. The present contributions did not produce conclusive assessments of the relationship between procedural and substantive integration. Policies that differ widely on substantive grounds, such as the regional and the environmental policy,[7] exhibit some form of procedural integration. Nevertheless, all contributions suggested that in the absence of substantive integration among policies, in terms of their theoretical and conceptual framing and value orientations, an instrumental, procedural-oriented approach to PI will not deliver PI, at least in the longer-term. Of course, because of the contextuality and contingency of policy problems, the ideal linkage between substance and process will be difficult to achieve in most cases. Moreover, even if a perfect PI scheme is designed and agreed at higher spatial/organizational levels,

---

[6] For example, lack of procedural rationality characterized until recently water resources management and the EIA directive's requirements. In some cases, developers were obliged to obtain and submit with the EIA tenths of different licenses from various administrative agencies with competencies on environmental and water management issues. Under the WFD, each MS will have to elaborate and implement effective cooperation and coordination mechanisms and fast, efficient and clear procedures in order to ensure the achievement of the aims and objectives of both Directives.

[7] See, Chapter 3, this volume.

substantive and procedural 'misfits' will arise during implementation at lower levels that are difficult to presage at the formulation stage.

Overall, it is not advisable to conclude that two or more policies are integrated if they are procedurally coordinated and produce immediate, short-term benefits. Even in this case, the possibility that other factors have been present and have produced the observed results should be investigated; in other words, the case of spurious effectiveness of procedural integration should be thoroughly investigated.

PI to Address Issue Areas for Which no Common EU Policy Exists

Chapter 8 on the role of PI for sustainable forest management suggests another function of PI that has not received attention until now; namely, its use to address issue areas for which no common EU policy exists. In addition to the issue of forest management that Chapter 8 covered, other issue areas include soil protection and tourism. The characteristic of these 'policy-orphan' areas is that they concern problems that are multi-dimensional, encompass diverse activities and associated actors, are affected by multi-level forces, are difficult to coordinate spatially and temporally, and historically no common EU policy has been proposed to address them. At the same time, however, several existing EU policies contain provisions related to particular facets of these issue areas but these provisions are not coordinated among them to address the associated problems holistically. Instead of introducing a new common EU policy, PI comes as a potential answer to fill this policy void in two forms. One approach is to systematically introduce issue-related concerns (e.g. tourism) into the objects, goals, actors, procedures and instruments of relevant EU policies; in other words, to produce integrated policies in the EPI fashion. The effectiveness of this approach is not guaranteed as it was argued before. The other approach is to attempt the integration of existing EU policies, issue-integrated or not as suggested before, guided by issue-specific requirements; e.g. sustainable forest management in the case of forests, sustainable management of soil (or, more broadly, land) resources in the case of soils, sustainable tourism development in the case of tourism, and so on. Evidently, the optimum approach is to combine both forms of PI but the EPI-like integration seems more viable at present. In any event, the case of 'policy-orphan' areas suggests that properly conducted PI can exploit the provisions of extant policies to address more than one issue areas thus adding value to the current stock of policies and achieving multiple benefits.

At what Level PI? From Above (EU) or from Below?

A recurring theme in most contributions, and in the broader literature, concerns the appropriate spatial/organizational level at which to carry out PI. Is it better to integrate policies at the level of the EU and then transpose PI to lower levels or is it better to delegate the task to lower levels where problems and the associated actors are more visible, concrete and solutions can be adapted to local conditions and needs? The answer to this question depends importantly on the contemporary socio-political landscape and the political will to promote PI.

EU policy making makes heavy use of the subsidiarity principle which is a double-edged sword in the present context, at least. On the one hand, the principle recognizes the right of lower levels to self-determination and decision making autonomy at the national and subnational level where problems arise mostly. However, contemporary problems are produced by the constant, context-specific interplay of forces acting from the macro to the micro level rather than by forces acting on a single level. Consequently, their resolution implicates more than one institutional levels and groups of actors. But as Hajer (2003, p.175) argues:

> ...policy making now often takes place in an 'institutional void' where there are no generally accepted rules and norms according to which politics is to be conducted and policy measures are to be agreed upon. More than before, solutions to pressing problems transgress the sovereignty of specific polities.

Within this context, subsidiarity encourages decentralization and the creation of dispersed and diffuse power contexts that, however, generate centrifugal forces countering centralization and, by implication, coordination and cooperation. Therefore, unfettered application of the subsidiarity principle opposes the quest for policy coherence and detracts from the achievement of both vertical and horizontal PI. As Detsis (Chapter 7, this volume) notes for the case of the EU biodiversity policy, in practice integration is an open option rather than an obligation of the actors involved.

The question of the possibility of PI amidst an institutional void, among other detracting factors, gives rise to other related questions such as whether it is possible to muster and sustain the required political will to promote the PI goal. Currently, all analyses of PI, including the present contributions, identify political will as the *sine qua non* of successful integration and its lack as the most important factor hindering the realization of PI. A reading of representative recent documents such as the Third Cohesion Report (EC, 2004) and the Cardiff Stocktaking report (CEC, 2004) reveals that the obstacle of insufficient political commitment still remains to be overcome. The decentralization trend and the devolution of decision making powers to lower levels mark a gradual retreat of the EU from matters of common interest and raises concerns as to the existence of genuine interest in PI.

This brings to the fore the related question of accountability in policy making for complex, horizontal, cross-cutting policy problems under a decentralized policy regime (Peters, 1998). Since the resolution of these problems implicates several decision levels, diverse policy areas and numerous actors and decision making power is dispersed, who will be held accountable for what policy outputs and outcomes? The answer is further complicated if it is taken into account that PI is not equally desirable in all policy contexts as the case of rural development policy suggests (Chapter 4, this volume). The absence of strong, central, political commitment to PI does not allow optimism as to the resolution of such dilemmas and the feasibility of effective accountability procedures to safeguard the delivery of PI outcomes.

Under these conditions, it is not surprising that procedural rationality in policy making rules and procedural integration dominate as they secure a minimum of effectiveness and contain, to some extent, the centrifugal forces that subsidiarity generates. The asymmetric desirability of PI among policy actors does not favour efforts towards substantive integration among policies especially at times of fiscal austerity and limited public resources.

Recapitulating, a state of generalized decentralization and strong subsidiarity seems to shrink the chances of achieving PI at higher levels. Even if the conditions were more favourable, however, PI designed at high levels, such as the EU, could be felt as too restrictive on local level autonomy and planning decisions; thus, the chances of its realization could be thin again. The solution may lie somewhere in between the two extremes, i.e. in a combination of EU-level coupled with lower level PI, a topic that is addressed below.

How to Promote PI?

The variety of multi-level, country-, sector- and problem-specific factors that condition the success of PI efforts makes unrealistic the prescription of broadly applicable PI recipes. Moreover, the preceding selective negotiation of important PI-related questions suggests that perhaps the current quest for PI may not be moving in the right direction. More specifically, policy problem complexity and the negotiation, and not unquestionable acceptance, of scientific knowledge in decision making (Hajer, 2003) preclude the *a priori* and exact description of the output of a policy integration process, be it an integrated policy or a coherent and smoothly functioning policy system. Therefore, it is not possible to specify particular ways and procedures to achieve a fuzzy or unspecified PI output. This has been recognized, for example, in water policy making (see, Chapter 6, this volume) where the logic of procedural regulation dominates in the WFD; processes that competent authorities should implement are prescribed instead of standards or measures. The assumption is that good water status will result from the proper implementation of these procedures. This logic can be extended to the PI issue as it is argued next.

Evidently, PI cannot be imposed because it is an emergent property of a smoothly functioning and well-connected policy system. Its particular manifestations will vary with the geographic and physical context, the socio-economic conditions and political regimes as well as the particular problem considered. Therefore, what is important to specify are the proper rules that should be followed within[8] and between policy domains to satisfy certain prerequisites for

---

[8] An important observation is that policies that are inherently incompatible cannot be integrated, even through procedural arrangements. Hence, they should be appropriately modified first (through vertical integration) before they are integrated with other policies.

substantive and procedural integration. Because of the overall uncertainty of the undertaking, an adaptive management mentality is suggested that is specifically suited to situations engendering high uncertainty (see, Chapters 1 and 9, this volume). An iterative process of designing, testing and revising PI schemes may provide a flexible mechanism to elaborate the requisite procedures and rules.

The Future Prospects of Integration of EU Policies

Only a few of the present contributions attempted an assessment of the future prospects of integration among EU policies; those that did it, found them dim! A continuation of present global and EU level socio-economic, political and policy trends do not lend much hope to the feasibility of essential PI that will meet the current demand generated by complex socio-environmental problems. Important changes have to take place to reverse the current trends and provide an environment conducive to policy dialogue and cooperation. Particular changes may be introduced at different spatial/organizational levels but their ultimate effectiveness will depend on the composite workings of the forces influencing PI. The discussion of global level socio-political conditions is beyond the present purposes. Although globalization and the attendant socio-economic and political developments may be difficult to reverse in the near future, long-term surprises may not be entirely ruled out given the non-linear, co-evolutionary nature of human-environment systems, some of which may prove favourable to PI.

At the EU and the national levels, reducing the departmentalization of policy making and allowing for more interaction among policy domains may facilitate not only procedural but also substantive integration. In the context of the Cardiff Integration process (CEC, 2004), the extension of EPI to policy domains such as social policy, tourism, education, etc. is an indication of willingness to broaden the scope of PI in general. Similarly, the Luxembourg and the Cologne processes, concerning employment and macro-economic policy issues respectively, and the hesitant introduction of sustainability criteria in the Broad Economic Policy Guidelines (BEPGs) signify a slowly changing rhetoric, at least, that may bring fruits at some point in the future.

Against this background, the Commission should take steps to improve the methods of consultation between its Directorates General, with a view to improving the overall coherence of the policy process and changing the current administrative culture (Avery, 2001). A parallel deepening of integration processes is also necessary involving changes in the theoretical/ideological and conceptual underpinnings of policies with the adoption a sustainability-informed paradigm where the economic, social and environmental dimensions of sustainable development are treated simultaneously, not in isolation but as indivisible and interdependent aspects of reality. The adoption of PI as an explicit policy goal at the political level will signify a serious intent to move ahead with this endeavor, rendering it thus a realistic expectation rather than an elusive goal.

As regards the particular case of combating desertification, it is difficult to offer a fully documented assessment because of the limited analysis provided and the bio-physical, socio-economic and institutional complexity of the phenomenon. However, it can be safely argued that because desertification is a complex policy problem demanding the coordination of several policies at and across various spatial levels, the weak and mostly procedural PI at present is not favourable for the long-term resolution and management of the issue. An integrated conception of the phenomenon is missing as it has been approached and analyzed from different, narrow disciplinary perspectives and traditions, where natural sciences approaches dominate social sciences or integrated approaches. Not surprisingly, policy prescriptions are similarly colored and biased, favoured by the extant sectoralized policy environment. Most of them concern the environmental dimensions of desertification despite the recent emphasis on its socio-economic and cultural dimensions. The connections between all dimensions are still thin and not well-elaborated, especially on a spatial basis, exposing the lack of a common substantive basis for PI to resolve the issue. If PI moves in the right direction as suggested above, it may be possible to assist desertification-sensitive areas develop sustainably, within their biophysical constraints. PI at the EU level is, however, still necessary to provide the enabling framework for the elaboration of particular PI schemes in the context of the National Action Programmes (NAPs) of each country or, better, of the regional plans of desertification-sensitive areas.

Future Research Directions

The PI theme is a relatively recent addition to the policy sciences research area, especially as construed in this book as integration among policies in terms of their objects, goals, actors, procedures and instruments. The limited attempt to explore it in this volume suggests a very rich theoretical and empirical future research agenda. The most important research directions are highlighted below to complement the proposals offered in the individual chapters.

Theoretical research should study in depth the substantive integration of EU policies, including those that were considered in this volume, for a variety of issue areas such as desertification, biodiversity protection, rural development, tourism development, etc. Of particular interest is the analysis of the complexity of the respective policy problems and the examination of how the integration of pertinent policies can address it. This necessitates the thorough study of the actor networks involved in the various EU policies both at the EU and at the national and subnational levels through suites of empirical studies covering a variety of country and regional situations.

To produce useful results that can inform policy change towards greater and effective PI, research should analyze thoroughly the objects and goals of EU policies, investigating their theoretical and conceptual underpinnings in relation to the actors involved in the respective policy domains. Suggestions for reconciling the theoretical framing and underlying value systems of EU policies around the sustainable development paradigm need to be developed to overcome the weak/strong sustainability division and move towards more operational approaches to PI.

Closely related to the study of substantive integration should be the theoretical and empirical exploration of the analytical dimension of PI focusing on alternative approaches to harmonize and/or coordinate the spatial and temporal frames of reference as well as the methodologies associated with EU policies. Particular emphasis should be placed on integrated methodologies used in common by all policies to provide compatible analyses of complex policy problems.

Although several proposals have been made with respect to the procedural integration of EU policies (e.g. Avery, 2001), research is needed on procedural arrangements meaningful for particular issue areas as well as on the linkages of procedural provisions to the substantive integration of policies. Again, this analysis necessitates the study of the actor networks involved in each case and the specific procedural needs to harmonize them. It would be interesting to see if common procedural arrangements can address the diversity of issue areas and thus simplify the institutional complexity of EU policy making.

The integration of policy instruments should be studied with reference to particular issue areas to suggest how to remove conflicts and inconsistencies as well as to provide guidelines for the development of suitable policy instrument mixes. The special case of integrative instruments should be re-examined in the light of the more essential need to promote the substantive in addition to the procedural integration of EU policies. Empirical research on the use of the variety of policy instruments in diverse socio-spatial contexts and issue areas is indispensable to carry out this research stream.

Lastly, the practical aspects of PI should receive special attention as they are the least considered and researched. Issues of data and information compatibility, consistency and availability should be explored, capitalizing on and carrying further related, on-going efforts. These are the most critical requirements for properly analyzing the substantive and procedural dimensions of PI and putting it into practice.

References

Avery, G. (2001), 'Policies for an Enlarged Union', Report of Governance Group 6, *White Paper on European Governance*, Area no. 6, Defining the framework for the policies needed by the Union in a longer-term perspective of 10-15 years taking account of enlargement, European Commission, Brussels.

Briassoulis, H. (2001), 'Policy-oriented integrated analysis of land use change: An analysis of data needs', *Environmental Management*, 26(2), pp. 1-11.

CEC (2002a), *Communication from the Commission on Impact Assessment*, COM(2002) 276 final, Brussels.

CEC (2002b), *Simplifying and improving the regulatory environment*, Action plan, Communication from the Commission, COM(2002) 278, final, Brussels.

CEC (2004), *Integrating environmental considerations into other policy areas - a stocktaking of the Cardiff process*, Commission Working Document, COM(2004)394 final, Brussels.

EC (2004), *Third Report on Economic and Social Cohesion*, COM(2004) 107, Office for Official Publications of the European Communities, Luxembourg.

Hajer, M. (2003), 'Policy without Polity; Policy analysis and the institutional void', *Policy Sciences*, 36, pp. 175-95.

Lafferty, W.M. and Hovden, E. (2002), *Environmental Policy Integration: Towards An Analytical Framework?* PROSUS, Centre for Development and the Environment, University of Oslo, Oslo, Report 7/02.

Lenschow, A. (2002), 'Greening the European Union: An introduction', in A. Lenschow, (ed.) *Environmental Policy Integration: Greening sectoral Policies in Europe*, Earthscan, London, pp. 1-21.

Liberatore, A. (1997), 'The integration of sustainable development objectives into EU policy-making: Barriers and prospects', in S. Baker, M. Kousis, D. Richardson and S. Young (eds), *The politics of sustainable development: Theory, policy and practice within the European Union*, Routledge, London.

Peters, G.B. (1998), 'Managing Horizontal Government: The Politics of Coordination', Canadian Centre for Management Development, Research Paper No. 21, Catalogue Number SC94-61/21-1998, ISBN 0-662-62990-6.

Reitsma, F. (2002), 'Measuring Geographic Complexity', On-line Paper, (http://www.glue.umd.edu/~femke/written/GeographicComplexityPaper.pdf).

Simon, H.A. (1979), 'From Substantive to Procedural Rationality', in F. Hahn and M. Hollis (eds), *Philosophy and Economic Theory*, Essay 5, Oxford University Press, Oxford.

Index